AF693853

UNDERSTANDING BIOTECHNOLOGY LAW

KF
3133
.B56
U53
1993

UNDERSTANDING BIOTECHNOLOGY LAW

Protection, Licensing, and Intellectual Property Policies

edited by
Gale R. Peterson
Cox & Smith
San Antonio, Texas

LAW LIBRARY
JAN 3 1 1994
GOLDEN GATE UNIVERSITY

Marcel Dekker, Inc. New York • Basel • Hong Kong

Library of Congress Cataloging-in-Publication Data

Understanding biotechnology law : protection, licensing, and
intellectual property policies / edited by Gale R. Peterson.
p. cm.
Includes bibliographical references and index.
ISBN 0-8247-8935-0
1. Biotechnology—United States—Patents. 2. Biotechnology-
-Research—Law and legislation—United States. I. Peterson, Gale
R.
KF3133.B56U53 1993
346.7304'86—dc20
[347.306486] 93-7101
CIP

This book does not and cannot give legal advice. It is written to inform readers generally of legal developments of interest. Laws, rules, and regulations are subject to frequent, sometimes dramatic, change. The information herein thus cannot serve as a substitute for current, competent, and reasoned legal advice, given by licensed attorneys in response to specific requests.

The publisher offers discounts on this book when ordered in bulk quantities. For more information, write to Special Sales/Professional Marketing at the address below.

This book is printed on acid-free paper.

Copyright © 1993 by Marcel Dekker, Inc. All Rights Reserved.

Neither this book nor any part may be reproduced or transmitted in any form or by any means, electronic or mechanical, including photocopying, microfilming, and recording, or by any information storage and retrieval system, without permission in writing from the publisher.

MARCEL DEKKER, INC.
270 Madison Avenue, New York, New York 10016

Current printing (last digit):
10 9 8 7 6 5 4 3 2 1

PRINTED IN THE UNITED STATES OF AMERICA

I would like to acknowledge my deepest and sincerest appreciation to the contributors who gave of their time and expertise in preparing this book. I would also like to express my heartfelt gratitude to Francess Gutierrez without whose able assistance this book would not have been possible.

Gale R. Peterson

Contents

Contributors

Reid G. Adler Director, Office of Invention Development, National Institutes of Health, Bethesda, Maryland

Joyce Brinton Director, Office for Technology and Trademark Licensing, Harvard University, Cambridge, Massachusetts

Katherine L. Chapman Associate Vice President for Legal Affairs and Technology Transfer, University of Texas Southwestern Medical Center at Dallas, Dallas, Texas

Bradford J. Duft Vice President, Amylin Pharmaceuticals, Inc., and Counsel, Lyon & Lyon, San Diego, California

William L. LaFuze Department of Intellectual Property, Vinson & Elkins, Houston, Texas

Edward T. Lentz Vice President and Director, Corporate Patents–U.S., SmithKline Beecham Corp., Philadelphia, Pennsylvania

Charles E. Lipsey Finnegan, Henderson, Farabow, Garrett & Dunner, Washington, D.C.

Peter E. Mims Department of Intellectual Property, Vinson & Elkins, Houston, Texas

Lita Nelsen Associate Director, Technology Licensing Office, Massachusetts Institute of Technology, Cambridge, Massachusetts

Gale R. Peterson Head of Intellectual Property, Cox & Smith, San Antonio, Texas

Amy L. Tsui Collins Division Patent Counsel, Chiron Corporation, Emeryville, California

1

Introduction to the Field of Biotechnology Law

Gale R. Peterson

Cox & Smith, San Antonio, Texas

I. THE BUSINESS OF BIOTECHNOLOGY

Since Robert A. Swanson and Herbert W. Boyer first formed the plans for Genentech in the mid-1970s, reportedly in a San Francisco tavern, companies focused on developing "biotech" products have spawned at a pace rivaling the replication rates for some of their products. Names such as Chiron, Centocor, Amgen, ImmunoGen, Cal Bio, Biogen, and Genentech, unknown a few years ago, are now familiar from Wall Street to Main Street. Moreover, as each day passes, new companies emerge and enter the field. It has been reported that more than 50 new companies made public offerings in 1991, and a similar number raised additional equity through the public markets.[1] The market for biotechnology products is also expanding. According to current estimates, present sales by the U.S. biotechnology industry of approximately $4 billion are expected to balloon to a forecasted $50 billion worth of biotechnology products on sale throughout the world by the turn of the century.[2] By anybody's measure, biotechnology will be a "growth" industry for the remainder of this decade and beyond.

Recognizing the international competitive advantage the United States currently enjoys, the government has been active in promoting biotechnology. In January 1992, for example, the Bush Administration launched the "Biotechnology Initiative" that promised a 7% increase in federal support in fiscal year 1993 for biotechnology research, thus raising the support to approximately

$4.03 billion annually.[3] The U.S. Department of Health and Human Services and the National Institutes of Health have requested $3.1 billion of that funding.[4] In late February 1992, the federal government also announced new ''ground rules'' for regulating biotechnology that would relax some of the strict regulatory controls over biotechnology products. Coupled with the 3-year, $3 billion Human Genome Project, the government's support for the biotechnology industry has been on the rise, and it should continue for the foreseeable future.

II. DEVELOPMENTS IN THE LAW FOLLOW BUSINESS

As with other burgeoning and dynamic fields, laws governing such fields to be effective must change, adapt, and evolve as the field changes, adapts, and evolves. If that was not so, laws would run counter to their intended purpose and would hinder economic and technological progress rather than foster such progress. And so it is with the field of biotechnology.

From the earliest days of Hammurabi, the law has served to establish individual rights and responsibilities and to settle disputes short of physical combat. Such laws have typically followed on the heels of an event or discovery or societal change that has compelled a need for regulation or governance. During the initial days of the California gold rush, for example, frontier law prevailed. Claims were asserted and protected with physical prowess, guns, and knives. Frontier law dealt ''justice'' to claim jumpers and trespassers swiftly and severely by generally irreversible means. More formal laws governing staking claims, abandonment, and loss of rights, though, inevitably and quickly followed. Such laws thus brought some semblance of law and order to the myriad of miners seeking their fortunes.

A similar development of laws necessary to address specific problems evolving from a discovery occurred a half century later with the advent of the oil and gas industry. Once again, extensive laws and regulations quickly emerged to govern rights and responsibilities in the ''oil patch.'' Oil and gas law was born. In far earlier times, merchants plying the trade routes from Phoenicia to what is now Europe sorely needed a system for dealing with trade accounts. The law of the Law Merchant was born, which became the progenitor for our current laws governing sales and commercial transactions.

In similar fashion, the law today is developing rules for the modern equivalent of staking and protecting claims (primarily through developments in patent law), that address rights and responsibilities, and that create a system for dealing with commerce in products of biotechnology. As biotechnology developments occur, and as businesses are formed to capitalize on those developments, our laws have struggled to keep pace with the rapid, and sometime unpredictable, changes in the field. In some cases, these laws have been evolutionary, lagging far behind scientific developments, whereas in others, the law has changed in revolutionary fashion. Indeed, as will become clear in the chapters that follow, some of the more significant developments in the law have occurred through a single court decision adapting long-standing principles to the ever-changing world of biotechnology. Rules on patentability and infringement, although broadly applicable to a wide range of technological developments, are now beginning to take on a new "gloss" in order to address the particular issues that arise from developments in the field of biotechnology. Such developments in this rapidly changing field have created (and will continue to create) dramatic changes in the law as the legal community becomes increasingly more familiar with the business of biotechnology. The result has been the emergence of a field of law generically known as "biotechnology law."

III. THE NEED TO UNDERSTAND THE LAW

The advent of laws specific for a field, however, itself can create disputes and controversies. Such laws can also work unintended injustices. A miner unaware, for example, that failing to work a claim diligently for 30 or more consecutive days would lead to a loss of rights to, or an abandonment of, that claim would be dismayed to learn, after an extended absence, that another had legally usurped his or her rights. The point is that such laws must be understood by the workers in the field. Miners failing to understand the mechanics of properly staking, registering, and working their claims, as well as the legal boundaries of those claims, could easily loose their rights.

The law of biotechnology is not significantly different. Just as the miners, wildcatters, and traders needed to understand the laws governing their respective fields, so too must the scientists, academicians, and technicians who work daily on innovations that will change our lives forever understand at least the fundamentals of the evolving law relating to developments in the field of biotechnology.

IV. THE AUDIENCE

This book is written for the above-mentioned scientists, academicians, and technicians. Specifically, this book is intended to explain, clearly and simply, the seeming labyrinth of rules, regulations, and procedures that are typically encountered by researchers in the development of their innovations. At this point, though, it should be explained that 'biotechnology law" actually encompasses a wide field of law ranging from contracts and licensing to food and drug law. Within that broad range there are laws dealing with such diverse topics as environmental issues, employee health and safety issues, tax, and corporate and securities issues.

The genesis of biotechnology law, though, lies in the identification and protection of the products and processes of biotechnology. That will be the focus of this book. Also, because many of these innovations are developed in university and nonprofit research settings, and are ''intellectual property'' protected under the patent laws, this book will especially focus on institutional intellectual property programs, forms of legal protection (with a particular emphasis on patent protection), inventorship, ownership, patent infringement, conflicts of interest rules, and technology transfer. Those are the fundamental concepts the authors believe should be understood by those working in the field.

Each of the chapter authors is a recognized authority on the topics covered in their respective chapters. The breadth of subject matter in each chapter, and the necessity to keep this book short of a legal treatise on each topic, however, required that this book be written as an overview rather than as a critical analysis. Each of the authors has, therefore, undertaken to eliminate jargon and legalese to the extent possible in an effort to present a full, yet clear, explanation of their respective topics. In doing so, it must be understood, however, that the authors' comments are necessarily general and should not be taken literally as the ''law'' that would be applicable to a particular fact pattern.

V. EVENTS GIVING BIRTH TO THE FIELD OF BIOTECHNOLOGY LAW

It should also be borne in mind that the field of biotechnology law is of fairly recent vintage.[5] Indeed, although ''biotechnology'' has been with humankind since the early Egyptians used yeast to ferment grain, and we have had a fundamental understanding of deoxyribonucleic acid (DNA) since the work of Drs. Crick and Watson, modern innovations in the field probably would not have been possible without Paul Berg's work on recombinant DNA in the early 1970s and Caesar Milstein's and George Kohler's work on hybridoma cell fusion and selection in producing monoclonal antibodies in the mid-1970s. As noted above, the ''law'' associated with such innovations tends to lag several years behind the actual innovations. Thus, it was not until 1980 that we began to see a rational scheme developing for protecting such innovations. In fact, events in the early 1980s, although individually significant at the time, have proven in hindsight to have been the origin and key to the field of biotechnology law as we know it today.

A. Federal Legislation in the 1980s

The first of those events was passage of the Bayh-Dole Act in 1980.[6] At that time, the federal government allocated somewhere in excess of $2.7 billion for research and development. In fiscal year 1992, the federal government was scheduled to allocate approximately $3.7 billion for research and development. That was and is a significant investment of public resources. The Reagan Administration recognized, however, that innovations must not only be discovered—they must be put to economically meaningful use. Prior to 1980, innovations developed with those federal moneys were governed by 26 different federal agency policies. The multiple policies governing innovations then in use frustrated efforts to commercialize those innovations. Thus, one of the avowed purposes of the Bayh-Dole Act was to establish a uniform federal government policy on the disposition of patent rights arising from government funded research. That legislation did so by adding a new chapter to the patent statute entitled "Patent Rights in Inventions Made With Federal Assistance."

In the words of the statute, the policy and objective of the legislation was "to promote the utilization of inventions arising from federally supported research or development" and "to encourage maximum participation of small business firms in federally supported research and development efforts."[7] The legislation sought to meet that objective by establishing a procedure that allowed nonprofit organizations (such as universities and nonprofit research institutions)[8] and small business firms[9] to elect to retain title to inventions developed with government funding. The legislation, however, also imposed certain disclosure, election, and reporting requirements that are covered in detail in Chapter 8.

Although many universities and nonprofit research institutions had intellectual property programs prior to the passage of the Bayh-Dole Act, the Act was largely responsible for the expansion of existing programs and the development of new programs in institutions where such programs did not exist. Also, those programs and the offices that administered them found added responsibilities. The federal budget deficit was rapidly getting out of hand. The budget ax was increasingly brought to bear on federal funding available for civilian research and development (R&D). In the biotechnology field, available research grants were being reduced and funds for funded research were being cut. As a result of that legislation, universities and research institutions, however, had the ability to elect to retain title to their innovations. That meant that such innovations could be licensed and the revenues realized from such licensing could be used to replace some of the moneys lost because of the reduced federal funding. Also, such licenses could be used to create a closer relationship between industry and the research community. The anticipated result was that industry could provide a possible source of continuing research support. Intellectual property offices at universities and nonprofit research institutions thus had the added responsibility of creating and administering active licensing and sponsored research programs.

In the late 1970s and early 1980s, many of the nation's research dollars were also going to support research at over 700 government laboratories employing one-sixth of the nation's scientists and engineers. Legislation paralleling the Bayh-Dole Act, the Stevenson-Wylder Technology Innovation Act, was also passed in 1980 that was intended to establish a link between those federal laboratories and industry. Indeed, that Act made the policy statement that

> It is the continuing responsibility of the Federal Government to ensure the full use of the results of the Nation's federal investment in research and development. To this end the Federal Government shall strive where appropriate to transfer federally owned or originated technology to State and local governments and to the private sector.

That Act was amended by the Federal Technology Transfer Act of 1986 in an effort to further foster the development of commercially useful technology in federal laboratories and to create a statutory mechanism for moving those technologies into the private sector. More recently, the Technology Competitiveness Act, passed as part of the Omnibus Trade and Competitiveness Act of 1988, and the National Institute of Standards and Technology Authorization Act for Fiscal Year 1989, provide enhanced statutory authority and assistance to that technology transfer effort. Indeed, the Technology Competitiveness Act charged the National Institute of Standards and Technology (NIST) (formerly the National Bureau of Standards) with assisting private sector initiatives to capitalize on advanced technology. The NIST was also charged with advancing, through cooperative efforts among industries, universities, and government laboratories, promising research and development projects that could be "optimized" by the private sector for commercial and industrial applications. Despite such efforts, however, in a recent report, the Congressional Office of Technology Assessment identified improving industry-university relationships, coordinating federal agencies, and protecting intellectual property as three of the seven broad policy areas that required further enhancement. Similarly, the Office of Science and Technology Policy recently identified technology transfer as one of the areas that required improvement. Thus, the current efforts at fostering even closer working relationships between the research community and industry may be expected to continue.

The increasingly close relationship between the research community and industry, however, has itself created some new issues; especially the issue of conflicts of interest. Specifically, a researcher's impartiality and independence could be open to question if the researcher held an ownership or equity interest in the company providing the funding for the research. Thus, rules on dealing with actual and potential conflicts of interest have become necessary. Such rules are still evolving. As Katherine Chapman points out in Chapter 8, such potential conflicts of interest, and the rules for dealing with such conflicts, do not necessarily imply any wrongdoing. Indeed, such conflicts are many times more a matter of perception than actual fact. Nevertheless, institutional intellectual property programs and offices have been called upon to formulate workable conflict of interest rules that reduce or eliminate even an appearance of bias or impropriety. The issue of conflicts of interest remains, however, as one of the more challenging and unresolved issues in the field.

B. The *Chakrabarty* Decision and Its Progeny

Another significant event in 1980 was the U.S. Supreme Court's decision in *Diamond* v. *Chakrabarty*.[10] Ananda Chakrabarty was the inventor of "a bacterium from the genus *Pseudomonas* containing therein at least two stable energy-generating plasmids, each of said plasmids providing a separate hydrocarbon degradative pathway." That man-made, genetically engineered bacterium had the ability to break down the several components of crude oil and thus was useful in controlling oil spills. The commercial potential, of course, was significant, and Chakrabarty's employer, the General Electric Company, filed a patent application on the invention. That application sought a patent covering the method of making the bacterium as well as the bacterium per se.[11] The U.S. Patent and Trademark Office approved the "method of making" claims,[12] but it refused to issue a patent on the bacterium per se, reasoning that living microorganisms are not patentable subject matter. At the time of that decision, it was unclear whether § 101 of the patent statute—providing that only "new and useful" processes, machines, "manufactures" (i.e., products), and compositions of matter[13] were patentable subject matter[14]—was intended by Congress to cover living organisms,[15] even though manmade.[16] The court hearing appeals from the U.S. Patent and Trademark Office at that time was the U.S. Court of Customs & Patent Appeals. On appeal, that court held that Chakrabarty's invention *was* patentable subject matter and reversed the decision of the U.S. Patent and Trademark Office.[17] In view of the potential importance of the issue, the U.S. Supreme Court agreed to decide the question whether living, manmade microorganisms were patentable subject matter.[18] In its decision in *Chakrabarty*, the Supreme Court, in a 5 to 4 decision, held that such organisms *were* patentable subject matter; thus paving the way for effective patent protection in the field of biotechnology.[19]

As noted at the outset, a single court decision can have a far-reaching impact on the development of a field of endeavor. *Chakrabarty* was one such decision. Furthermore, in this case, it was the decision of a single justice that created the 5 to 4 majority balance in favor of patentability.

Following the decision in *Chakrabarty*, the Patent and Trademark Office began granting patent protection to a wide variety of inventions in the biotechnology field. Hybridomas, polyclonal and monoclonal antibodies, DNA transfer and expression vectors, and specific DNA sequences are now considered patentable subject matter.[20] Indeed, the door was opened to patenting multicellular organisms by a decision, *Ex parte Allen*,[21] of the Patent Office's internal appellate body, the Board of Patent Appeals and Interferences, that an polypoid oyster constituted patentable subject matter. That decision ultimately led to the grant, in 1988, of a patent to Harvard Medical School on a transgenic mouse used in studying cancer. Shortly thereafter, the U.S. Patent & Trademark Office announced: "The Patent and Trademark Office now considers non-naturally occurring non-human multicellular living organisms, including animals, to be patentable subject matter within the scope of 35 U.S.C. § 101."[22] The result has been that a literal flood of biotechnology patent applications have been filed. According to figures released by the U.S. Patent & Trademark Office, filings of biotechnology patent applications have risen from just 30 in 1978 to over 11,000 in 1991,[23] causing the Office in 1988 to establish a new examining group, Group 180,[24] to handle those applications. One consequence of those large-scale filings has been an increased time of pendency; namely, the time between when an application is filed and when it is issued as a patent or abandoned. Currently that time is approximately 2+ years.[25]

These developments have been controversial. There have been efforts in Congress, for example, to restrict patent protection for multicellular organisms, and the Patent and Trademark Office suspended granting patents on multicellular organisms for 1 year following the *Allen* decision, but that voluntary suspension ended on April 12, 1988. Although such Congressional efforts at curbing or limiting patent protection for multicellular organisms are continuing, currently the only express limitation is that patents will not be granted on modified humans under the theory that doing so would create a property right in humans violating the Thirteenth Amendment—the antislavery amendment—to the U.S. Constitution.[26]

C. Creation of the Federal Circuit

Another significant development in the 1980s was the creation of the U.S. Court of Appeals for the Federal Circuit in 1982. Prior to 1982, patent infringement suits were tried in federal district courts and appealed to one of 11[27] regional or circuit U.S. Courts of Appeal. Appeals could be taken from one of those courts to the U.S. Supreme Court, but such appeals were only by permission—i.e., by issuing a *writ of certiorari*, or *cert*. Few patent cases were heard by the U.S. Supreme Court. As a result, the law for a number of years was developed primarily by the 11 regional U.S. Courts of Appeal. That law, in many respects, developed inconsistently from circuit to circuit. Furthermore, the law developed in one circuit was not binding on another circuit. Yet, with the exception of the U.S. Court of Appeals for the District of Columbia, those circuits covered several states and relatively vast territories. "Bad" law developed in one of those circuit courts could thus have widespread impact.

Some of those circuit courts evidenced a virtual hostility toward patents[28] or, at least, that was the perception. Patents in such circuits were frequently held invalid.[29] Other circuits were perceived as favoring (or at least tolerating) patents, and patents in those circuits were more frequently upheld. Thus, a practice called "forum-shopping" developed. Patent owners attempted to obtain jurisdiction for their patent suits in "friendly" circuits, whereas the alleged infringers attempted to move those suits to "hostile" circuits. All of that created substantial uncertainty, and patents rapidly began loosing their value. Companies and individuals spending several thousands of dollars obtaining a patent could find their investment evaporate overnight if they were caught in a "hostile" circuit. As a result, many elected to forego obtaining a patent owing to that diminished value and the uncertainty of protection.

Additionally, even if the validity of a patent was upheld, damages for patent infringement were many times meager. Thus, patent infringement was treated by some as an acceptable business risk. That attitude too contributed to an increasingly general perception that patents ultimately were not worthwhile.

All of that changed in 1982. Prior to that time, two courts, the U.S. Court of Customs and Patent Appeals and the U.S. Court of Claims, shared a building on Layfayette Plaza in Washington, D.C., across the street from the White House. Both courts had nationwide, but specific subject matter, jurisdiction and both courts had developed substantial experience in patent cases. The U.S. Court of Customs and Patent Appeals, as mentioned above, heard appeals in patent and trademark matters from the U.S. Patent and Trademark Office in addition to appeals from the U.S. Customs Court and the International Trade Commission. The U.S. Court of Claims handled claims brought against the government, including claims for patent infringement.

In 1982, the Federal Courts Improvement Act combined both of the U.S. Court of Customs and Patent Appeals and the U.S. Court of Claims into a single U.S. Court of Appeals for the Federal Circuit. In addition to the combined jurisdiction of the two former courts, the new "Federal Circuit" was given *exclusive* nationwide appellate jurisdiction in patent infringement cases. As a result, since 1982 appeals in patent infringement cases from all of the federal district courts go to this single court. With a single stroke of the pen, the inconsistency between the 12 regional courts of appeal in patent cases was eliminated.

Further, the U.S. Court of Appeals for the Federal Circuit began its existence with a judiciary trained in patent matters. Several of the original judges of the Court, including Chief Judge Howard Markey,[30] were former patent attorneys. Circuit Judge Rich, previously on the U.S. Court of Customs and Patent Appeals, in particular had extensive patent experience and was one of the co-authors of the 1952 Patent Act. Additionally, the court maintained (and continues to maintain) a staff of "technical advisors," many of whom also had (and have) strong patent law backgrounds. Some are former patent examiners from the U.S. Patent and Trademark Office.

The result has been a substantial improvement in the consistency of decisions on appeals in patent matters. Also, the Federal Circuit has issued rulings that have greatly increased the financial risk of patent infringement. Sizable damage awards in patent infringement cases have been sustained by the court. As a result, patents today have infinitely more value than a mere 10 years ago. That enhanced value generally means that there are more frequent licensing opportunities and more favorable licensing terms than were available previously.

D. Patent Protection for Computer Software

The final factor, patent protection for computer software, is only tangentially related to biotechnology law, but nevertheless it has had an effect on the issues addressed in later chapters. Patent protection for computer software has had a turbulent past. The principal issue has been whether computer software was statutory subject matter under § 101 of the patent statute; i.e., the same issue that initially prevented patent protection for microorganisms. In the 1950s through the early 1970s, the U.S. Court of Customs and Patent Appeals and the U.S. Patent and Trademark Office struggled with the issue. Then, in 1972, the U.S. Supreme Court decided *Gottschalk* v. *Benson*.[31] That decision has been the subject of a substantial body of critical comment, including various opinions on what the holding of that case was. Nevertheless, in brief, the case probably stands for the rule that a mathematical formula—an algorithm—though implemented on a digital computer, is not patentable subject matter. The Court, however, left the door open for programs using such formulas or algorithms.

Following the *Benson* decision, the U.S. Court of Customs and Patent Appeals decided a number of cases in which they attempted to square the language of the *Benson* decision with the court's view of what the law should be. The issue was again presented to the U.S. Supreme Court in 1978 in *Parker* v. *Flook*.[32] Although the issue is not entirely free from doubt, the Supreme Court seemed to support the proposition generally that software could be patented but held that the software at issue *in that case* was not patentable as claimed. In subsequent cases, the U.S. Court of Customs and Patent Appeals again tried to develop a consistent analysis to determine whether particular computer software was patentable subject matter. But the court was hampered and somewhat constrained by the U.S. Supreme Court's decisions in *Benson* and *Flook*.

In 1981, however, the U.S. Supreme Court considered the issue a third time in *Diamond* v. *Diehr*.[33] The third time was the charm. Now the Supreme Court had a new panel because Associate Justice William O. Douglas had left the bench in 1980. In *Diehr*, in a 5 to 4 decision, the Supreme Court held that computer software *could* constitute patentable subject matter and opened the door for "software patents." Although, the "rules" governing the patenting of software are still evolving, the important point, for the present discussion, is that some, but not all, software is subject to patent protection.

Computer software is also copyrightable subject matter and can be protected under the Copyright Act, administered by the U.S. Copyright Office under the Library of Congress. The U.S. Copyright Office began accepting registrations for computer programs in May 1964 by construing such programs as "books" under the provisions of the 1909 Copyright Act then in effect. In 1976, a new copyright statute was passed (the "1976 Act"), and fully implemented as of January 1, 1978, that expressly included computer data bases and computer programs within the subject matter available for copyright protection. Copyright protection for computer software was further clarified by the Software Protection Act of 1980 that, among other things, added a definition of "computer programs" to the Copyright Act and a clarification of what constituted infringement of a copyright in a computer program. Most recently, the Software Lending Act has enhanced the scope of copyright protection available for computer programs.

It is now clearly established that computer programs in either source code or object code are copyrightable subject matter. Operating systems and "utility-type" programs alike have been held to constitute copyrightable subject matter. The exact scope of protection, however, is an issue that is still evolving. Also, ownership and authorship issues, especially for software produced by independent contractors as "works made for hire," are evolving.

In the research setting, those dual forms of protection for computer software created additional issues that had to be addressed by institutional intellectual property policies and the offices that administered those policies. As Lita Nelsen points out in Chapter 2, universities originally handled computer software in the same manner as other copyrightable works such as journal articles and books. The authors usually retained title. More recently, however, universities have taken the view that computer software is more in the nature of an "invention," and it should be treated like other "inventions" regardless of the method of protection used. That, in turn, brings up issues such as whether software, e.g., sequencing software, developed using government funding should be made available royalty free to other universities and nonprofit research organizations or, indeed, whether such software should be dedicated to the public and made available for use without restriction by industry and academia alike.

VI. THE IMPACT ON THE FIELD OF BIOTECHNOLOGY

The above developments have had a profound impact on the field of biotechnology. Not only is effective protection available under the patent laws for biotechnological developments, but the validity of resulting patents is much more likely to be upheld, and substantial remedies exist to deter infringements. Companies, individuals, and institutions now have the legal tools available for protecting their substantial and continuing investments in the technology. Companies, individuals, and institutions that create properly designed programs for protecting their "intellectual property rights" also have additional "assets" that can attract investment capital, licensing opportunities, and research funding. Indeed, the property rights created or confirmed, depending on how one looks at it, by those developments have virtually necessitated the creation of administrative offices and personnel capable of initiating and administering programs for identifying, reporting, and evaluating research results for proper protection and possible licensing. As noted at the outset, such offices and personnel have become particularly important in universities and nonprofit research institutions.

VII. OVERVIEW

This book is arranged, in general, to follow the stages of innovation identification and development. Thus, in Chapter 2, Lita Nelsen describes the attributes of an effective intellectual property program for research drawing on her expertise with the program at the Massachusetts Institute of Technology and similar programs at other institutions. As she points out, an effective program should encourage the dissemination of research results, provide faculty incentives, provide an avenue to increase industrial research sponsorships, provide needed funding for discretionary programs, and respond to prodding by federal and local governments to use our national resourcefulness in research in enhancing our competitiveness in world markets. Nelsen provides an excellent introduction to the fundamentals of an intellectual property program, including policies on ownership, disclosure, and licensing, staffing and personnel considerations, and an explanation of exemplary invention reporting and evaluation procedures.

In Chapter 3, Reid Adler discusses the forms of legal protection available for biotechnology-related inventions and offers a number of suggestions for choosing the appropriate form of protection. First, however, Adler appropriately sets the stage by tracing the development of "biotechnology law" in parallel with developments in biotechnology, dividing the topic into developments prior to discovery of the genetic code, developments between discovery of the genetic code and the advent of recombinant DNA technology, and developments after the appearance of recombinant DNA technology. With that background, Adler discusses utility patents and the statutory requirements for such patents. He covers the topics of patentable subject matter, novelty, utility, nonobviousness, and enablement. Importantly, Adler gives examples of patented biotechnology-related products that can serve as useful guides in making initial decisions on whether to consider seeking patent protection and, if so, on what. Adler also discusses utility patents on microorganisms, plants (other than under plant patents), and animals, as well as the related topics of plant patents and protection under the Plant Variety Protection Act. Although, as Adler points out, utility patent protection is by far the most prevalent means for protecting biotechnology-related inventions, copyright, trade secret protection, and trademarks are used as well. Adler discusses those forms of protection and gives a number of suggestions for selecting the appropriate form (or combination of forms) of protection.

Once the decision is made to proceed with patent protection, proper drafting of a patent application becomes vitally important. Poorly or improperly drafted applications will prove difficult to navigate through the patent prosecution process in the U.S. Patent and Trademark Office and, even if issued, may prove to be of little value or worth. On the other hand, as Brad Duft explains in Chapter 4, a properly drafted and prosecuted application can be of enormous value, and may become one of the principal assets a company or institution may own.

The inventor, as Duft points out, plays a critical role in the proper preparation of a patent application. Nascent inventors should understand that at the time they prepare their original "disclosure" of their invention, they know more about that invention than probably anyone on the face of the earth. Skilled patent professionals, particularly those with experience in the biotechnology field, will, of course, have some background in the technology. Nevertheless, each "invention" by definition possesses, or should possess, some degree of novelty. It is incumbent on the inventor to provide the patent professional with sufficient "instruction" in the invention (and the field of technology in general) that the professional can bring his or her expertise to bear on preparing a proper patent application. For those inventors that believe that patent applications are simply "legalese," nothing could be further from the truth. Although the actual preparation and prosecution of an application requires a thorough understanding of patent law and practice, patents are intended to be read and understood by individuals of ordinary skill in the art. If the inventor does not understand the application, then there has been a lack of communication somewhere in the process that must be identified and corrected if the system is to work as intended.

Chapter 4 thus provides an excellent description of the patent application process from the time of discovery of an invention to issuance of the patent. In the almost 200 years since passage of the first patent statute in 1793, a variety of rules, regulations, and procedures have developed that can prove daunting to someone new to the field. Duft, however, has parsed those rules, regulations, and procedures to their elemental form, and he has clearly and simply explained those that are fundamental to appreciating the patent application process. First-time patent applicants in particular should review this chapter in detail.

U.S. law has historically placed an emphasis on the rights of individuals, and the laws of intellectual property are no different. Indeed, true innovation has historically occurred frequently (and perhaps most often) in the small business sector. If the issuance of patents is any indication of innovation, 1984 figures prepared by the National Science Foundation reported that 60% of the patents issued since the turn of the century have been to small businesses or individuals.

The Senate Report accompanying the Small Business Innovation Act of 1982 reported that, based on figures for 1953 to 1973, firms with fewer than 1000 employees accounted for approximately one-half of the major U.S. inventions.

Consistent with that rationale, from the very beginning, U.S. patent law has required that the true inventor or inventors be identified on the patent application. Indeed, the inventor or inventors must individually review and sign-off on an application before it is filed, subject to special procedures that are available if the inventor is deceased, cannot be located, or refuses to sign. In all cases, however, the individual inventor or inventors are the patent "applicants," although, of course, the application may be assigned to a company or other legal entity that actually directs and pays for the preparation and prosecution.

The same is not true in most foreign countries. Most foreign countries permit the company or entity employing the inventor to apply for a patent, and in some cases the inventor need not even be named.

In the United States, however, determining proper inventorship is critical for the reasons Ed Lentz explains in Chapter 5. Specifically, a failure to name the proper inventor or inventors can lead to invalidation of an issued patent. Also, and many times of equal importance, determining proper inventorship will affect other aspects of the patenting process, such as the "date of invention," the ability to obtain legal benefits coming from the filing date of an earlier application, and ownership of the invention covered by the application in addition to the ownership of the application itself. Unfortunately, determining proper inventorship is not always easy and is rarely "black and white," as Lentz points out.

Lentz, however, explains the legal attributes of an "inventor"; joint inventorship; how inventorship is determined (including several examples); the effects of the inventorship determination on validity, "date of invention," priority, and ownership; and concludes with an explanation of how improper inventorship may be corrected. After reading Lentz's chapter, researchers should understand, that the rules for naming proper inventors on a patent application are substantially different than naming authors on a journal article. It would be the unusual case, for example, to find that all of the several authors typically identified on a journal article describing a new development, in fact, would meet all of the criteria to be properly named as "inventors" under the law.

An issue somewhat related to inventorship is ownership, although the two are separate issues. In Chapter 6, Bill LaFuze reviews the law governing ownership of laboratory discoveries and work products under several scenarios. First, LaFuze provides an overview of the historical common law[34] rules on ownership. As LaFuze points out, the issue of ownership usually begins by determining the property right in question. There are property rights, or at least potential property rights, in the physical or tangible property typically developed in research laboratories, e.g., cells, as well as property rights in the "intangible" scope of rights the law associates with innovations and discoveries. As an example, when one purchases a book, one acquires a property interest in the tangible book. If the book is stolen, the owner can sue to recover the book and obtain any damages that might have been suffered as a result of the theft. The purchase of the book, however, does not include the "intangible" property rights associated with the book; i.e., the copyright. Thus, the owner of the book does not enjoy the rights of the owner of the copyright and may not lawfully make copies or reproductions of the book, prepare translations, or produce other derivative works based on the book or otherwise take actions that would infringe the rights of the copyright owner. Similarly, one may purchase the "intangible" property right—the copyright—without ever purchasing or acquiring an actual book or the printing plates used to produce the book.

LaFuze also explains the common law rules of ownership applicable when there is no written agreement between the employee and employer. As LaFuze points out, the purpose for which the employee was hired can prove pivotal. LaFuze also explains how written agreements may be used to vary the common law rules and, indeed, clarify the ownership issues. Oral agreements are also covered and, as LaFuze points out, some, but not all, oral agreements may be enforceable. In cases in which the law requires a "writing," however, informal writings, e.g., an exchange of correspondence, can provide sufficient foundation to enforce such an agreement.

Ownership rights of independent contractors, i.e., nonemployees, are also covered in Chapter 6, as are issues of joint ownership and the rights of joint owners. LaFuze addresses those issues in the context of those typically encountered in common biotechnology research agreements. Lastly, LaFuze discusses the issues of ownership under federal legislation applicable to innovations developed using federal funding, by government contractors, under federal procurement contracts, and by government employees.

As noted above and in Duft's Chapter 4, one of the measures of the value of a patent is its ability to cover "infringements" and its ability to withstand challenges to validity after suit is brought. As also noted above, the creation of the Federal Circuit in 1982 has been instrumental in the significant increase in the worth of patents in general due both to increasingly consistent treatment in the courts and to the substantial damage awards that have been sustained. In Chapter 7, Charles Lipsey and Amy Collins provide a thorough explanation of the topic of patent infringement as it applies to the field of biotechnology.

Claim interpretation and the issues involved in determining both "literal infringement" and "infringement under the doctrine of equivalents" are some of the more controversial and complex topics in the field today. The law is continuing to evolve, and the application of rather traditional infringement "rules" in the field of biotechnology is sometimes not all together clear. Lipsey and Collins provide an exceptionally clear discussion using actual patents and cases as examples. Under the law, patents are considered as items of personal property that can be bought and sold, leased or licensed, or abandoned. Lipsey and Collins begin with that precept and use analogies to everyday events to describe the sometimes difficult to understand intangible property rights granted by a patent.

One of the more challenging tasks in determining infringement is claim interpretation. Claim interpretation is subject to a variety of "canons of construction," i.e., various "rules," that are discussed in Chapter 7. Specifically, however, claim construction requires an understanding of the parts of a patent, the specification and claims, and the interplay between the two. Second, claim construction requires an understanding of claim structure—the preamble, the transitional phrase, and the body. Lipsey and Collins provide a clear explanation together with numerous examples illustrating the various forms claims can take. Next, Lipsey and Collins discuss common types of claims for products, processes, and products that are defined by the process used to produce the products (product-by-process claims). The authors also explain how choosing the form of a claim impacts on the ultimate scope of protection. In addition, there are "special" forms of claims, and the authors discuss those special claim forms, namely "Jepson-type" claims, "Markush" claims, and "means-plus-function" claims. It is further typical to draft a series of claims in "independent," "dependent," and "multiple-dependent" format. The authors provide a thorough explanation of each.

Next, the authors in Chapter 7 discuss how, once the claims are properly interpreted, infringement is determined. The authors first discuss literal infringement, including claim interpretation based upon literal or dictionary meanings, based upon the specification, based upon other claims (so-called "claim differentiation"), based upon prosecution history, based upon expert testimony, and other factors. Second, the authors discuss infringement under the "doctrine of equivalents" and the limitations on that doctrine, including limitations imposed by prior art, prosecution history estoppel, and "the reverse doctrine of equivalents." In each instance, Lipsey and Collins provide case examples and illustrations.

Not all uses, however, create liability for infringement. Lipsey and Collins thus discuss the common law and statutory exemptions from infringement for experimentation (which as the authors point out is a exceedingly narrow exemption) and for uses reasonably related to the development and submission of information under federal law regulating the manufacture, use, or sale of drugs or veterinary biological products, typically submissions to the U.S. Food and Drug Administration (FDA).

Lastly, the authors discuss the remedies available for patent infringement including injunctions and damages. Damages, however, can take several forms, and the authors discuss damages in the form of recovering lost profits, including the types of evidence necessary to show entitlement to lost profits. The patent statute provides that the patent owner, assuming that infringement is shown, is at least entitled to a "reasonable royalty." Therefore, the authors discuss the proof requirements for showing a reasonable royalty. The authors close with a discussion on the personal liability of corporate officers, prejudgment interest, recovery of attorney's fees, and increased or treble damages.

In Chapter 2, Lita Nelson introduced the topic of, and explained the necessity for, clear institutional policies on intellectual property and conflicts of interest as part of an effective intellectual property program. In Chapter 8, Katherine Chapman provides an in-depth explanation of intellectual property policies, research agreements, consulting agreements, and conflicts of interest. Although she draws on her experiences at The University of Texas Southwestern Medical Center at Dallas, Chapman also has wide experience in the field and draws on that wide experience in describing features common to intellectual property policies across the United States. As a result, Chapter 8 conveys a firm background for understanding such policies as well as their rationale.

As noted above, industry-supported research at universities and nonprofit research institutions is becoming increasingly more common. It is thus becoming increasingly more important for scientists and researchers to understand the primary features of research and consulting agreements in use today. Contract issues relating to confidentiality, rights to future developments, and restrictions on publications can have a profound impact on daily routine in the research laboratory. Agreeing to accept a company's information "in confidence" and to maintain certain levels of confidentiality may, for example, require separate and locked research laboratories, coolers, and safes to meet those contractual obligations. In extreme cases, contractual obligations could preclude the use of graduate students or technicians in performing the research and result in overwhelming obligations on the research scientist. Broadly or inartfully worded agreements could be subsequently interpreted to preclude further research in a given area even after industry stopped supporting the research. In short, Chapman's chapter is "must reading" for those contemplating a research or consulting arrangement with an industry sponsor.

Lastly, Chapman provides a superb discussion of conflicts of interest—including a discussion of when such conflicts will most likely arise as well as the types of conflicts that are possible. As Chapman points out, some conflicts are unavoidable and minor. Others, however, can have serious and far-reaching consequences. Indeed, the policies themselves can have unintended consequences. For example, broad policies restricting scientists from working on technologies owned by companies in which they or a family member may have an interest would potentially remove significant sources of continued research funding. Accepting a research or consulting contract with industry under some of the broader policies initially proposed by the National Institutes of Health that Chapman discusses could lead to a loss of federal funding on related projects. Again, Chapter 8 should be carefully considered by scientists contemplating a continuing relationship with an industry sponsor.

Joyce Brinton's discussion of biotechnology licensing in Chapter 9 completes the circle. As mentioned at the outset, events beginning in the early 1980s have shaped the field of biotechnology law for the past decade and are likely to continue to shape the field until at least the turn of the century. One of those events is the continued reduction in civilian research moneys available despite the recently announced "biotechnology initiative". Universities and nonprofit research organizations have been forced and will continue to be forced to capitalize on their innovations to replace that reduced funding either through income realized from licensing or from industry-sponsored research incident to licensing.

The field of technology licensing is exceptionally broad and demands a variety of unique skills. The license negotiator must have a command of fundamental contract law in addition to an understanding of intellectual property law, the rules governing inventorship and ownership discussed above, and an appreciation for the merits, scope, and possible uses of the technology. Too, the licensing professional must have good negotiating skills, the ability to persuade, and the many times unique ability to keep the transaction in perspective, refusing to get bogged down in largely insignificant minutiae. As Brinton points out, the license should be the start of a long-term and hopefully fruitful relationship between the parties. The license negotiations should therefore not create lingering animosities or other impediments to a continuing relationship. Brinton brings a wealth of experience to bear on the topic, and she provides a concise yet complete explanation of licensing in the biotechnology field.

License agreements, of course, can take a variety of forms. However, modern license agreements tend to follow a now widely accepted format and include similar topics, although those topics can, of course, be addressed in a variety of ways varying from strongly weighted in favor of the licensor to being strongly weighted in favor of the licensee. In Chapter 9, Brinton discusses each of the principal terms encountered in biotechnology licenses, and she explains those provisions from both the licensor's and licensee's viewpoints. The result is a balanced explanation of a typical license, as well as an explanation of why licenses can be the subject of extensive negotiations.

Brinton also discusses issues unique to licensing by universities and nonprofit research institutions, including the goals of such licensing programs, finding potential licensees, evaluating potential licensees, and negotiating the license. Lastly, Ms. Brinton discusses the importance of developing and culturing the relationship between a licensor and licensee that began with such a license.

VIII. CONCLUSION

Once again, the intent of this book is to provide the research community with a source that explains at least the intellectual property aspects of biotechnology law from initial identification and reporting through licensing. In closing this Introduction, it should be noted that the authors of the following chapters have selflessly taken valuable time away from work, friends, and family to prepare these materials because they share the view that a book devoted to explaining biotechnology law is sorely needed and long overdue. On behalf of all of the authors, we hope that mission has been accomplished.

NOTES

1. BUSINESS WEEK at 66 (March 2, 1992).
2. Idem.; Statement of President Bush on February 24, 1992.
3. The President's Budget for fiscal year 1992 proposed approximately $3.7 billion in funding. That represented an 11% increase over the $3.4 billion allocated in fiscal year 1991. 12 GENETIC ENGINEERING NEWS at 5 (March 1, 1992).
4. The other agencies involved, together with their funding requests, are the Departments of Agriculture ($167.7 million); Commerce ($13 million); Defense ($86.6 million); Energy ($242.7 million); Interior ($5 million); Justice ($2.3 million); Veterans Affairs ($88.4 million); Agency for International Development ($30.7 million); Environmental Protection Agency ($18.3 million); NASA ($44.7 million); and the National Science Foundation ($206 million).
5. Reid Adler in Chapter 3, however, provides an overview of how current ''biotechnology law'' has actually evolved over a number of years with advances in science. His chapter provides a unique insight into how the law as developed in response to changes in science.
6. The need for a change was recognized 2 years earlier, in 1978, during the Carter Administration. In May 1978, President Carter had called for a major policy review of industrial innovation. That resulted in the formation of an Advisory Committee that issued a 300-page report in 1979, a key segment of which contained recommendations on government patent policy. The Bayh-Dole Act was a direct result, albeit during a new administration.
7. 35 U.S.C. § 200.
8. The regulations under the Act define a ''nonprofit organization'' as ''universities and other institutions of higher education or an organization of the type described in section 501(c)(3) of the Internal Revenue Code of 1954 . . . or any nonprofit scientific or educational organization qualified under a state nonprofit organization statute.''
9. The regulations under the Act define a ''small business'' by reference to regulations adopted by the Small Business Administration. 37 C.F.R. § 401.3(g). Those vary by industry. 13 C.F.R. § 121.5.
10. 447 U.S. 303 (1980).

11. A patent covering the product, the bacterium per se, was potentially more valuable than a patent covering only the method of making the bacterium. A patent covering the bacterium per se would be infringed by someone making, using, or selling the bacterium covered by the patent regardless of what process was used to make the bacterium.
12. As pointed out in detail in Chapters 4 and 7, the "claims" of a patent define the invention. Here Chakrabarty had multiple sets of "claims," some of which were drawn to the method of making the bacterium and some to the bacterium per se.
13. That portion of the statute has not changed substantively since the first Patent Act of 1793 authored by Thomas Jefferson. Jefferson defined patentable subject matter as "any new and useful art, machine, manufacture, or composition of matter, or any new and useful improvement thereof." Act of Feb. 21, 1793, ch. 11, § 1, 1 Stat. 318. The current 1952 Act substituted the term *process* for *art*, but that was not a substantive change. The Committee Reports of Congress accompanying the 1952 Act indicated that Congress intended in § 101 to cover "anything under the sun that is made by man." S. Rep. No. 1979, 82d Cong., 2d Sess., 5 (1952); H.R. Rep. No. 1923, 82d Cong., 2d Sess., 6 (1952).
14. Court decisions have established that there are inherent, although unstated, limits on the breadth of patentable subject matter under § 101. For example, the courts have held that the laws of nature, physical phenomena, and abstract ideas are not patentable subject matter. Parker v. Flook, 437 U.S. 584 (1978); Gottschalk v. Benson, 409 U.S. 63 (1973). Nevertheless, products that *use* the laws of nature or physical phenomena *are* patentable subject matter. O'Reilly v. Morse, 15 How. 61 (1853); LeRoy v. Tatham, 14 How. 155 (1852).
15. Those arguing that Congress did *not* intend the patent statute to cover living organisms pointed out that Congress had passed two separate statutes, the Plant Patent Act of 1930 (covering asexually reproduced plants) and the Plant Variety Protection Act of 1970 (covering certain sexually reproduced plants, but excluding bacteria), specifically dealing with "living" materials.
16. Although the statute does not specifically require that an invention be "manmade," it has long been the law that naturally occurring products, i.e., not manmade, are not patentable subject matter. Funk Seed Co. v. Kalo Co., 333 U.S. 127 (1948). On the other hand, naturally occurring products that are changed from their natural state by man *may* be patentable subject matter depending upon the circumstances. American Fruit Growers, Inc. v. Brogdex Co., 283 U.S. 1 (1931).
17. This was not the first case, however, in which the courts had addressed the issue. The U.S. Court of Customs and Patent Appeals (CCPA) had previously held in *In re* Bergy, 563 F.2d 1031 (CCPA 1977), that living microorganisms were patentable subject matter. After that decision, the Supreme Court issued its decision in Parker v. Flook, 437 U.S. 584 (1978), a case dealing with the patentability of computer software. The issues involved in deciding whether computer software was patentable subject matter also arise under § 101 of the patent statute. Thus, in *Bergy*, the Supreme Court agreed to hear the case (i.e., granted a petition for certiorari), but then vacated the CCPA's decision and remanded the case for consideration of that case in light of what the Supreme Court had said in *Flook*. After remand, the CCPA

decided the *Bergy* and *Chakrabarty* cases together. Ultimately, however, the *Bergy* case was dismissed as moot.

18. Here it should be understood that neither the U.S. Court of Customs and Patent Appeals nor the U.S. Supreme Court actually decided whether Chakrabarty's invention was, indeed, patentable. The sole question being decided was whether inventions such as Chakrabarty's were the *types* of inventions that *could* be patented; i.e., whether such inventions constituted patentable subject matter. Chakrabarty and other inventors must still show that their inventions are novel, unobvious, and otherwise meet the overall requirements for patentability. Those requirements are discussed by Reid Adler in Chapter 3 and by Brad Duft in Chapter 4. In particular, an invention must represent something more than what would have been considered obvious to one of ordinary skill in the art at the time the invention was made. 35 U.S.C. § 103. One consequence of the advance of technology and the experience the examiners in the U.S. Patent & Trademark Office have gained with the technology, is that rejections based on "obviousness" are now more frequently encountered. It is expected that such rejections will become even more frequent as the predictability in the biotech arts increases.
19. One of the more significant aspects of the decision was that biotechnology products per se could be protected without regard to the method or process used to produce those products.
20. Even prior to *Chakrabarty*, the U.S. Patent & Trademark Office had issued, perhaps unwittingly, patents on living organisms. In 1873, a patent was issued to Louis Pasteur on "yeast, free from organic germs of disease, as an article of manufacture." In 1967 and 1968, two patents were also issued on living microorganisms. Reply Brief of United States, at 3, and n. 2, in *Chakrabarty*.
21. 2 U.S.P.Q.2d 1425 (Bd.App.&Int. 1987).
22. 1118 O.G. 19 (1987).
23. On an historical note, the Patent Office, as it was called before its name was changed to the U.S. Patent & Trademark Office in the 1970s, originally had but three examiners, who were the Secretaries of War and State and the Attorney General of the United States. All applications were read and discussed by these three men. The first patent was issued in 1790 to Samuel Hopkins for a method of making potash used in making soap. Times have changed. Currently, the U.S. Patent & Trademark Office is part of the Department of Commerce, is headed by an Assistant Secretary of Commerce/Commissioner of Patents, and has a patent examining corps of over 1000. Practice in patent matters before the Office is restricted to registered patent attorneys and patent agents and is governed by a number of rules of practice, set out in Title 37, Code of Federal Regulations (37 C.F.R.) and in a voluminous *Manual of Patent Examining Procedure* (MPEP).
24. The patent operations of the U.S. Patent & Trademark Office are divided into three main categories: electrical, mechanical, and chemical. Within each category, there are examining "Groups," headed by a Group Director, that have responsibility for a broad range of technologies. Chemical groups are numbered in the 100 series, electrical groups are numbered in the 200 series, and mechanical groups are numbered in the 300 series. Within each group there are several "Art Units," headed

GOLDEN GATE UNIVERSITY LAW LIBRARY

by a Supervisory Primary Examiner, or SPE, that have a narrower range of technologies that they are responsible for. Prior to establishing Group 180, cases involving biotechnology were scattered across several groups and art units. One of the reasons for creating a new group having specific responsibility for biotechnology was to more effectively manage biotechnology applications and to increase the experience of examiners with such cases. Group 180 (now Group 1800), consisting originally of approximately 70 examiners, was initially organized into five Art Units having the following areas of responsibility:

Art Unit 181: Chemical apparatus such as analyzers, reactors, and sterilizers; processes of chemical and clinical analysis, sterilizing and preserving; immunology and liquid purifications or separations by living organisms

Art Unit 182: Clinical chemistry, microbiology, immunology and enzymology, purification, and chemical engineering

Art Unit 183: Peptide and carbohydrate chemistry, and drugs, bioeffecting and body-treating compositions containing peptides, carbohydrates, antibodies, antigens, enzymes, or animal or plant extracts of undetermined constitution

Art Unit 184: Multicellular organisms (animal/plant), molecular genetics, cell culture, nucleic acid assays, immunology, hybridomas, molecular biology, microbiology, fermentation, and chemical engineering

Art Unit 185: Molecular genetics, catalysts, enzymology, microbiology, and chemical engineering

At the time of this writing, the Group has been substantially reorganized into 14 Art Units (1801 through 1814).

25. That figure is probably conservative. Under the rules of practice of the U.S. Patent & Trademark Office, the second action by an examiner in an application is normally made "final." In response, a "continuation" application (many times a "file-wrapper continuation application") is filed and the parent application is abandoned. Thus, the application is continuing to "pend," despite the abandonment of the parent, albeit under a new serial number. It is not atypical for "biotech" applications to "pend" a number of years through one or several "continuation" applications.

26. The Office announced:

> A claim directed to or including within its scope a human being will not be considered to be patentable subject matter under 35 U.S.C. 101. The grant of a limited, but exclusive property right in a human being is prohibited by the Constitution. Accordingly, it is suggested that any claim directed to a non-plant multicellular organism which would include a human being within its scope include the limitation "non-human" to avoid this ground of rejection.

1118 O.G. 19 (1987).

27. There were originally 10 regional circuits having jurisdiction over several states plus the D.C. Circuit that had jurisdiction over the District of Columbia. The Fifth Circuit was split in the early 1980s and the Eleventh Circuit was formed thus creating a total of 12 regional circuits.

28. Patents were frequently denounced as "monopolies" that were therefore subject to close scrutiny. Although the U.S. Supreme Court in Northern Pacific Railway Co. v. United States, 356 U.S. 28 (1958), clearly explained why patents were *not* evil monopolies; nevertheless, that perception seemed to pervade decisions by a number of circuit courts.
29. I. Kayton, The Crisis of Law in Patents (1970).
30. Chief Judge Markey was also formerly Chief Judge of the U.S. Court of Customs and Patent Appeals.
31. 409 U.S. 63 (1972).
32. 437 U.S. 584 (1978).
33. 450 U.S. 175 (1981).
34. In the United States, courts follow the rule of *stare decisis*, meaning that court decisions in one case involving one fact pattern may establish law that would be followed by other courts in other cases having similar fact patterns or issues. "Common law" refers to such "judge-made" law as contrasted with statutory law that is created by legislative bodies. In some cases, rules or law developed by court decisions have become so firmly established over time that they become widely adopted by courts as "black letter" law roughly equivalent to statutory law. In other cases, however, the common law developed in one case may have no true applicability in a subsequent case due to the necessarily different fact patterns. In all cases, judge-made or common law is subject to being overruled or limited by subsequent statutory law.

2

Identifying, Evaluating, and Reporting Innovative Research Developments at the University

Lita Nelsen

Massachusetts Institute of Technology, Cambridge, Massachusetts

I. INTRODUCTION

In the United States, university laboratories (and related nonprofit institutions) are the primary source of basic research findings and the basic inventions that arise from them. The challenge to universities and to the government agencies that fund much of the universities' research is to get these discoveries effectively transferred to industry and the market so that the public can benefit. The key to this transfer is intellectual property.

The key position of intellectual property in exploitation of university technology derives from the early, undeveloped nature of university discoveries. At the time the researchers are "finished" with them, most university inventions are still a long way from products. They will require substantial investment of time and money—often at considerable risk of failure—before they are ready for the market. Industrial developers will demand intellectual property protection (and exclusivity) before they will undertake the risk of development. Frequently, the earlier and more basic the technology, the higher the risk, and the more that intellectual property protection is required.

On the other hand, direct dealings between the university and industry ultimately aimed at product development carry with them the risk of conflict of interest. The objectives of the university—discovery of new knowledge and open dissemination for the common good—may be in conflict with the proprietary and profit objectives of companies. Further, dealings with industry may motivate the university to redirect its research to more "practical" (and inevitably short-term) ends, and away from longer range discovery that is the university's primary mission. Finally, even in the absence of any change in behavior of the university, much less any unethical behavior, overzealous watchdogs of government or private agencies may perceive a "smoking gun" at the university based simply on the fact of university/industry dealings.

II. WHY DOES A UNIVERSITY WANT AN INTELLECTUAL PROPERTY PROGRAM?

University intellectual property and licensing programs are proliferating rapidly. Within the past 5 years alone, the effort devoted to such programs has increased probably 10-fold or more. However, only a few such programs are yet profitable, and of those few yielding more than a million dollars a year net to the university, the majority are based on single "blockbuster" patents that arose serendipitously at the university. New licensing programs cannot, of course, expect to model themselves on these exceptions.

Why then is so much effort being spent on this activity (particularly in light of the conflict of interest hassles that may result)? The reasons are many, in addition to the university's perception of its obligation to the public to encourage development of research findings. These include the following reasons.

A. Prodding by Local and Federal Governments

The federal government has identified basic research as one of the major competitive strengths of our nation in world markets and is looking for ways to translate this strength into product competitiveness by increasing the incentives for universities to exploit their intellectual property. The passage of Public Law 96-517 (the Bayh-Dole Act) was a milestone in this effort. Under this Act, title to university inventions arising from research programs funded by the U.S. Government is given directly to the university. Prior to this act, title to such inventions belonged to the government, which had a preference for nonexclusive licensing and lacked internal systems for effective technology transfer. Universities could petition to assume title to selected inventions, but this was usually a long and uncertain process. Each federal agency had different "guidelines" regarding patent clauses in its funding agreement, further complicating the process. A few universities obtained blanket agreements with the government granting them title to all inventions arising from federally funded research, but most universities simply ignored technology transfer opportunities.

Now, under PL 96-517 (and subsequent guidelines and executive orders during the 1980s that removed many of the restrictions on granting long terms of exclusivity that still remained in the Bayh-Dole Act), universities are relatively free to exploit their inventions as they wish, consistent with public interest and providing that preference is given to U.S. manufacturers. This has led to a rapid growth in size and expertise of university technology transfer operations. At the same time, various federal agencies have made it clear that effective intellectual property and technology transfer programs are among the criteria on which they may judge universities in awarding grants, further encouraging the trend.

At the local level, universities are beginning to be seen as a potential source of regional economic development. A recent study by the Bank of Boston, for example, identified at least $10 billion per year in income and 300,000 jobs in Massachusetts attributable to companies formed by alumni of the Massachusetts Institute of Technology. (Although it should be noted, in caution and modesty, that much less than 1% of these stemmed from technology acquired directly from MIT's own intellectual property program.)

Local, regional, and state governments throughout the country are setting up programs aimed at similar technology-based economic development centered around local universities. These programs may include industrial parks, technology incubators, funds for intellectual property programs, direct research and development grants to the university, and so forth. The first step in all of these is development of an effective technology transfer program at the university.

A further federal influence on universities' interests in technology transfer has been the threatened decline in federal funding of research. Particularly in those fields thus far funded primarily by the defense agencies, universities are seeking industrial support of research as a substitute for federal funding. Although it is unrealistic to expect that royalties from intellectual property themselves will replace any significant fraction of research funding, sponsored research funds from industry can be an important part of the university research budget, and getting such funds is often dependent upon the university having an effective policy for transferring intellectual property.

B. Encouraging Dissemination of Research Results

Paradoxically, a good intellectual property policy at a university rather than inhibiting the dissemination of research findings can actually encourage it. Once the information is protected by patents or copyright, the researcher can freely publish it without fear that the ideas will be stolen and exploited by others.

Although it is true that a pending patent application cannot be used legally to prevent a competitor from using the invention, the time scale for development of university "research stage" inventions into products is such that the patents will probably issue before the invention can be brought to market. It may also be advantageous for the university to publicize the fact that it has filed a patent in order both to encourage potential licensees to come forward and to warn off potential infringers from investing in the technology.

Achieving the balance between protecting intellectual property and free discussion of research, however, depends on the university being able to decide *quickly* whether it wants to protect a particular technology and its ability to get the patent application filed quickly. A good intellectual property program will also make it clear who the owners of the technology are and will provide clear paths for licensing and exploitation of the technology.

C. Faculty Incentives

An effective intellectual property program allows the faculty member to share in the economic potential of his or her discoveries without having to leave the university. (Most university programs share between 15 and 50% of the net earnings from patents with the inventors, with the most common fraction being one-third of net.)

In certain fields, of which biotechnology is one, "world class" researchers are also interested in participating in the industrial fruits of their research, frequently through participation in new companies formed around their research results. The more sophisticated faculty have learned that for this process to be successful, their university must have an effective licensing program. The presence of such a program can be a decided plus in competing to attract academic stars to a university.

D. Industrial Research Sponsorship

As government research funds become increasingly hard to obtain, universities are looking to industry to make up the difference. Industrial sponsors usually demand access to intellectual property coming from the research that they sponsor. Unless the university is willing to give ownership of such property directly to the sponsor (which is seldom necessary), the university must have a good intellectual property policy and technology transfer program in place to make the relationship effective and profitable for both sides.

In addition, new industrial research funds may come from intellectual property that the university already owns. These inventions are often in only an embryonic state and require substantial further basic research to understand and develop their potential. Often this research is best done in the university laboratory from which the invention originated. As federal funds become less available, the university may be dependent upon industrial funding to move the invention along. Typically, the company will be granted an exclusive option or license to the original invention under favorable terms if it agrees to sponsor research on the invention at the university. Any "improvement" patents which come out of the research will also fall under the sponsor's option.

Such a relationship between the company and the university is likely to be of long duration. Once the university research stage is completed, the company will undertake product development at its own facilities, often with consulting help from the inventor. Royalties will flow to the university once the product reaches the market.

At MIT we find that even though only a small minority of our licenses are tied into continuing research sponsorship, the total "additional industrial sponsorship funds" that come in through our licensing program are about equal to our total licensing royalty income each year. Though such funds comprise only a few percent of our total research budget, they are often critical in advancing particular inventions beyond the "concept" stage.

E. Income

All university intellectual property programs hope to make a profit in the long run. Almost all universities know the history of the major inventions at Stanford (gene splicing), MIT (synthetic penicillin and magnetic core memory), University of Wisconsin (warfarin and vitamin D), the University of Florida (Gatorade), and elsewhere that have made universities tens of millions of dollars. And many (including MIT) have stories of "the one that got away"—very valuable technology that was never protected (for lack of an effective intellectual property monitoring program) but that ultimately became widely exploited without profit to the university. All intend to be alert when the next big invention comes into their office!

Realistically, universities can expect that, on the average, even a fully developed licensing program will yield royalty income no more than 2 to 5% (at most) of their total annual research budget. However, this income is among the most valuable types of money for the university in that it is "unrestricted funds" not dedicated to any single project. Some of the money goes to the inventors (providing incentives to researchers to stay at the university), whereas the rest may be used at the university's discretion. And, in addition, all can hope that the "big one" will come through soon.

III. WHAT'S PROTECTABLE AND LICENSABLE?

A range of different types of intellectual property arise from biotechnology research and lend themselves to licensing by the university. These include the following.

A. Patents

Patents are by far the most frequently licensed form of intellectual property and are usually the most valuable. As the number of patents filed in the biotechnology area have grown, it becomes increasingly clear which types of innovations will be awarded patents. The strengthening of patents in the United States through the establishment of the Federal Circuit Court of Appeals and the unequivocal ruling that patents are allowable on transgenic animals have both increased the potential value of university biotechnology patents. The change in the law that made it an infringement to import into the United States products made abroad whose process of manufacture fell under U.S. process patents has greatly increased the value of those process patents, and has greatly increased the incentive for universities to file such patents.

B. Biological Materials

Deoxyribonucleic acid (DNA) for new genes, expression promoters, and enhancers, DNA probes, hybridomas, antibodies, plasmids, vectors, cell lines with high expression of certain genes or that display particular receptors, transgenic animals, and so forth, are all examples of physical property which can be licensed. Access to the physical property, even when its description is published, can save researchers, including industrial researchers, many months or years of laboratory effort that would be required to duplicate the material. Sometimes (as in the case of unique antibodies with unusually high binding characteristics and low cross reactivity) the material may be almost impossible to duplicate.

These materials therefore have substantial value in themselves, whether or not patented. Universities are aware of this, and are protecting them as "trade secrets." Even those universities that refuse, in the name of free dissemination of information, to restrict publication of trade secrets in the form of data or other written information have shown a willingness to protect these "physical trade secrets," and there is a considerable amount of licensing of such materials now going on.

There is another paradox concerning licensing of these materials that is troubling many universities. For many of the materials, the licensing value may decrease as the number of people having access to them increases, because the objective is to get a time advantage over competitive researchers. At the same time, current norms in the biological research community (enforced by several of the best scientific journals and by the National Institutes of Health on its grantees) state that new biological materials must be made available to other "basic researchers" in order to allow them to duplicate and build upon published results. It is generally assumed, also, that such materials should be available at little or no cost, at least to other academic researchers.

This paradox is being addressed in a number of different ways at present, and the university community is struggling to evolve standards. Following are some of the interim solutions currently being practiced. None are universally subscribed to. All are usually effected through one form or another of a Material Transfer Agreement. They include:

1. Requiring all recipients except specific licensees to use the material transferred only for "noncommercial" research.
2. Distinguishing between research at nonprofit institutions and even publishable, basic research at commercial institutions. The latter is taken as "commercial" research, and the university may charge a much higher price for the material used in such research or may refuse altogether to release the biological material to the "commercial" investigator.
3. Requiring that recipients agree not to file any patent applications on any results coming from use of the biological material without the permission of the university (or its exclusive licensee).
4. Requiring that recipients agree that any inventions made using the invention be accessible to the university owner or its exclusive licensee. This access may range from assignment of any patents or an exclusive royalty-free license (thereby negating the value of the invention to the inventors) to some sort of option rights or rights of first refusal to royalty-bearing licenses.
5. In some cases, the owners of the technology demand ownership of "derivatives" of the biological property made by the recipient but may admit that "new materials" belong to the inventors (with perhaps option rights to the owner or licensee of the original material).

Crafting of agreements defining these arrangements takes considerable technical knowledge and ingenuity on the part of the negotiators on both sides to arrive at acceptable and unambiguous terms. Often definitions are the most difficult part of the negotiation. *Derivatives* in particular is an important but ambiguous term. Sometimes it is used to mean a modification of the original material and sometimes it means, instead, "products made through the use of but not containing any part of" the original material.

The fairness of the various arrangements made in transferring biological research materials are currently under intensive debate throughout the university community with considerable input from the National Institutes of Health and other government funding agencies. Two interim Universal Biological Materials Research Agreements have been devised by a committee of university and National Institutes of Health (NIH) technology transfer experts under the leadership of Joyce Brinton of Harvard University (see Appendix). These are expected to serve as model agreements between universities and between a universities and industrial research laboratories, simplifying the materials transfer process "under ordinary circumstances." It is recognized by all committee members, however, that these model agreements are, despite their name, far from universal and that particularly valuable materials will still require individually crafted agreements.

An additional issue in the protection of biological property is the requirement of many governmental patent offices that samples of the biological property be deposited at the time the patent is filed or when it issues. The materials then become available to the public within a specified amount of time. In some cases, this availability seriously decreases the value of the biological materials. As a result, some companies and some universities are choosing not to try to patent the materials but to protect them as trade secrets.

C. Software

A surprising amount of software is developed in biological laboratories, and this can represent considerable value to the university. Most of this software can be protected very inexpensively through standard copyright procedures. When the software embodies not only good code but unique mechanisms for handling data, it may also be patentable. This latter is far more expensive but can be important to protect very valuable property.

The following simple guidelines may be used when deciding whether to use only (inexpensive) copyright protection—currently only $20 dollars in the U.S.—or whether to invest in a (perhaps 100-fold more expensive) patent: (1) if the primary value of the software is in the specific code (i.e., the specific embodiment of the program), then copyright protection may be adequate; however, (2) if the value of the software lies in the algorithm (i.e., the calculational methods used to obtain the results), such that different programmers might use the same underlying calculational methods and arrive at identical results through very different, independently written code, then patenting may be advisable.

Universities originally handled software as they did other copyrightable works (such as journal articles and books) written by their faculty, with ownership resting in the authors. In the past few years, however, many universities have decided that software "inventions," whether copyrighted or patented, should be handled just like patents, with title resting in the university.

Software presents many of the same problems associated with biological property when it comes to the trade-off between value from limited or exclusive access and free dissemination to other researchers to build on the results. (It has been pointed out that there are also many physical similarities between biological property and software: They both reproduce themselves; they both can be derivatized, with the derivatives moving farther and farther from the originals, so that over time very little of the original contribution is left; and they can both get "bugs and viruses"—but perhaps that's going too far.)

The dilemma in choosing between free access and exclusivity has been resolved somewhat more easily for most software property. Universities routinely deliver copies of their software, unsupported and in a raw, "as is" state, to other universities at nominal cost (typically, $200 or less), sometimes with the understanding that it is not to be given to third parties. Access for commercial concerns may be limited. One or more distributors may be licensed, who will debug and support the software and sell it to commercial customers (and those non-profit users who need more fully developed and supported product).

There is a great deal of debate now going on (led by the Free Software Foundation and others) as to whether patenting software will ultimately redound to the public good (by allowing universities and their inventors to profit from their work and by providing incentive for commercial investment), or whether it is a pernicious trend that will discourage creative hackers from improving current software. Most universities currently find it appropriate to file patents on software when the technology is sufficiently valuable and unique to warrant the expense.

D. Trade Secrets and Know-How

There is a great deal of know-how developed in biotechnology research, much of which may not fit the usual perception of the breakthroughs of academic research. This know-how can range from the fact that a particular brand of glass fibers makes cells grow better to the apocryphal "left-handed stirring" that makes the reaction go. Generally, such know-how is developed through trial and error, frequently as a side product of more traditional research and over long periods of time. Seldom is know-how fully documented or published, and often it resides only in the researchers' heads. Yet sometimes it is very valuable indeed.

There are a number of ways that universities and companies can take advantage of the know-how arising from university research. The simplest and perhaps most effective way for the company to gain access to such know-how is through people transfer: hiring the faculty member as a consultant (see Chapter 8) or the graduating student as a full-time employee; or assigning a company employee to work in the university laboratory as a "visiting scientist." The close interaction that develops between university and company researchers in a well-run sponsored research program is also a very effective means of know-how transfer, and this transfer is often seen by the company as one of the most valuable products of sponsoring research at the university.

The university may seek to capitalize on its know-how by making it a trade secret, subject to licensing fees. However, a trade secret (in the information sense, not physical property) is only valuable if it is not published. Once publicly available, its licensable commercial value is nil. There is therefore an inherent contradiction between establishing ownership in trade secrets at a university and encouraging the "free and open" dissemination of scientific results.

Many universities, therefore, will not attempt to establish trade secrets and even deny recognition of them in their intellectual property offices. Such universities will also refuse to hold in confidence data developed under commercial sponsorship (except for a month or so in order to get patents filed before publication). Though these universities recognize that they are giving up potentially valuable rights, they believe that their duty to publish and to encourage open exchange of information within the university is more valuable.

IV. FUNDAMENTALS OF A UNIVERSITY INTELLECTUAL PROPERTY PROGRAM

A. Clear Policies

Before beginning any technology transfer/intellectual property program, the university *must* set in place clear policies for such a program. Although policies will change over time as the program evolves, the major issues must be decided in advance. Otherwise, a new program is likely to stall or fail altogether in a morass of committee indecision and policy ambiguity. A particular license deal simply cannot be negotiated simultaneously with evolution of university policies concerning individual clauses of the license agreement!

Among the policy issues that must be decided in advance are discussed below.

1. Who Owns the Technology?

At MIT our intellectual property policy states that the university owns all intellectual property that arises under research grant funding or from significant use of MIT facilities. In some cases, MIT may waive its title in favor of the inventors, but this is understood to be an exception.

Inventions made without significant use of university facilities, not under any research grant, and not as "work for hire" (e.g., software programs written for a specific task by a hired programmer) are assumed to be the property of the inventor. This is the typical Sunday-afternoon-in-the-cellar invention, or the invention made by a student not supported by a grant who makes her invention in the library. Upon request MIT will provide such an inventor with a "nonassertion of rights" form for the invention, making it clear that the university claims no ownership in the invention. Alternatively, as a member of the MIT community, the inventor may choose to have MIT elect title in the invention, handling it as we would any MIT-owned invention and sharing royalties with the inventor. Many inventors choose this latter route because MIT (if it chooses to elect title) will support the patent costs and will market the invention and negotiate the license agreements. MIT cannot accept inventions from inventors not affiliated with the university.

Enforcement of such university policies on ownership of inventions is not altogether straightforward. At MIT, faculty researchers and visiting scientists (including scientists on industry payrolls who are assigned to MIT for a limited period of time) are supposed to sign an Inventions and Proprietary Information Agreement (see Appendix) prior to their beginning work at MIT. We also hold that in the absence of such a signed form (which may have been overlooked in the initial procedures when the scientist came aboard), the policy nevertheless prevails as a condition of being allowed to perform research at the university.

It is highly recommended that universities have such intellectual property forms, and that they require that all researchers sign them as a condition of employment or visiting scientist status. Unfortunately, the case law on who owns the invention in the absence of such a signed invention is not altogether clear for universities because it appears that a researcher's relationship to a university may be somewhat different from an employee's relationship to a company.

2. Is Disclosure of an Invention to the University Mandatory and Who Monitors It?

Most research grants (both government and industrial) require that the university report new inventions to the sponsor. Federal reporting requirements are delineated in 37 C.F.R. SS 401.8-401.14(c). Some sponsorship agreements with companies require that the investigators give draft manuscripts to the sponsor in advance of publication so that the company can identify new inventions and ask that patents be filed. The amount of time by which publication can be delayed is subject to negotiation. (MIT will usually allow no more than a 30- to 60-day delay.)

Compliance with any of these reporting requirements, however, is usually in the hands of the investigators, since unless they report new inventions to the university, no one will know about them. It has been suggested by some that universities implement a review procedure for all publications and other public presentations to prevent inadvertent disclosure of inventions. Some have suggested that the department head perform such reviews, whereas others would leave it to the technology transfer office.

Although such prepublication review is common in industry, it would appear to be impractical in any major university. The immense volume of publications (thousands of manuscripts and hundreds of presentations per year from major research institutions—most of which do not contain anything of commercial value) would require a squad of reviewers. The inevitable delay in publication that would result would be resented by the faculty as an infringement of their academic freedom, and the university would be accused of inhibiting the flow of knowledge to protect its commercial interests.

The only practical way to bring inventions from academic research to light is through education of the researchers. If they understand the potential value of their inventions, believe that the technology transfer office can effectively commercialize such inventions with minimum hassle to the inventors, and understand their obligations to research sponsors, most will remain alert to identifying valuable intellectual property and will bring it to the attention of the technology transfer office.

3. Will the University Allow Publication to Be Restricted (or Banned Altogether) to Protect Intellectual Property?

This is the acid test of the open university. MIT's answer is "under no circumstances will we ban or significantly delay publication to protect intellectual property." Clearly, then, "confidential research" is also prohibited.

As stated above, we will delay publication for up to 60 days by agreement with a sponsor to allow patents to be filed. In the absence of such a sponsorship agreement, our Technology Licensing staff will not ask any delay in publication, but we will aim to get the patent filed "immediately" (in 1 or 2 days if absolutely necessary) to predate an imminent publication.

4. Who Gets to Decide on Licensees and Licensing Terms?

At MIT, this is done by the Technology Licensing staff, with the approval of the Vice President of Research when equity is involved. We will almost always consult the inventor(s) as to the suitability of a particular licensee, and, all things being equal, will often follow the inventor's preference *if* there is no conflict of interest. This deference to inventor's preference is clearly informal and not required by policy, but it is frequently advisable since the inventor is often familiar with the industrial players in their field and the inventor's cooperation in the informal part of the technology transfer is often key to the effectiveness of the whole deal.

The actual terms of the license, however, are negotiated without the inventor's participation.

5. How Will Licensing Revenue Be Shared?

We recommend a firm, cut and dried policy, with a simple formula and very few exceptions. Historically, special deals have usually caused us trouble and resulted in inequities.

MIT's current policy gives the inventor(s) one-third of net royalties (after taking 15% for administration and any unrecovered patenting costs for the case). The remaining funds are shared between the academic departments and the university general fund under a formula involving patenting costs for unlicensed cases that is too complex to document here.

6. How Will Disputes Be Resolved and Who Is the Adjudicator if Informal Procedures Does Not Work?

There are no clean cases. All new technology licensing officers soon learn this maxim. Almost all inventions reported into our office at MIT have at least one messy complication that must be resolved before the technology can be licensed. The complications may include a dispute over co-inventorship (with a student, another faculty member, a sponsor's employee, even another co-inventor's spouse!). Another university may claim co-ownership because its property or funds were used. There may be an outstanding option to a company as a result of initial conception under a sponsorship agreement that long ago expired. The professor's own company may also claim ownership. Et cetera, et cetera—there is a new complication every day and none of them is straightforward to resolve.

Fortunately, most of these disputes can be resolved informally, given skilled licensing officers, particularly if the attempts at resolution are made early—before the technology is licensed and there is money on the table. In the case of inventorship disputes, we will often call in an outside patent attorney to investigate the facts and make a nonbinding recommendation to the disputants and to our office. Frequently, an explanation by the attorney of the difference between contribution of *effort*, which may deserve recognition in the form of coauthorship of a paper or even a share of royalty income, and the intellectual contribution necessary for inventorship, is enough to resolve the issue.

Sometimes, however, whether the issue is over inventorship or something else, informal resolution is not possible. Success (defined as a relatively *quick*, straightforward compromise that all sides can accept as reasonable and that enables the technology to be licensed) then depends on a clear chain of command for resolution. Ideally, single individuals make the decision at each level, and the case comes to a committee only in extremis.

At MIT, the Vice President for Research (who is also the Associate Provost), in consultation with individual department chairs, is highly instrumental. Within the past 5 years, fewer than 5 cases (out of more than 3000 active cases) have had to go further.

B. Staffing an Intellectual Property Program

There are three components to a university intellectual property program staff: patent attorneys, marketing and licensing personnel, and financial and support staff.

1. Patent Attorneys

Universities vary in whether they use in-house patent attorneys or hire outside firms. In-house staff is certainly cheaper, that is, if the attorneys can be isolated such that they are able to devote most of their time to patenting, and not get diverted into everything from a professor's private consulting contract to the university's dispute with the local sewerage authorities.

For multidisciplinary universities, the use of outside patent attorneys has the advantage that different attorneys can be chosen for different technologies (few can handle both molecular biology and aeronautical inventions effectively!). Also, there are no fixed costs if the volume of work declines. Unfortunately, the per-hour costs are much higher.

In using outside attorneys, university technology administrators seek people who are technically skilled and up-to-date on the technology, responsive to the short time frames of a university (particularly when publication looms), and able to deal diplomatically with the sometimes quirky demands of faculty members. Attorneys working for universities should also be ever-conscious of the limited funds of universities, seeking ways to save money in the patenting process. Finally, highest on the "list of angels" are those attorneys who recommend against filing if they believe that the claims likely to issue will be too narrow to be commercially useful.

2. Marketing and Licensing Personnel

Effective technology transfer personnel possess a peculiar combination of skills. They are cross cultural in that they understand and identify with the different constraints, objectives, and even ideals of both the university and industry. They are fascinated by science, yet at the same time are most interested in practical applications and the pragmatism of the market. They are motivated to make deals but are willing to walk away if the deal runs counter to the long-range idealistic objectives of the university. They are inherently compromisers (as you must be if you are going to complete license agreements), yet have a strong, almost absolute sense of ethics that they will use effectively in "gray areas." They are self-starting "closers," who want to make things happen but have the patience to keep prodding a company who may take more than a year to get moving on a license. They must be able to deal with complexity and come up with original solutions when others are stymied. Such personnel have to be able to juggle many assignments at once without confusion.

Above all, the successful technology transfer person is a great communicator. He or she can get to the heart of the issue within a couple of sentences—whether it concerns how a new device works, or why university policy will not allow a concession on the definition of a particular term.

University technology transfer personnel come from many different professions and backgrounds: law, science, engineering, marketing, academia, and industry. No single background is most successful but at least some in-depth technical training and experience is helpful.

3. Financial and Support Personnel

The need for support personnel in a technology transfer office is frequently underestimated (and underbudgeted). This is a paperwork-intensive business and must be recognized as such. Without adequate support staff, the professional staff will spend much of their time word-processing, filing, and looking up records—activities that are neither motivating nor cost effective for such people.

It is also important to recognize the price of success. Licenses are ongoing agreements that must be monitored and nurtured. Licensees make frequent requests of their "partner" (the university) and expect to be attended to. And revenues from licenses must be carefully monitored, with revenue shares to inventors, departments, other universities, and so forth, carefully calculated according to sometimes very abstruse formulas. As the number of licenses increases into the dozens and hundreds (if the university program is successful), these license-maintenance functions become more and more burdensome, requiring professional management and adequate staff.

In our office, the ratio of support staff to licensing officers is slightly more than one-to-one. We do not expect it to decrease despite a new project to make our computer system more effective.

C. The Invention Reporting and Evaluation Procedure

Invention reporting in most universities is voluntary. Individual researchers, perhaps with prodding from senior investigators, identify a potentially important invention and report it into the intellectual property office. If no invention report is filed, the invention may enter the public domain and be unrecoverable.

This invention reporting process is fragile: The incentives for such reporting (patents and royalty income) are all tenuous and long range, and university researchers see this process as "not really their job"—in contrast to teaching, research, and publication. If the disclosure, evaluation, and patenting process are perceived as a hassle interfering with the academic process or "never resulting in anything," researchers may simply choose not to report their inventions.

The results of perceived differences in incentives for invention reporting can be seen by comparing the number of new technology disclosures received in different university intellectual property offices. They range from as high as one disclosure per million dollars of research to less than one-fiftieth of that rate—in universities of comparable size and distinction. The rate of invention reporting correlates well with researchers' views of the effectiveness of their universities' intellectual property programs.

The key to encouraging invention reporting are *simplicity* and *feedback*: Keep the procedure as simple as possible, and keep the researcher informed as to what is happening (even if the results are negative).

1. Timing: Protecting Patent Rights from Prior Disclosure

a. U.S. Patent Law on Prior Disclosure United States Patent Law is more lenient than that of Europe and Japan. It allows a 1-year grace period after publication during which a patent application may be filed. Oral disclosure at any time is not a bar to patenting. However, most university researchers are not aware that a slide presentation at a meeting or a simple informal poster presentation outside the university will count as *written* disclosure and start the clock ticking.

Similarly, few researchers are aware that handing out even two or three copies of a preprint prior to publication without agreement by the recipients to hold the material in confidence qualifies as "publication." MIT recently lost all possibility of patent rights on what may have been a fairly important invention when it was discovered that a few members of our Industrial Liaison Program had been given preprints 5 months before the article was published. Even a single copy of a handwritten document in an obscure library—if it is indexed so that it is at least theoretically accessible to outsiders—can count as a publication.

And finally, it should be noted that publications by others more than a year before patent filing, even if these others were not the inventors, will bar patenting by the inventors. This is particularly important in academic research when a researcher describes his or her results to colleagues who then may, with all good intent, describe and credit the original researcher's results in a publication.

What counts? Seminars and chalkboard sessions outside the university, poster sessions, abstracts, and other informal and formal descriptions of the invention all count as public disclosure if the description is *enabling*: if someone of "ordinary skill in the art" could replicate the invention from the description. Sometimes a very detailed description is required (such as the sequence of a gene), and sometimes the mere title of the invention can give it away.

On the positive side, it has been established that material submitted for publication or grant proposals submitted to government agencies are formally held in confidence by the reviewers, and the submission itself does not count as public disclosure. (However, once the proposal is accepted and the grant awarded, the material may become publicly available and count as a publication.)

In disputes between parties as to who should be awarded a patent, U.S. law awards patent rights to the *first to invent*. Timing of disclosure and of filing of a patent is therefore, at least in theory, less important than in foreign patent law, if the actual dates of conception and reduction to practice of the invention are well documented and if no written public disclosure has taken place. In practice, however, proceedings to establish the first inventor in such disputes are usually long, tedious, and expensive—often as expensive as infringement proceedings.

b. Foreign Patent Law on Prior Disclosure European patent law does not allow a grace period. A patent must be filed in at least one country (which may be the United States, Japan, a European country, or one of a number of other countries) prior to any *written or oral* public disclosure. In the absence of prior public disclosure of the invention by anyone (including the inventors), European patent law awards the patent to the *first to file* a patent application.

Japanese law is similar to European law on these issues except that, under certain circumstances, it will allow a 6-month grace period after publication by the inventor in which a patent application may be filed.

c. Importance of Worldwide Patent Coverage In today's increasingly global marketplace, and particularly in biotechnology where the cost of developing a product is much higher than in most other fields, worldwide patent coverage is usually required to justify investment in the product. Also, start-up biotechnology companies frequently use foreign patent rights in their products as a mechanism for raising product development capital from large foreign companies in return for giving the foreign company marketing rights for the products. Yet, one of the most frequent criticisms of university technology transfer offices heard from biotechnology companies and venture capital firms is that the universities usually do not file foreign patents on their inventions.

Universities should adopt the view that if it is worth filing a patent at all, it is worth preserving foreign patent rights. This means that, *whenever possible, the original (usually U.S.) patent should be filed before the first public oral or written disclosure.*

The excuse for not doing this is often that foreign patents are too expensive for a university in the absence of a licensee to pay for them. This is true *if* one considers the entire cost of the foreign patents over the life of the patents. However, the possibility of obtaining foreign patents can be preserved for several years for no more than a few thousand dollars through the Patent Cooperation Treaty (PCT) mechanisms, and during those years the university can determine whether it is likely to find a licensee to the technology. The procedure for using the PCT process is as follows:

1. The initial patent application is filed in a single treaty country prior to the first public oral or written disclosure. This country is usually the United States for U.S. universities.
2. The patentee is then allowed 12 months to decide whether to file in other PCT countries for a minimal fee per designated country (and a number of "regional offices" that further reduce the designation fee; e.g., the "European" region allows most of the major countries of Europe to be designated for a modest fee.)
3. Later expenditures are significantly delayed according to a preset schedule.

With the awareness of the importance of foreign patents has come a sensitivity about disclosure and an unsureness by researchers as to when and to whom they can describe new inventions. "Can I discuss it with my colleagues at our university?" "What about someone at another university?" "What about publication of a thesis?" "Can I give a poster session? Or a seminar?" "Do abstracts count?" And running counter to this caution is the desire of the university to remain a place of open discussion and free dissemination of information—its primary mission.

Although the patent laws in the United States and in foreign countries are far from completely clear on the issue, we use the following guidelines at MIT to strike this balance.

1. Disclosure within MIT (including seminars open only to MIT staff and students) is encouraged. If the information has not yet been published or a patent filed on it, the speaker should inform the audience that the results should be kept in confidence.
2. Disclosure outside of MIT of inventions that have not been published and on which a patent application has not yet been filed should be done under a confidentiality agreement. We keep ours short and simple, and since we expect fairly prompt publication and/or patent filing, the term of confidentiality need not be overly long.
3. U.S. courts have held that theses that have not been catalogued are not "publicly available." We can therefore arrange that a thesis describing a new invention be held uncatalogued by our library for a limited period of time until a patent application is filed.
4. We will not delay publication for more than a couple of months. Even this brief delay will not be asked if the technology is "hot" and priority of publication is important. If necessary, patent applications can be written and filed within a couple of weeks—or even a couple of days. Mercifully, the latter time-frame is seldom necessary unless the inventor "forgot" to tell us about the manuscript at the time of submission.
5. Once a patent is applied for, the inventor is encouraged to publish and discuss his or her research freely. The inventor is alerted, however, to the fact that the mere filing of a patent application does not prevent third parties from practicing the invention or even selling it. Only an *issued* patent offers that protection. If the patent does not issue, and the invention has been disclosed publicly (or in the case of foreign patent applications, it is published as part of the patenting process), the invention will fall into the public domain. But this is the chance that the university should take to balance its desire to protect its intellectual property against its more important responsibility to publish the results of its research.

2. Protecting University Patent Rights: "Good Hygiene" Lectures

A few simple rules conveyed to all researchers at the university will go a long way in protecting patent rights. These include

1. Keep good notebooks. Document inventions as they occur—including the dates. Have the notebooks read and witnessed by someone who can understand the implication of the invention but who is not an inventor. This is very important if the U.S. patent is disputed by someone claiming that another party was the first to invent. Document both the conception of the invention and diligent and continuing effort to reduce it to practice. (A recent interference in a "hot" area of research was lost because of a few weeks' interruption in diligence.) If the invention is likely to be particularly

important, the researcher may want to have the notebook pages notarized within a few days of conceiving the invention and again upon its full reduction to practice. (Most universities have notaries on campus; if they do not, they should.) Note, however, that notarization will establish only the date and the signature—it will not provide a witness to the invention except in the unlikely event that the notary has the technical ability to understand the invention.

2. It is never too early to report the invention to the university's intellectual property office. It may be too early to consider filing a patent application, but at least the report is documented. The intellectual property office may find it useful to set up a reminder system on early-stage invention reports, contacting the inventors every 6 months or so to see whether the invention has matured enough to file a patent application on it.
3. Researchers should routinely contact the intellectual property office before submitting a publication (or at least simultaneously with submission) if the publication describes a new invention.
4. Beware informal disclosure before publication. Any disclosure outside the university can endanger patent rights. A simple confidentiality agreement with the recipients can protect them. (The intellectual property office should have such agreements readily at hand, and should be willing to negotiate quickly acceptable changes in wording with recipients. Many companies have peculiar, but not particularly significant, variations on such agreements that can tie up communications for months if no one is available to negotiate compromise.)

3. The Invention Reporting Process: Keeping It Simple

Remembering that invention reporting is a largely voluntary process, it is important to keep the process simple. We use a one-page Technology Disclosure form that asks for

1. Title
2. Name of the inventors and where they can be reached plus addresses and social security numbers
3. Sponsors of the research work (and grant numbers so that we can trace the contract)
4. Description of the invention (frequently fulfilled by an appended manuscript)
5. Dates of first public disclosure (past or future)

This is signed by the inventors and witnessed by someone who can read and understand it.

A Practical Note: We find that the best witness for a Technology Disclosure, if he or she is not an inventor, is the principal investigator. In addition to being able to understand the significance of the invention, the principal investigator will often help to identify any problems concerning proper credit to other contributors to the invention, issues with sponsors, and so forth. If the principal investigator is an inventor, the laboratory director or department head may be chosen as the witness.

4. Involving the Inventors

Throughout the process of invention reporting, patent filing, and licensing, the inventor is key. He or she may have little or no industrial orientation or, conversely, may be very entrepreneurial. In either case, however, it is unlikely that a university-stage technology will get successfully developed in industry without the enthusiastic participation of the inventor.

First of all, since the invention reporting process is largely voluntary, the inventor must first identify the invention and contact the intellectual property office. (To be precise, most government and industrial research sponsorship require the university to report all new inventions to the sponsor. In reality, however, only the researchers themselves are usually capable of making such identification, so their cooperation is key.) How does the intellectual property office cause this to happen? There are a few things it can do; namely

1. Educate researchers about the existence of the office and how to interact with it. A few well-publicized examples of licensing success will spur many new invention reports.
2. Keep it simple. Make the invention reporting process as simple as possible.
3. Keep it relatively hassle free. The inventors should not be made to feel that they must justify having filed an invention report. If literature searches must be run by the inventors, make the searches as painless as possible. (We used trained information specialists from MIT's computerized literature search service to work with the inventors in such searches—and we have the charges billed to our office rather than the researchers' accounts. Such little gestures can make big differences.)
4. Keep the inventors informed. If the decision whether to file a patent application seems to be coming out negative, let the inventors know that—and let them know the reasons behind it. Usually, given the same evidence, the inventors will agree with the decision. Sometimes they have additional evidence that will reverse the decision.
5. Keep the patent filing process responsive. Researchers should not have to run after the patent attorneys to keep the filing process moving, although frequently the reverse may be true.
6. Involve the inventors in seeking licenses. The researchers are often the best source of leads to identify companies interested in the technology. And, if no licensees can be found, work with the inventors to determine why not.

Sometimes a little experimental effort to further demonstrate the utility of the invention is all that is needed. Sometimes, of course, the invention turns out to have no real commercial utility, and patenting efforts should be abandoned. Again, the inventor should be kept informed—so as to keep him or her enthusiastic about filing an invention report on the next (perhaps much more important) invention.

V. TO FILE OR NOT TO FILE: EVALUATING THE COMMERCIAL POTENTIAL OF A UNIVERSITY INVENTION

University intellectual property offices encourage reporting of new inventions. Yet each of these invention reports presents a major, sometimes nerve-wracking, decision: Should we file a patent on this? The risks are great, no matter what the decision.

If the answer is no, the university may later find that it gave up a very valuable invention that could have made it millions of dollars.

If the answer is yes, the university is committing itself to patenting expenditures that may total \$20,000 to \$50,000 or more. (Multiply that number by several hundred new invention reports per year in some major universities and the totals become very worrisome indeed.)

Some universities retreat to a somewhat lower-risk position: As a matter of policy, they will not file patents unless they already have a licensee lined up who will assume all patenting expenses.

This position is conservative but results in the university potentially losing many valuable inventions. The great majority of university inventions will not have found licensees at the time they are reported. The university must then either delay publication while it seeks licensees (a process that can take years and therefore delay the dissemination of knowledge) or must let the invention go into the public domain. Many seminal inventions, particularly in biotechnology, may have to "ripen" for 2 to 5 years before their commercial potential is adequately recognized.

Most intellectual property offices must therefore base the decision whether to patent on less easily answered questions: Is the invention patentable? If it is patentable, will it be commercially attractive so that someone will want to license it?

Unless the invention is an exact duplicate of something already known, the answer to the first question will too frequently be positive, because patent law allows patents to be granted to an astonishing array of minor improvements so long as they are "novel, nonobvious and useful." The first question should be rephrased as: Is it likely that this invention will get broad enough claims that it can be protected against competitors?

If the answer to the first question (usually found after a search of the literature and a discussion between a patent attorney and the inventor) is still positive, the second question must be answered; it can be rephrased as: If the claims that seem possible are received, would anyone pay to practice them?'' Answering this question often means predicting the future: of science, of technology, and of business. The problem can be broken down into a series of questions that help in the evaluation:

1. What need in the market might this fill? Is it a minor need or a major one?
2. Is the market miniscule or large (e.g., a few academic astrophysicists or the general public?)
3. How is this market need being filled now?
4. Is this market need recognized now or will it have to be developed?
5. Is anyone investing in that market. Is it a dying field or an unprofitable one or one that is otherwise out of fashion?
6. How does this invention differ from present means of attacking the same problem? And is it not only different but better? In what way is it better? How much better?
7. How sure are we that this invention will work for the purpose intended? How difficult will that be to demonstrate? (Some basic biotechnology patents, for example, are based on laboratory findings whose efficacy in actual medicine will take years or perhaps decades to demonstrate.)
8. What is the nature of likely patent claims? Are they composition-of-matter claims or process claims? If they are process claims, can you prove that someone is using them? (An example of an invention on which MIT did not file patents is a finding that certain cells grow better if the pH is confined within a small range. Since the findings were published and could be practiced by simply adding a few drops of acid or base, we felt that it would be virtually impossible to identify infringers, and therefore the patent would not be practically defendable.)
9. Is this invention so early that any patents obtained from it will expire before products come to market? (This has happened to us with several electronics patents and may someday happen with several of our biotechnology patents.)
10. Is the field moving so quickly that by the time a patent issues the technology will be obsolete. (This is sometimes true for software patents, for example).
11. Is this patent likely to be dominated by someone else's patent? Will the other party be interested in a license or at least grant licenses to the dominating patent to our prospective licensees?

And then there are the other questions concerning the inventors. These are usually unwritten but are of similar importance to the more "objective" questions in predicting the success of a new university invention: Are the inventors recognized in their fields? As leaders or kooks? Do they have a good understanding of the commercial needs in their field or are they inventing solutions to nonproblems? Have their previous inventions been successful? Why not? Are they cooperating in their interactions with potential licensees or do unrealistic expectations, overbearing egos, or just general uncooperativeness prevent any licenses from coming about? Is this invention in a field in which the inventors are actively working, so that it will get further developed, or is this just a flash in the pan?

Usually, none of the evaluation questions are easy to answer. Some may take many months of work to attack and others will be impossible to answer. The newer the technology, the less likely there are to be answers. And meanwhile, manpower for evaluation is scarce (no university can devote months of labor to evaluating a single invention in the disclosure stage) and a publication date is coming, sometimes within 6 months, sometimes within a few weeks or less. What is an intellectual property officer or patent committee to do?

There is no straightforward answer. At MIT, we base our decisions on educated guesses. Our licensing officers are technically trained and have worked in the fields in which they are evaluating inventions. They understand the current markets. Over time in our office, they also develop a sense of which type of inventions are licensable and which are not. (Although this sense must be updated continually. Things that are licensable at one time may later become too common to be worth much. New expression systems in prokaryotic cells are an example of technology that has become progressively less important with time.) Our licensing officers also develop contacts in their fields and can sometimes get a technology confidentially and quickly evaluated by a company in the field, helping us to decide whether to file. Finally, over time we get a better and better fix on which investigators are likely to have licensable inventions and which will be helpful in the licensing process.

Because we have a fairly large budget for patent expenditures, we look at patent filing decisions statistically. It turns out that we decide to file on about 45% of the disclosures sent to us each year. Of those on which we file, about a third get licensed. The remainder are either abandoned or languish in our files.

As mentioned above, only one-third of those inventions on which we file ultimately get licensed. Numerically, then, we can be seen as making the wrong decision two times out of three (and perhaps more frequently if we count those "good" inventions that were never filed.) However, revenue from the one-third that does get licensed more than compensates for the "wasted" patents, so the system seems to work.

VI. FINAL NOTES ON EVALUATION CRITERIA

A. Market Size

It will be noted that no attempt was made to define market size in the evaluation procedure, other than to decide whether the market was miniscule or fairly large. Further precision is seldom useful in making the filing decision. In the licensing process itself, somewhat better precision is needed by the licensor (in order to assess upfront fees and other early considerations), but even there the licensor does not need a fine definition or market size. This is because the major value of the invention to the licensor will presumably come in the form of running royalties based on a percentage of sales rather than a single sum. The licensor with a 5% royalties on sales, for example, is protected whether the ultimate sales are $50 million or $500 million per year.

Licensees, of course, are in a different situation. They must have at least a ballpark estimate of the market size before they can decide whether to invest in the technology, be it the initial license fees or, more importantly, the development funds required to bring the university invention to market. But licensees are usually much better equipped to make such market size estimates because they may be actively working in the market under study or at least have the resources to conduct a full study.

B. Patent Committees

Many universities use patent committees to decide whether to file on a disclosure rather than individual technology licensing officers. Such committees share the burden of the risky decisions and can exert tighter control on patent expenses. They appear to work well when

1. They can meet frequently and can delegate decisions when they are not in session, so that urgent decisions are not delayed pending the next committee meeting.
2. Committee members are knowledgeable in the particular fields of the technology under discussion. (This is relatively easy in a medical school or biological research institute; it is a nightmare in a multidisciplinary university, where a professor of aeronautical engineering may be asked to vote on the commercial potential of an invention in immunology.)
3. There is adequate representation by people knowledgeable about the commercial aspects of the technology.
4. It is clear that the committee is judging the commercial potential of the inventions. This is not a peer review committee judging the academic worth of the technology or its inventors.

If all of these criteria are not fulfilled, however, patent committees may be far less effective than technically trained, market knowledgeable individual licensing officers.

C. Taking Time on the Big Ones

Because of the pressure of patenting expenses, university offices are frequently anxious to get inventions licensed quickly so that the licensee can assume the increasing patent costs. For the ordinary invention, this haste can be beneficial since it also leads to early investment in commercial development of the invention and therefore quicker introduction to the market and earlier running royalties.

For the really "blockbuster" patent, however, and particularly the core technology patent that may form the basis of an entire new industrial segment, such haste can be very dangerous. It can result in tying up the technology with a company incapable of adequately developing the technology or incapable of defending it against infringers.

What is needed in university intellectual property offices is some mechanism for identifying these core technology inventions and purposely handling them differently, viewing them as an investment for the university—and potentially for the public at large. Every invention should be subject to an initial scrutiny that asks: Is this a single product (or product line) invention, or is it a basic technology that could underlie an entire industry? Very few will fall in the second category—perhaps only one every few years in even the largest universities—but those inventions that *might* do so should be treated with care. The university may want to wait a year or two (or maybe more) before committing to any licenses. This will give it time in which to monitor the external world's reaction to publication of the technology, get some feedback from U.S. and foreign patent offices as to the breadth of claims likely to issue, let the science behind the invention ripen a bit, identify some of the potential applications for the invention, and begin discussions with potential licensees.

During this time, the intellectual property office can address such questions as

1. Would it be appropriate from a public responsibility point of view to license this technology exclusively or would it result in too little exploitation? Would exclusivity likely enhance or retard the speed of development of products?
2. Is this technology so basic that it will be developed whether or not intellectual property protection is available to the developers? Will any attempt at granting exclusivity therefore just result in massive infringement and/or endless litigation?

3. Even if exclusivity can be justified from a public policy view, is it fiscally wise for the university, or would a broad nonexclusive licensing program likely lead to wider investment in the technology and more royalties to the university?
4. If exclusivity still seems advisable, are the current bidders capable of exploiting the full potential of the technology? Are they scientifically strong enough? Are they fiscally and managerially strong enough both to exploit the technology and to defend the patents?
5. If exclusivity is granted, should sublicensing be compulsory? Will the licensee agree to mandatory sublicensing provisions that are readily enforceable by the university?
6. If exclusivity is granted, can the university and the licensee agree on strong but realistic development milestones that can be clearly interpreted so that if the licensee does not meet them, the license can be given to someone else?
7. If exclusivity is not justified, can the university support the promotional and administrative costs of a broad nonexclusive licensing program, including defending against infringers? If not, should it seek an agent to carry out the program? Under what terms and how can the university assure performance?
8. Should the university seek the highest royalty rates possible, or will lower royalty rates result in more licensees, more development of the technology, fewer infringers, and ultimately more revenue to the university?

None of these questions can be answered quickly, and while the answers are being developed, patent costs will be mounting. Such a strategic approach to licensing must therefore be confined to a few, exceptionally broad and promising inventions. But for those, the costs and time can be justified by the potential revenue—and by the potential losses if a strategic approach is not taken.

The experience of Stanford University with their strategic crafting and management of licenses to the Cohen-Boyer gene splicing patent should be instructive to all university licensing officers. Stanford chose a very broad, nonexclusive licensing program. Initial fees and annual maintenance fees were low, and royalty rates were very low (one-half of 1%) to early licensees. Early licensees were also given substantial credits for their upfront payments against future running royalties. Royalties were subject to change for those who delayed becoming part of the licensing program. Stanford also made it clear that it was building a legal "war chest" and would pursue infringers aggressively after the patent issued.

After approximately 10 years of this program, Stanford now has over a hundred licensees. Although some practicers of the technology expressed an initial reluctance to take a license, all of those who have been approached by Stanford have eventually agreed to a license, and none has yet had to be taken to court. Stanford estimates that royalties paid on this patent will ultimately total about $100 million. This amount might have been doubled if Stanford had filed for a patent before the initial publication of the technology and thus have been able to get foreign patent protection.

An appendix follows.

APPENDIX

Massachusetts Institute of Technology
Inventions and Proprietary Information Agreement

Name (Please print or type): ______________________________
Social Security No. ______________________________

In consideration of:

* my present or future employment at the Massachusetts Institute of Technology (MIT); and/or
* my participation in research at MIT; and/or
* opportunities made or to be made available to me to make significant use of MIT funds or facilities; and/or
* opportunities to share in royalties and other inventors/author rights outlined in the "Guide to the Ownership, Distribution and Commercial Development of M.I.T. Technology" dated May 24, 1989 (the "Technology Policy Guide"), I agree:

A. to disclose promptly and assign to MIT all rights to all inventions, copyrightable materials, computer software, semiconductor maskworks, tangible research property and trademarks ("Intellectual Property") conceived, invented, authored, or reduced to practice by me, either solely or jointly with others which:

(i) are developed in the course of or pursuant to a sponsored research or other agreement in which I am a participant as defined in Paragraph 2.1.1 of the Technology Policy Guide; or

(ii) result from the significant use of MIT funds or facilities as "significant use" is defined in Paragraph 2.1.2 in the Technology Policy Guide; or

(iii) result from a work-for-hire funded by M.I.T. as defined in Paragraph 2.1.3 of the Technology Policy Guide; and

B. to execute all necessary papers and otherwise provide proper assistance, at MIT's expense, during and subsequent to the period of my MIT affiliation, to enable MIT to obtain, maintain or enforce for itself or its nominees, patents, copyrights or other legal protection for such Intellectual Property; and

C. to prepare and maintain for MIT adequate and current written records of all such MIT Intellectual Property; and

D. to deliver promptly to MIT when I leave MIT for whatever reason, and at any other time as MIT may request, copies of all written records referred to in Paragraph C. above as well as all related memoranda, notes, records, schedules, plans or other documents, made by, compiled by, delivered to, or manufactured, used, developed or investigated by MIT, which will at all times be the property of MIT; and

E. not to disclose to MIT or use in my work at MIT (unless otherwise agreed in writing with MIT):

(i) any proprietary information of any of my prior employers or of any third party, such information to include, without limitation, any trade secrets or confidential information with respect to the business, work or investigations of such prior employer or other third party; or

(ii) any ideas, writings, or Intellectual Property of my own which are not included in Paragraph A. above within the scope of this Agreement (please note that inventions previously conceived, even though a patent application has been filed or a patent issued, are subject to this Agreement if they are actually first reduced to practice under the circumstances included in Paragraph A. above).

This Agreement replaces all previous agreements relating in whole or in part to the same or similar matters which I may have entered into with MIT. It may not be modified or terminated, in whole or in part, except in writing signed by an authorized representative of MIT. Discharge of my undertakings in this Agreement will be an obligation of my executors, administrators or other legal representatives or assignees.

I represent that, except as identified on the reverse side hereof, I have no agreements with or obligations to others in conflict with the foregoing.

________________________ ________________________

WITNESS SIGNATURE
(To include first name in full)

____

DATE

Note: A detailed discussion of MIT's policy related to intellectual property is outlined in the "Guide To the Ownership, Distribution and Commercial Development of M.I.T. Technology" referenced above. This document is available upon request from the Technology Licensing Office.

(To be made out and signed in triplicate. Distribution: original copy to the signatory's personnel file; second copy to the signatory; and third copy to the Technology Licensing Office.)

SIMPLE LETTER AGREEMENT FOR BIOLOGICAL MATERIAL TRANSFER FROM NON-PROFIT TO NON-PROFIT

To: ______________________________ (RECIPIENT)
Address: ______________________________

From: ______________________________ (PROVIDER)
Address: ______________________________

Re: Biological Material identified as ______________________________
____

In response to RECIPIENT's request for the above-identified Biological Material, PROVIDER's institution asks that you and your Institution agree to the following before you receive the Biological Material:

1. The above Biological Material is the property of PROVIDER's institution and is made available as a service to the research community.

2. The Biological Material will be used for research purposes only.

3. The Biological Material will not be further distributed to others without my written permission; however, so that others may replicate or verify your research, I agree, upon request, to provide the Biological Material (subject to its availability) or enabling information to appropriate investigators for that purpose.

4. You agree to acknowledge the source of the Biological Material is any publications reporting on your use of it.

5. The Biological Material is experimental in nature and IT IS PROVIDED WITHOUT ANY WARRANTIES, EXPRESS OR IMPLIED, INCLUDING ANY WARRANTY OF MERCHANTABILITY OR FITNESS FOR A PARTICULAR PURPOSE. You and your Institution agree to assume all liability for damages which arise from your use, storage or disposal of the Biological Material.

6. You agree to use the Biological Material in compliance with all applicable statutes and regulations, including, for example, those relating to research involving the use of human and animal subjects or recombinant DNA.

Please sign and have your Institution sign both copies of this letter and return one signed copy to me. I will then forward the Biological Materials.

PROVIDER

____
(signature)

RECIPIENT: ______________________________
(signature) (date)

Institutional Approval: ______________________________________

(authorized signature) (date)

Name:

Title:

Address:

NON-PROFIT TO NON-PROFIT
UNIFORM BIOLOGICAL MATERIAL TRANSFER AGREEMENT
(UBMTA)

Definitions:

PROVIDER: Institution providing the Original Material. (Enter name and address here:______________________________________)

Provider's Scientist: (Enter name and address here:____________________
______________________________________)

RECIPIENT: Institution receiving the Original Material. (Enter name and address here:______________________________________

Recipients Scientist: (Enter name and address here:____________________
______________________________________)

Original Material: (Enter description):____________________
______________________________________)

MATERIAL: Original Material plus Progeny and Unmodified Derivatives. The MATERIAL shall not include (i) Modifications or (ii) other substances created by the RECIPIENT through the use of the MATERIAL which are not Progeny or Unmodified Derivatives.

Progeny: unmodified descendant from the MATERIAL, such as virus from virus, cell from cell, or organism from organism.

Unmodified Derivatives: substances created by RECIPIENT which constitute an important unmodified functional sub-unit of the Original Material. Some examples include: subclones of unmodified cell lines, purified or fractionated sub-sets of the Original Material when those were known to PROVIDER, proteins expressed by DNA/RNA supplied by Provider, DNA/RNA sequences for a protein with an activity known to or suspected by PROVIDER, monoclonal antibodies secreted by a hybridoma cell line, sub-sets of the Original Material such as novel plasmids or vectors.

Modifications: substances created by RECIPIENT which contain/incorporate the MATERIAL (Original Material, Progeny or Unmodified Derivatives).

Terms and Conditions of this Agreement

1. The MATERIAL is the property of PROVIDER and is to be used by RECIPIENT solely for research purposes at RECIPIENT's institutional facilities only and only under the direction of the Recipient's Scientist. PROVIDER [has/has not] filed patent applications claiming the MATERIAL or uses thereof.

2. The Recipient's Scientist agrees not to transfer the MATERIAL to anyone who does not work under his or her direct supervision at RECIPIENT's facilities without the prior written consent of PROVIDER. Recipient's Scientist shall refer any request for the MATERIAL to PROVIDER. To the extent supplies are available, PROVIDER or Provider's Scientist agrees to make the MATERIAL available under a UBMTA to other scientists (at least those at non-profit or governmental institutions) who wish to replicate Recipient's Scientist research.

3. (a) RECIPIENT shall have the right, without restriction, to distribute substances created by RECIPIENT through the use of the MATERIAL only if those substances are not Progeny. Unmodified Derivatives, or Modifications.

(b) Upon notice to PROVIDER and under a UBMTA, RECIPIENT may distribute Modifications to non-profit or governmental organizations for research purposes only.

(c) Upon written permission from PROVIDER, RECIPIENT may distribute Modifications for commercial use. It is recognized by RECIPIENT that such commercial use may require a commercial license from PROVIDER and PROVIDER has no obligation to grant such a commercial license. Nothing in this paragraph, however, shall prevent RECIPIENT from granting commercial licenses under RECIPIENT's patent rights claiming such Modifications.

4. (a) Ownership of tangible property as between PROVIDER and RECIPIENT is defined in Attachment A.

(b) RECIPIENT is free to file patent applications claiming inventions made through the use of the MATERIAL but agrees to notify PROVIDER if it files patent applications claiming Modifications or uses of the MATERIAL.

5. (a) Except as expressly provided in this Agreement, no rights are provided to RECIPIENT under any patents, patent applications, trade secrets or other proprietary rights of PROVIDER. In particular, no rights are provided to use the MATERIAL or Modifications and any related patents of PROVIDER for profit-making or commercial purposes, such as sale of the MATERIAL or Modifications, use in manufacturing, provision of a service to a third party in exchange for consideration, or use in research or consulting for a for-profit entity under which that entity obtains rights to research results.

(b) If RECIPIENT desires to use the MATERIAL or Modifications for such profit-making or commercial purposes. RECIPIENT agrees, in advance of such use, to negotiate in good faith with PROVIDER to establish the terms of a commercial license. It is understood by RECIPIENT that PROVIDER shall have no obligation to grant such a license to RECIPIENT, and may grant exclusive or non-exclusive commercial licenses to others.

6. The provision of the MATERIAL to RECIPIENT shall not alter any preexisting right to the MATERIAL. If PROVIDER has granted any rights to a third party (other than the customary rights granted to the federal government or non-profit foundations) which would affect RECIPIENT, those rights are listed in Attachment B.

7. PROVIDER shall inform RECIPIENT of any toxicity, health risks, etc. associated with the MATERIAL which are known to PROVIDER. Recipient's Scientist shall inform PROVIDER of any toxicity, health risks, etc. discovered through the use of the MATERIALS.

8. Any MATERIAL delivered pursuant to this Agreement is understood to be experimental in nature, and PROVIDER MAKES NO REPRESENTATIONS AND EXTENDS NO WARRANTIES OF ANY KIND, EITHER EXPRESS OR IMPLIED. THERE ARE NO EXPRESS OR IMPLIED WARRANTIES OF MERCHANTABILITY OR FITNESS FOR A PARTICULAR PURPOSE, OR THAT THE USE OF THE MATERIAL WILL NOT INFRINGE ANY PATENT. COPYRIGHT, TRADEMARK, OR OTHER RIGHTS.

9. RECIPIENT assumes all liability for damages which may arise from its use, storage or disposal of the MATERIAL. PROVIDER will not be liable to RECIPIENT for any loss, claim or demand made by RECIPIENT, or made against RECIPIENT by any other party, due to or arising from the use of the MATERIAL by RECIPIENT, except to the extent caused by the gross negligence or willful misconduct of PROVIDER.

10. This Agreement shall not be interpreted to prevent or delay publication of research resulting from the use of the MATERIAL or Modifications. Recipient's Scientist agrees to provide appropriate acknowledgment of the source of the MATERIAL in all publications and, if requested, agrees to send PROVIDER a copy of any such publications at the time of submission for publication.

11. RECIPIENT agrees to use the MATERIAL in compliance with all applicable statutes and regulations, including, for example, those relating to research involving the use of human and animal subjects or recombinant DNA.

12. (a) This Agreement will terminate on the earliest of the following dates: (1) when the MATERIAL becomes generally available from third parties, for example, through reagent catalogs or from a repository under the Budapest treaty, or (2) completion of RECIPIENT's current research with the MATERIAL, or (3) on thirty (30) days written notice by either party to the other, or (4) on the following date ________________. Paragraphs 8 and 9 shall survive termination.

(b) If termination should occur under 12 (a)(1), RECIPIENT shall be bound to the PROVIDER by the least restrictive terms applicable to MATERIAL obtained from the then-available sources.

(c) Except as provided in 12 (d) below, on termination of this Agreement under 12 (a)(2), (3), or (4) above, RECIPIENT will discontinue its use of the MATERIAL and will, upon direction of PROVIDER, return or destroy any remaining MATERIAL. RECIPIENT will also either destroy Modifications or remain bound by the terms of paragraphs 4 and 5 as they apply to Modifications.

(d) In the event PROVIDER terminates this Agreement under 12 (a)(3) other than for breach of this Agreement or with cause such as an imminent health risk or patent infringement, PROVIDER will defer the effective date of termination for a period of up to one year, upon request from RECIPIENT to complete research in progress.

13. The MATERIAL is provided [with/without] a fee which [is/is not] solely to reimburse PROVIDER for its distribution costs. The amount of the fee, if any, is $________ .

AGREED:

PROVIDER

Institution: ______________________________

Address: ______________________________

____

____

Authorized Official: ______________________________

Title: ______________________________

Signature: ______________________ Date: __________

Provider's Scientist

Name: ______________________________

Title: ______________________________

Signature: ______________________ Date: __________

RECIPIENT

Institution: ______________________________

Address: ______________________________

____

____

Authorized Official: ______________________________

Title: ____

Signature: _______________________ Date: __________

Recipient's Scientist

Name: ____

Title: ____

Signature: _______________________ Date: __________

Attachment A

Belonging to PROVIDER

MATERIAL
- Original Material
- Progeny
- Unmodified Derivatives

Belonging to RECIPIENT

Modifications (however PROVIDER retains ownership rights to any form of the MATERIAL included therein)

Those substances created through the use of the MATERIAL or Modifications, but which are not Progeny, Unmodified Derivatives or Modifications (e.g., do not contain the Original Material or Unmodified Derivatives).

Attachment B

[PROVIDER describes any preexisting obligations that PROVIDER has to third parties (other than the federal government or non-profit foundations) which would affect RECIPIENT.]

3

Choosing the Form of Legal Protection

Reid G. Adler

National Institutes of Health, Bethesda, Maryland

I. INTRODUCTION

An October 1991 study by the Congressional Office of Technology Assessment noted that "many see protection of intellectual property as a paramount consideration when discussing a nation's competitiveness in industries fostered by the new biology."[1] Intellectual property rights, particularly patents, have been especially critical to the development of the U.S. biotechnology industry for reasons of more fundamental importance simply than suing infringers. Such rights are essential business tools; e.g., for fund raising, identifying strategic marketing partners, and recruiting and retaining "intellectual capital" in the form of scientists to conduct research and development.

As a general proposition for biotechnology-related inventions, the selection of intellectual property protection reduces to the question of whether or not to file a patent application. Understanding the history and current trends as well as anticipating future developments in intellectual property law as applied to the field of biotechnology is vital to effective use of the patent system. Science today is entering new frontiers in biomedical research, clinical medicine, and agriculture. The entrepreneurs who transform research results into developed products also must appreciate the strengths and limitations of patent systems internationally, as described in this and several other chapters, because many of these products also will push back the frontiers of biotechnology patent law.[2]

The United States generally is considered to have the most progressive patent system. Because biotechnological and pharmaceutical products are marketed internationally, however, relative deficiencies and imbalances in foreign patent laws create uncertainties that undermine the ability of the U.S. biotechnology industry to plan for and to conduct research and development. Without adequate patent protection in major markets, competitors may be able to copy existing products cheaply rather than develop new products that require a more substantial investment. Thus, patent law also has become an important component, for example, of international trade negotiations. Enjoying the benefits of international patent laws while avoiding their hazards also has become a fact of life for the biotechnology industry.

For some companies in limited circumstances, trade secrecy is a preferable alternative, although this option is not available as a practical matter to universities or government laboratories. For example, the National Institutes of Health (NIH) has a statutory mandate to disseminate knowledge resulting from its biomedical and behavioral research programs. Greater flexibility than patents is afforded by law to plant inventions; however, patent protection is preferable in many circumstances. This chapter describes the various modes of intellectual property that are applicable to inventions and discoveries in the field of biotechnology and provides suggestions about which mode of protection might be favored under various circumstances.

II. THE COEVOLUTION OF "BIOTECHNOLOGY" AND "BIOTECHNOLOGY LAW"

Microorganisms have been applied for millennia to industrial purposes such as baking and fermentation.[3] Plants and animals have been domesticated and bred for human use even longer. Humankind's ability to knowingly manipulate genetic material, though, is of much more recent origin. The granting of exclusive rights as a way of encouraging innovation also is an ancient practice. For example, the doges of Venice and kings of England granted monopolies (from the Greek *monopolion*, right of exclusive sale) to inventors as an inducement for them to trade new articles in commerce that would be of benefit to society.

A less beneficial practice of the English sovereigns involved the granting or selling of monopolies to favored subjects, thereby conferring an exclusive right to sell already-traded commodities (such as salt, iron, vinegar, or playing cards) or to engage in a trade (such as the transportation of beer or the importation of Spanish wool). Monopolies to inventors or merchants were awarded in the form of "letters patent" (from the medieval Latin *litterae patentes*, open letters) addressed to the public at large. The latter use of such patent monopolies, however, restrained professions and commerce in times that previously had been freely enjoyed by the public, typically at lower prices. Eventually, Parliament enacted the Statute on Monopolies to make all monopolies illegal except those awarded to inventors.

The roots of modern biotechnology, however, are much more recent than the patent law. Both the science of genetics and modern agricultural breeding techniques stem in large part from the work of an eastern European monk, Gregor Mendel, whose pea plant research was reported in 1865.[4] Mendel concluded that characteristics varying from individual to individual within a species were transmitted as distinct, inherited traits. Deoxyribonucleic acid (DNA) itself was discovered in 1869,[5] but it was not identified as an agent of heredity until 1944.[6] The discovery of the structure of DNA by Watson and Crick in 1953[7] led to an understanding of, and an ability to manipulate, the DNA genetic code.

A. Developments Prior to Discovery of the Genetic Code

From its creation in 1790, the American patent law reflected the purpose of stimulating the manufacturing industry to provide sorely needed tools and machines for a nation with both unlimited land and shortages of manpower. The general patent law of the United States evolved over time with occasional statutory codification of court-made law. Most recently, the patent law was itself recodified and somewhat revised in the Patent Act of 1952 (Patent Act) but certain provisions from 1930 creating plant patents were not significantly changed. Although these requirements are discussed in greater detail below, the Patent Act requires that an invention—to be patentable—be of appropriate subject matter, novel, useful, and nonobvious. Additionally, a patent application must provide sufficient description and guidance to skilled workers in the field of an invention to "enable" them to make and use a patented invention.

Developments in the law of intellectual property with respect to protecting rights of invention in living organisms has paralleled technological developments in biology, with progressively decreasing lag times. The Convention of Paris for the Protection of Industrial Property (which the United States joined in 1883)[8] defined industrial property[9] to extend "not only to the products of industry in the strict sense but also to agricultural products (e.g., wines, grain, fruit, cattle) and mineral products that are put into trade.[10] Subsequent revisions to the Paris Convention and various European statutes and court decisions addressed the need for protection of agricultural advances and the propriety of the patent system. No uniform practice among the nations was apparent.

By 1906, some members of Congress felt it desirable for U.S. agriculture, which had entered into a scientific industrial phase, to receive the same benefits from the patent system as did industry.[11] Two obstacles were perceived. First, plant varieties were thought generally to be unpatentable as products of nature,[12] notwithstanding the time, expense, and application of human intellect required to produce a novel plant variety. Second, plant inventions were thought not to be capable of description in writing, so that an inventor could not comply with the fundamental quid pro quo of enabling workers to make the invention for which patent protection was sought. To circumvent these perceived limitations, Congress ultimately enacted the Plant Patent Act of 1930[13] (hereinafter the Plant Patent Provisions), by which Congress expressly extended the patent system to the plant agricultural sector:

> No one has advanced a just and logical reason why reward for service to the public should be extended to the inventor of a mechanical toy and denied to the genius whose patience, foresight and effort have given a valuable new variety of fruit or other plant to mankind. . . . This Bill is intended not only to correct such discrimination, but [also to stimulate] invention. . . . [14]

The foregoing reasoning is equally valid for today's agricultural sector, which is poised on the brink of a new era of productivity driven in turn by the research and development efforts of science and industry.

As Congress noted in 1930:

> Today, the plant breeder has no adequate financial incentive to enter upon his work. A new variety once it has left the hands of the breeder may be reproduced in unlimited quantity by all. The originator's only hope of financial reimbursement is through high prices for the comparatively few reproductions that he may dispose of during the first two or three years. After that time, depending upon the speed with which the plant may be asexually reproduced, the breeder loses all control of his discovery. Under the bill the originator will have control of his discovery during a period of 17 years, the same as industrial patent. If the new variety is successful, the breeder or discoverer can expect an adequate financial reward. . . . It is hoped that the bill will afford a sound basis for investing capital in plant breeding and consequently stimulate plant development through private funds.[15]

The same federal interest—but with greater direct federal involvement—in stimulating private enterprise applies today,[16] as do the concerns recognized in 1930 of making new technology widely available at lower initial market prices.

The Plant Patent Provisions modified existing patent law by designating asexually reproduced plants (i.e., reproduced by grafting, budding, cuttings, layering, and division) as patentable.[17] The right granted by a section 161 plant patent (the provisions are now codified at sections 161–64 of the Patent Act) is "the right to exclude others from asexually reproducing the plant or selling or using the plant so reproduced."[18] A Section 161 plant patent need not comply with the stringent "enablement" requirements of 35 USC § 112 because the "claim" is extremely limited, being restricted to the plant that is "shown and described" by photograph appended to the patent.[19]

The Plant Patent Provisions are silent as to the possibility that the identical subject matter could be patented under Section 101 of the Patent Act because, as discussed above, it was thought that Section 101 could not apply to plants because plants are products of nature and because Section 112 enablement requirements could not be met. Sexually reproduced seed plants were excluded from the Plant Patent Provisions because it also was believed that plants could not be stably reproduced by seed for commercial purposes.[20] Therefore, protection against unauthorized sexual reproduction of plants was not considered important. By court decision in 1940, bacteria were excluded from protection under the Plant Patent Provisions because bacteria were not plants as understood or contemplated by Congress.[21]

Between 1940 and 1950, various European countries had adopted differing approaches to protecting plant-related inventions. For example, Czechoslovakia (1921), Holland (1946), and Austria (1946) enacted plant breeders' protection laws or plant variety registration systems; France (1922) and Germany (1934) enacted breeders' rights laws but also issued patents on plants; other countries, such as Italy (1951), issued patents for plants without adopting a special breeders' rights law; and Sweden, Hungary, and Japan have also issued plant patents.

The U.S. Patent Law, as noted above, was recodified and revised in part by the Patent act of 1952 without significantly changing the Plant Patent Provisions of 1930. Contemporaneously, the desirability of an international accord to protect plant inventions was being debated by Western European nations. Ultimately, an international conference was convened, resulting in the creation of the Union for the Protection of New Varieties of Plants and the adoption of the International Convention for the Prevention of New Varieties of Plants (referred to by their French acronym as UPOV and the UPOV Convention). The UPOV Convention became effective in 1968.

Also in the late 1950s and early 1960s, European nations were negotiating a treaty that culminated in the establishment of a European patent system. Because of the difficulties in reconciling the varied national protection schemes for plants, the resulting Strasbourg Convention on the Unification of Certain Points of Substantive Law on Patents for Inventions (1963) permitted contracting nations to "refrain if they chose, from granting patents on plant or animal varictics or essentially biological processes for the production of plants or animals."

During the unification process, a committee of experts on patents recommended that European patent harmonization efforts defer to the UPOV for the development of a separate convention relating to plant varieties. Efforts to incorporate patent protection for living organisms into the European Patent Convention of 1973 were abandoned in the interest of expediency in achieving a harmonized regional patent accord. European national patent laws, revised since 1973, exclude living organisms from patentability in an apparent effort to harmonize with the European Patent Convention, although it never decided the issue. Current trends in European patent law, however, generally reflect a desire to catch up to the United States in expanding the scope of patent protection to provide maximal support for biotechnology industries.

1. Developments Between Discovery of the Genetic Code and the Advent of Recombinant DNA Technology

On the technical front, the complete genetic code was determined in the 1960s, thereby providing the Rosetta stone for biotechnology. At the same time, the commercial importance of sexually reproduced (seed) plants became appreciated. Congress and the president then began to contemplate providing statutory protection for the breeders of sexually reproduced plants.

The President's Commission on the Patent System (hereinafter the Patent Commission) issued a report in 1966 that favored protection for plant inventions but recommended that all plant protection be deleted from the Patent Act and instead be provided for by an alternative protection scheme. The Patent Commission acknowledged the benefit of breeders' efforts, but it concluded that patent applications for Section 161 plant patents could not be examined appropriately for compliance with the novelty, utility, and nonobviousness criteria of the Patent Act. This viewpoint echoed the conventional wisdom of the time that no living organisms could satisfy patent requirements.

Ultimately, a bill to expand Section 161 was proposed in 1968. The bill was not adopted owing to a perceived significant difference of opinion over the propriety and feasibility of extending the patent system to seed-reproduced plants as opposed to enacting an alternative mechanism to stimulate private development efforts. The UPOV Convention had been negotiated by this time, although the United States would not join it for over a decade. A major dissatisfaction with widening the scope of Section 161 centered on the distinction between sexually and asexually reproduced plants. Sexually reproduced plants are not genetically or phenotypically (i.e., in appearance) identical from generation to generation. Asexually reproduced plants, however, are genetically identical because they are reproduced from pieces of the parent stock. Thus, an additional objection to expanding Section 161 may have been that the protection offered by Section 161 for plants shown and described by a photograph would be rather limited.

After the concerned trade associations and other interested parties reached agreement on proposed legislation, the Plant Variety Protection Act (PVPA) of 1970 was enacted to extend patentlike protection to sexually reproduced plants. The PVPA authorizes the issuance to plant breeders of variety certificates that are analogous to the patents issued to inventors. Although applications for Section 161 plant patents benefit from a relaxed enablement standard, those claims must still satisfy the novelty and nonobviousness requirements of the Patent Act. However, the PVPA further relaxes the breeder's burden. For example, varieties protectable under the PVPA must be novel (i.e., distinct, uniform, and stably reproducible) but need not be nonobvious as required by the Patent Act. Following enactment of the PVPA, plant breeders enjoyed the option of applying for a variety certificate or, when the nonobviousness standard could be met, a Section 161 patent. However, the scope of protection under the PVPA also differs from patent protection under Section 161. A variety certificate protects against unauthorized selling, reproducing, importing, or exporting but does not preclude the use of a protected variety for development of other varieties. However, okra, celery, peppers, tomatoes, carrots, and cucumbers were expressly excluded from coverage by the PVPA. The reasons given were that (1) the vegetable industry conducted its own research on these varieties and there would be no useful purpose by covering them because these varieties were not sold in commercial channels; and (2) the large soup companies felt that prices of these six canning vegetables would increase as a result of protection.

2. Developments After the Advent of Recombinant DNA Technology

By 1970, a restriction enzyme capable of cutting DNA was isolated and, in 1972, a DNA ligase, an enzyme capable of joining DNA fragments, was first used to produce recombinant DNA molecules. In 1973, foreign DNA fragments were inserted into a cell via a plasmid,[22] and in 1974, the first gene from a foreign species was expressed in bacteria. Microorganisms per se were still considered unpatentable, although microorganism-related inventions such as antibiotics or other metabolic products were patentable.[23] Concurrently, Chakrabarty's patent application, claiming protection for a bacterium modified to digest petroleum, was working its way from the U.S. Patent and Trademark Office (PTO) to the U.S. Supreme Court.

As early as the 1950s, it had become evident that other scientist with access to the necessary cell cultures could reproduce microorganism-related inventions from patent specifications. The deposit practice was approved by the Court of Customs and Patent Appeals in 1970. To facilitate the patent process for these inventions, the United States, in 1978, joined the Budapest Treaty on the International Recognition of the Deposit of Microorganisms for the Purposes of Patent Procedure (hereinafter the Budapest Treaty). This treaty established uniform international procedures and criteria for repositories of biological material.

In 1980, the United States acceded to the UPOV Convention and subsequently amended the PVPA, in part to include the vegetables previously excluded, although discussions of recombinant DNA technology apparently did not enter into the process. Also in 1980, the Supreme Court decided *Diamond* v. *Chakrabarty*, holding that a living microorganism was encompassed by the categories of patentable subject matter set forth in Section 101 of the Patent Act. This decision came 10 years after the German Federal Civil Supreme Court's holding in 1969 that animals could be patented if the process for their production could be repeated, and after Hungary's legislation in 1969 providing that animals could be patented if they were "distinguishable, novel, homogenous and stable." The Canadian Patent Appeal Board followed after *Chakrabarty*, stating in 1982 that living organisms could be patented. Higher life forms also would be patentable in Australia.

Congress by 1980 had not contemplated the patent-related aspects of modern genetic engineering technology. In its *Chakrabarty* opinion, however, the Supreme Court noted that Congress had plainly contemplated that the patent laws would be given "wide scope" by wording 35 USC § 101 with such "expansive terms as 'manufacture' and 'composition of matter,' modified by the comprehensive 'any.' " The Supreme Court rejected arguments that enactment of the Plant Patent Provisions and the PVPA evidenced a Congressional understanding that Section 101 excluded living things. Rather, the Court noted that the original impetus for the plant-specific acts was the then-conventional wisdom that plant inventions could not satisfy the requirements of the patent law. The Court in *Chakrabarty* held that the broad congressional goals sought to be achieved by the Patent Act required a broad construction of Section 101. Thus, the Supreme Court concluded that Congress intended statutory subject matter to "include anything under the sun that is made by man."

By statutory definition, all patentable inventions must be novel and nonobvious. There is no indication that Congress intended the Commissioner to determine that some technological inventions were nevertheless unpatentable owing to the degree of their novelty or possible social impact. In fact, as long as the other requirements were satisfied, an applicant, according to 35 USC § 102 is *entitled* to a patent if the invention is novel. Recognizing these circumstances, the *Chakrabarty* opinion concluded that the broad language of Section 101 was selected precisely because such inventions were often unforeseeable. The Court specifically stated that

> A rule that unanticipated inventions are without protection would conflict with the core concept of the patent law that anticipation undermines patentability. . . . [T]he inventions most benefitting mankind are those that "push back the frontiers of chemistry, physics, and the like."[24]

By the early 1980s, developments in biological technology were accelerating. The first human recombinant DNA pharmaceutical (insulin) was approved in 1982 by the U.S. Food and Drug Administration (FDA). A foreign gene was first expressed by a transgenic plant in 1982, and a transgenic mouse expressing a rat growth hormone gene was also reported in that year.

Notwithstanding the intention of Congress (stated as early as 1930) that plant inventions should receive the same benefits from the patent system as mechanical inventions, as well as the broad import of the *Chakrabarty* decision, the PTO adopted a conservative approach to patentability of living organisms under Section 101. Although the Supreme Court held that the Plants Patent Provisions and the PVPA did not implicitly preclude the patentability of microorganisms under Section 101, the PTO determined that Section 101 patents for plant inventions had been preempted by these plant-specific statutes. The Commissioner expressly awaited judicial approval before issuance of Section 101 plant patents. Quasi-judicial authorization was granted by the PTO Board of Appeals and Interferences (the PTO Board) in *Ex Parte Hibberd*, a Section 101 patent application involving modified maize plants. The PTO Board, relying largely on the *Chakrabarty* opinion, flatly rejected the Commissioner's narrow construction of Section 101.

Despite *Chakrabarty* and *Hibberd*, the Commissioner continued to refuse patent applications on animal inventions until the PTO Board decided *Ex Parte Allen*, which ruled that polyploid (i.e., containing multiple sets of chromosomes (oysters) were patentable subject matter. Several days later, on April 7, 1987, the Commissioner announced that animals thereafter could be patented. This decision was extremely controversial and raised a firestorm of protest from certain interest groups. The issues of concern, however, do not generally relate to patenting per se but rather to animal welfare, environmental risks, ethics, and changing demographics in the agricultural sector where small family farms, unable to acquire leading-edge technologies, are not competitive with larger industrialized operations.

As genetic research developed over the past century, congressional efforts followed to enfranchise agricultural technology through patents or analogous protection schemes. Enactment of the Plant Patent Provisions and the PVPA was motivated by the then-contemporary understanding that agricultural technology could not accommodate the patent system's novelty and enablement requirements. The march of technology has since made that accommodation possible. Thus, the United States' accession to the UPOV Convention and the Budapest Treaty evidenced an awareness by Congress of advances in agricultural and biological sciences as well as reflecting a legislative initiative to extend the benefits of the patent system to such developments. Extension of patent protection to animal inventions is consistent with the statutory construction noted in *Chakrabarty* and the expressed intention of Congress to stimulate agriculture innovation.

III. THE FORMS OF INTELLECTUAL PROPERTY

The term *property* generally is associated with physical objects, such as household appliances or with real estate, for which ownership and associated rights are protected by law. "Intellectual" property, on the other hand, is intangible, encompassing the right to exclude others from making, copying, or using a product without permission. Intellectual property has incredible commercial value in many instances; e.g., Amgen's patent on erythropoietin and the Coca-Cola

Bottling Corporation's internationally famous trademark. The most widely utilized forms of intellectual property are discussed in more detail below and include patents, plant patents, plant variety protection certificates, trade secrets, copyrights, and trademarks.

The most useful and generally utilized form of intellectual property protection for the products of biotechnology, except for plants, is that provided by the general patent law through "utility" patents. As discussed above and detailed below, special modes of protection exist for plants.

A. Utility Patents

Under U.S. law, a patent represents a federal grant of the "right to exclude others from making, using or selling" an invention or discovery for a 17-year period, issued through the U.S. Patent and Trademark Office (PTO) of the Department of Commerce. A patent holder (patentee) may exercise this right by suing an infringer to obtain an injunction against continued infringement and to recover compensatory damages. Note that a patent does not convey an unconditional affirmative right to make, use, or sell an invention. Other federal or state laws may restrict the practice of an invention; for example, a patented pharmaceutical is still subject to the regulatory purview of the FDA before its clinical use is permitted. Similarly, a genetically engineered microbial pesticides would require Environmental Protection Agency permits under the federal Insecticide, Fungicide and Rodenticide Act before they are disseminated, and the physical location of companies manufacturing genetically engineered products may be regulated by municipal zoning authorities.

1. Statutory Requirements

By law, a patent contains a statement of the invention's title and of the grant of rights to the patentee, to which a copy of the specification is attached. The patent specification resembles most scientific journal articles in that it usually contains an abstract, a general discussion of the problem that the invention solves, related developments in the field, a "preferred embodiment" of the invention, and a written description that includes specific examples setting forth materials, methods, and conditions necessary to recreate the invention. The specification is also the patent application as originally filed, together with any amendments made to it during "prosecution"—the period in which an application is examined by the PTO to ensure compliance with statutory provisions.

The specification concludes with one or more "claims." These are various legal descriptions of the invention defining subject matter over which the patentee is entitled to assert his or her property right. A claim is analogous to a description in a deed of the location and dimensions of a tract of land. In short, it defines the scope of protection afforded by the patent. Selected claims from several key biotechnology patents are reproduced in Table 1.

Table 1 Selected Claims from Key Biotechnology Patents

1. The "Cohen and Boyer" U.S. Patent No. 4,237,224, issued on December 2, 1980 and is entitled "Process for Producing Biologically Functional Molecular Chimeras."

> Claim 1: A method for replicating a biologically functional DNA, which comprises: transforming under transforming conditions compatible unicellular organisms with biologically functional DNA to form transformants; said biologically functional DNA prepared in vitro by the method of:
>
> > (a) cleaving a viral or circular plasmid DNA compatible with said unicellular organism to provide a first linear segment having an intact replicon and termini of a predetermined character;
> >
> > (b) combining said first linear segment with a second linear DNA segment, having at least one intact gene and foreign to said unicellular organism and having termini ligatable to said termini of said first linear segment, wherein at least one of said first and second linear DNA segments has a gene for a phenotypical trait, under-joining conditions where the termini of said first and second segments join to provide a functional DNA capable of replication and transcription in said unicellular organism;
>
> growing said unicellular organisms under appropriate nutrient conditions; and isolating said transformants from parent unicellular organisms by means of said phenotypical trait imparted by said biologically functional DNA.

2. The Amgen "Erythropoietin" U.S. Patent No. 4,703,008, issued on , 1987 and is entitled "____."

> Claim 2: A purified and isolated DNA sequence consisting essentially of a DNA sequence encoding a human erythropoietin.

3. The "Chakrabarty" U.S. Patent No. 4,259,444, issued on March 31, 1981, and is entitled "Microorganisms Having Multiple Compatible Degradative Energy-Generating Plasmids and Preparations Thereof."

> Claim 1: A bacterium from the genus *Pseudomonas* containing therein at least two energy-generating plasmids, each of said plasmids providing a separate hydrocarbon degradative pathway.

The claims and specification are first judged by the PTO during the examination of the patent application, known as "prosecution," against the standards set forth in the patent statute, the basic provisions of which are excerpted in Table 1 and discussed in greater detail as follows. Far greater than 90% of all patent applications initially are rejected by the PTO as not satisfying the statutory standards for patentability. During the examination process, which involves a series of exchanges of views between the PTO and the applicant, issues are clarified and, frequently, amended claims are allowed. Note that patents carry only a rebuttable presumption of validity, so that their allowance by the PTO may be challenged and effectively second guessed through litigation. Recent statistics from the U.S. Court of Appeals for the Federal Circuit indicate that almost one-third of patents challenged are held invalid!

a. Patentable Subject Matter Section 101 of the patent statute sets for three requirements for patentability: utility, novelty, and statutory subject matter. The evolution of patentable subject matter, first to encompass living microorganisms and then plants and animals, has been described above and is summarized again below. Defining the limits of statutory subject matter as applied to biotechnological products has been problematic and required judicial action. The requirement for utility has been broadly interpreted and essentially excludes only things that are "mischievous or immoral," as discussed below. The requirement of novelty, although mentioned here, is applied under Section 102, which is discussed below.

b. Novelty Novelty is a criterion of Section 102 of the patent law that details situations in which the novelty of a claimed invention within the defined classes of statutory subject matter may be determined. The purpose of this criterion (and that of the nonexcerpted provisions of Section 102) is to prevent the granting of patents on applications claiming subject matter that already belongs to the public and to encourage the prompt filing of patent applications after their contents have been disclosed through publication in scientific journals or conference presentations. Any claim in a patent application that encompasses (patent attorneys say "reads on") something already in the public domain (known as "prior art")—whether machines, organisms, processes, conference abstracts, or published articles—is *anticipated* and thus unpatentable, subject to a 1-year grace period in the United States.

c. Utility Utility generally is considered to be a threshold requirement, understood to be a relatively low hurdle to leap so that subjective concerns about how useful is the invention do not unreasonably constrain applicants for patent rights. The seminal Supreme Court case in this area of law is Brenner v. Manson, 148 USPQ 689 (1966). In that case, Manson sought patents on a method for making steroid analogues, but the analogues themselves had no known use and were of interest largely because of their similarities to compounds with anticancer activities. The Court held that

> . . . unless and until . . . specific benefit exists in currently available form, there is insufficient justification for permitting an applicant to engross what may prove to be a broad field.

This decision has been loosely quoted as a proscription against the patenting of "research" reagents, although this characterization is misleading. Although the compounds in *Manson* were characterized as being useful only for research purposes, in that case the research was necessary to determine some function for the compounds themselves.

Note that proven biological activity in humans is not necessary to demonstrate the "specific benefit" required by the *Manson* decision. For example, in several decisions of lower courts, including Nelson v. Bowley, 206 USPQ 881 (1980), the U.S. Court of Customs and Patent Appeals accepted the use of in vitro tests to demonstrate utility for a potential pharmaceutical product. The court reasoned, in part, that "it is crucial to provide researchers with an incentive to disclose pharmacological activities in as many compounds as possible," so that "we conclude that adequate proof of any such activity constitutes a showing of practical utility." These cases that interpret *Manson* were not considered by the Supreme Court; thus, there is a risk that on appeal the *Manson* test for utility might be expanded in ways that overrule the post-*Manson* lower court holdings.

This issue of utility is particularly relevant to the biotechnology industry because much of the modern biotechnology research (that is publishable) is conducted at early stages of investigation that predates the determination of biological utility or arguably fails to satisfy a strict interpretation of *Manson*. For example, the National Institutes of Health filed a patent application in 1991 claiming complementary DNA (cDNA) sequences that correspond to human genes. Complementary DNAs are manmade compounds that correspond to the messenger ribonucleic acid (RNA) normally found in cells, the latter being an edited analogue of a gene. These cDNAs are useful as probes to identify various chromosomes and have potential therapeutic and diagnostic utility, but their biological function is largely unknown. Related issues pertain to the process of rational drug design where numerous structures, e.g., of antiviral agents, can be designed by software routines based on a knowledge of, e.g., a virus' structural configuration. At what point will sufficient testing or a demonstration of biological activity be sufficient to satisfy the requirements of utility? Answers from the PTO and the courts may take years; landmark patent decisions invariably deal with the technology of 5 to 10 years earlier.

As a more routine practice, most patent attorneys will encourage the filing of a patent application at an early stage of research, particularly when basic findings will be published. Then subsequent related applications will be filed that continue describing subsequent research and add additional disclosure and evidence about function and uses. Known as ''continuation-in-part applications'' (CIPs), the issue of utility may arise if an intervening publication can be asserted as prior art against a CIP and the original case is not deemed sufficient to satisfy the utility requirement or related statutory requirements that a patent application teach skilled workers how to make and use an invention.

d. Nonobviousness Section 103 (see Table 1) sets forth the criterion of nonobviousness by which useful and novel inventions are determined to be patentable. This provision is intended to preclude patent protection for those inventions that are so closely related to the prior art that their production is within the skill of a hypothetical person of ordinary skill in the pertinent field.

In a 1989 Supreme Court decision, the obviousness requirement was described as expressing a congressional determination that the purposes behind the patent law are best served by free competition and exploitation of that which may be readily discerned from publicly available material.[25] As further interpreted by the Supreme Court, Section 103 requires that the scope and content of the prior art be determined and the differences between this prior art and the claimed subject matter be ascertained. If the invention as a whole, notwithstanding these differences, would have been unobvious to an ordinary worker in the pertinent field at the time the invention was made, then the claimed invention is patentable. The determination of unobviousness is a subjective one, but the statute provides the procedures and the standard by which the decision must be made. As an illustration, the prior art references cited against the claims for producing recombinant DNA in the Cohen and Boyer patent include several articles from the *Proceedings of the National Academy of Sciences* from the early 1970s in which the making of recombinant DNA was described. Collectively, they include 15 articles and 1 issued patent, and represent the closest prior art considered by the PTO. The allowed claims in the Cohen and Boyer patent were found to have been unobvious to an ordinary worker aware of this prior art.

What the law considers to be obvious changes over time—often as quickly as new fields of research become passé. It is thus very important for scientists to realize that laboratory advances that were revelatory just yesterday have become legally mundane today. For example, the creation of recombinant DNA in 1972, for which Paul Berg received a Nobel Prize in 1980, may be accomplished by workers of ordinary skill in the art with a reasonable expectation of success. See *In re O'Farrell*. Similarly, the production of monoclonal antibodies in 1975 by Caesar Milstein and George Koehler was rewarded with a 1984 Nobel Prize but may not generally require the deposit of a hybridoma for enablement purposes, see discussion below, and often can be reproduced by skilled workers without undue experimentation.

The need to file for patent protection before an invention becomes obvious creates a tension with the need to delay filing until after an adequate showing of utility can be made. See *In re Wands*. As later chapters in this book describe, much still is unclear and unsettled about biotechnology patent law; e.g., with respect to the scope of claims adjudged during infringement litigation.

If other technologies are a guide to the future of biotechnology patent law, legal developments consistently will lag behind advances in technology. For example, the burgeoning chemical industry after World War II created a flood of patent applications, yet basic questions of prima facie obviousness (i.e., if a new structural formula looks like an old one, is the new one unpatentable) were answered in the 1960s. Also, questions of enablement (i.e., how much "teaching" need be present in a patent application) were not answered until the 1970s. One of the most important functions of a patent attorney is to identify issues and trends in their resolution and, as a practical matter, to err on the side of filing a patent application if the underlying technology will be important to his or her clients.

e. Enablement The first paragraph of Section 112 (see Table 1) includes the enablement provision of the statute. Its purpose is to require that the specification contain sufficient "written" information at the time it is filed to enable "any person skilled" in the pertinent art (not the general public) to make and use the claimed invention without undue experimentation. Experimentation that is "undue" is not necessarily time consuming, expensive or laborious but rather requires independent acts of invention to effect. Enablement makes the patent understandable by those in its field to avoid infringement; provides the public with the ability to practice the invention after the patent expires; and secures the continuous disclosure of novel, useful, and unobvious technical advances. Section 112 also requires the disclosure of the "best mode" that is contemplated by the inventor(s) for practicing their invention at the time that an application for patent is signed; thus, proscribing concealment of an inventor's preferred embodiment. The best mode requirement is satisfied by not withholding information. Satisfying the enablement requirement is more difficult and represents a particular area of uncertainty for patents on multicellular organisms.

A legally sufficient disclosure of a machine or chemical composition may easily be made through figures and text. In contrast, it is difficult to explain the production and use of a genetically engineered organism or a method for using a novel pure culture of a rare but naturally occurring microorganism unless the organism itself is available. One way to satisfy Section 112 is to deposit a microorganism, before the patent application is filed, with independent depositories such as the American Type Culture Collection. The United States is a signatory to the Budapest Treaty, which specifies deposit, maintenance, and distribution standards for patent purposes for depositories in member nations. These repositories will provide cultures to requestors after a patent is issued; thus, rendering the organism, equivalent to a "stock reagent" that may be used as described in writing. The following are examples of several kinds of patented biotechnology-related products from various fields of research:

1. Cellular and molecular biology: transformed cells, isolated proteins, mutant proteins, and active-site portions of antibodies
2. Genetic engineering: transgenic plants and animals
3. Hybridomas and monoclonal antibody technology: methods of preparing antibodies, monoclonal antibodies with various specificities against bacterial endotoxin, cancer cell antigens, various hormones, and the acquire immune deficiency syndrome (AIDS) virus (HIV)
4. Diagnostic products: test kits for Lyme disease and hepatitis, and pregnancy tests and antibodies bound with various detectable tags for whole-body scans
5. Drugs, therapeutics, and biologics: recombinant vaccines against hepatitis and various growth hormones

f. Patents on Microorganisms Since the *Chakrabarty* decision in 1980, the patenting of living cells and microorganisms has been routinely permitted by the PTO and has not raised substantial issues of patentability. A recent patent issued to Systemix Corporation that claims a "stem" cell that may be of great commercial importance because it is an excellent candidate for many purposes, including genetic therapy.

g. Patents on Plants In the case of *Application of Hibberd*,[26] the PTO's Board of Appeals and Interferences (Board) determined that the classes of patentable subject matter set forth in Section 101 of the Patent Act also include plant material that is otherwise protectable under the plant patent provisions or the PVPA. The PTO has adopted the holding of *Hibberd* in a reversal of its previously announced policies. The kind of claims that are permitted under the *Hibberd* case include the following example:

> A maize plant capable of producing seed having an endogenous free tryptophan content of at least about one-tenth milligram per gram dry seed weight, wherein the seed is capable of germinating into a plant capable of producing seed having an endogenous free tryptophan content of at least one-tenth milligram per gram dry seed weight.[27]

h. Patents on Animals Although animals are considered by the PTO to be patentable following the decision in the *Allen* case, only one transgenic animal patent has been issued by the PTO, the Harvard Mouse patent. As of August 1992, however, the PTO had not yet issued a second animal patent, although spokespersons for the PTO have indicated that other cases were allowed but not yet issued while the applicants pursued certain issues of claim scope. The patenting of life forms and of transgenic animals, in particular, has been very sensitive politically.[28] Nevertheless, the issuance of other animal patents in the United States and in Europe seems likely even though the official positions of various patent offices are subject to change and second thoughts about the permissible scope of animal patents. Various offices of the Commission of the European Communities also are studying the appropriate scope of animal-related patents.

B. Section 161 Plant Patents

The following is an example of a Section 161 plant patent claim:

> A new and distinct variety of chrysanthemum plant, substantially as herein shown and described, characterized by its very large, bright yellow blooms, its excellent production of well formed flowers, flowering with a very even eleven-week response and producing very few culls.

Pan-American Plant Co. v. Matsui, 198 USPQ 462, 464 (N.D. Cal. 1977). As a general proposition, Section 161 plant patents are useful for plants that will be reproduced through traditional asexual means. However, for plants that will reproduce by seed or by single cell–culturing techniques, the scope of protection afforded by utility patents would be preferable if the enablement requirements can be met.

C. Plant Variety Protection Act Certificates

As discussed above, the Plant Variety Protection Act (PVPA) authorizes the issuance of certificates that grant to a breeder the rights to exclude others from selling, reproducing, importing, or exporting the protected variety or using it in producing (as distinguished from developing) a different variety.[29] This mode of intellectual property protection generally should be considered for plants that reproduce by sexual means, but, as discussed above, the scope of protection is relatively limited in comparison with the patenting of plants under the Patent Act. Where it is difficult to satisfy the requirements of patentability under the Patent Act, PVPA certificate provide a valuable fall-back position.

D. Copyright

A copyright protects the expressed form of an idea but not the idea itself. This differs from patent and trade secret protection, which can encompass the substance of the idea behind a particular object. The purpose of the copyright law is to secure public benefit by encouraging the efforts of those who create "original works of authorship" (see Section 102 in Table 1). The copyright owner is granted the exclusive right to reproduce and distribute the work; however, the actual use of a copyrighted work is not protected. To be copyrightable, the work must be "fixed in a tangible medium of expression, now known or later developed," such as print, painting, or other media, including those in which the work cannot be directly perceived by the human senses, such as video tape. Copyright has evolved with the growth of technology since the invention of movable type; for instance, Congress amended the copyright statute a few years ago to recognize explicitly the copyrightability of computer software.

It has been suggested that copyright protection extends to original DNA sequences—works that arguably may embody creative, and indeed artistic, expression. This argument is made by analogy: A cell's DNA is a compilation of instructions to cellular machinery like the computer instructions embodied in software. The analogy does not fit precisely, however. Alternative computer programs can accomplish the same result through equivalent but different instructions. In contrast, the limited redundancy in the genetic code permits fewer ways to specify a particular amino acid. This might preclude protection under copyright because the judiciary is reluctant to protect works in which the underlying idea is capable of expression in only a few ways. Furthermore, the value of a copyrighted DNA sequence would be minimal. Once a gene has been disclosed, it would be a relatively simple matter to prepare an analog, without copying, to express the underlying idea by taking advantage of the code's wobble.

A work may also be denied copyright protection if its appearance has "an intrinsic utilitarian function." Because particular codons may ultimately determine the higher order structure necessary for effective transfer RNA functioning during translation, DNA sequences may be inherently useful and noncopyrightable for this reason as well. Computer data bases, photomicrographs of DNA, or instruction manuals related to biotechnology can be copyrighted, as is this book. But again, copyright protection is limited. This book cannot be reproduced, but all of the ideas in it may be freely used.

E. Trade Secrets

Trade secrets encompass private proprietary information or physical materials that afford a competitive advantage to the owner. The classic example of a trade secret is the Coca-Cola branch syrup formula. Biotechnological trade secrets might include hybridization conditions, cell lines, corporate merchandising plans, or customer lists. Unlike patents, trade secrets have a potentially unlimited duration, are primarily regulated by state law rather than federal law, and need not satisfy the more stringent requirements for patentability.

Courts will enjoin disclosure of a trade secret and compensate the owner for its unauthorized use when the secret has been discovered by improper means, such as breach of confidence. The trade secret, however, is no longer protectable when it becomes public knowledge through independent discovery, reverse engineering, or through disclosure, for example, at an open session of the National Institutes of Health Recombinant DNA Advisory Committee.

Biotechnology is characterized by certain unique aspects that affect reliance on trade secrets. Because present and former university researchers consult with or are employed by companies to develop the basic research necessary to commercialize biotechnology, the ability to publish research results is usually expected by them and is often an important aid in recruiting and retaining individuals from a limited talent pool. Additional risks to the secrecy needed to protect this type intellectual property include the natural tendency of "academic" scientists to share information at conferences during informal gatherings, the mobility of graduate students and employees, and perhaps even the computer theft of trade secret data bases containing sequencing or restriction map data. Some disclosure is also required by the federal government and may not remain as well-guarded as secrets in other fields that are totally controlled within a company. Thus, because environmental impact statements may become required for some genetic engineering projects, and certain recombinant DNA experiments are reviewed by the RAC, there is an enhanced risk of leaks or disclosures, and potential access to some information about trade secrets by competitors may be had through the Freedom of Information Act to obtain documents from the government.

Reliance on trade secrets, however, may be prudent notwithstanding the risks of disclosure. The rapid pace of development in biotechnology suggests that much patentable work may become outdated before a patent would be issued (generally about 3 years or longer after an application is filed). Moreover, certain types of products might be more appropriately protected in this manner. Trade secret protection may be more actively utilized for sequence data bases rather than for plasmids or other tangible creations.

F. Trademarks

A trademark is a word or symbol "adopted and used by a manufacturer or merchant to identify his goods and distinguish them from those manufactured or sold by others." Just as sports equipment or frozen foods sold under widely recognized trademarks have become known to their respective purchasing populations, laboratory equipment useful in biotechnology research and development or in physicians' offices for testing purposes bear trademarks that are well known to workers in these fields. Certain vectors useful in recombinant DNA or antibody-related research also have been trademarked; e.g., the "recombo-zap" vectors of Stratocyte, Inc. This area of intellectual property has not presented any particular new legal or business issues but is an important and valuable member of an industrial portfolio of intellectual properties.

IV. HOW TO SELECT A MODE OF INTELLECTUAL PROPERTY PROTECTION

Biotechnology processes and products are protectable by the various intellectual property rights described above. These properties can be bought, sold, and licensed like any other type of property, such as condominium apartments that can be rented, time shared, and sold. The worth of any particular biotechnology product once protected by intellectual property rights depends on what the market will bear and on the value placed on protection from competition by the property owner or licensee.

Selection of the most appropriate mode of protection is a business judgment for companies based on several factors that differ from case to case, as discussed in greater detail below. For government laboratories and the academic recipients of government grant funding, selection of intellectual property is somewhat a matter of economics but also is a matter of policy. For example, the Patent and Trademark Act Amendments of 1980 (also known as the Bayh-Dole Act, P.L. 96-517, 35 USC § 200) emphatically encourage the use of patenting as opposed to, e.g., trade secrecy:

> It is the policy and objective of the Congress to use the patent system to promote the utilization of inventions arising from federally supported research or development; to encourage maximum participation of small business firms in federally supported research and development efforts; to promote collaboration between commercial concerns and nonprofit organizations, including universities; to ensure that inventions made by nonprofit organizations and small business firms are used in manner to promote free competition and enterprise; to promote the commercialization and public availability of inventions made in the United States by United States industry and labor, to ensure that the Government obtains rights in federally supported inventions to meet the needs of the Government and protect the public against nonuse or unreasonable use of inventions and to minimize the costs of administering policies in this area.

This pro-patent sentiment carries over into subsequent federal legislation noted above that is applicable to government laboratories, particularly the Federal Technology Transfer Act of 1986.

The threshold question in many cases is whether or not to pursue intellectual property protection in the first place or simply publish the results of a particular research project. As a general proposition, pursuing intellectual property rights is not advisable when it is not cost effective or contrary to public policy. For example, the commercial value of an invention or discovery would be relatively low where numerous alternative or even superior products are already on the market; e.g., another expression vector system when dozens are already available through biotechnology catalogues. If a discovery is made by a university receiving federal funding, under circumstances in which no commercial interest is manifested on the part of industrial research sponsors, publishing the results of research alone may be advisable. Note however, that certain reporting requirements pertain to grant recipients that require them to offer patent rights back to the funding institution if the grant recipient elects not to pursue them.

Relevant decision-making factors include the pace of technological development. In rapidly developing fields, pursuing patents will not be cost effective because the technology will advance before patents would be expected to issue (3 to 4 years at present). Projected revenues also must be considered, particularly, e.g., if the costs of pursuing patent applications worldwide may exceed the anticipated revenues from licensing by a university). It would be difficult to say too much about anticipating costs, as even the costs of secrecy may exceed those of obtaining a patent but perhaps not the costs of trying to enforce it. Costs, however, are but one factor, and the need for small businesses to demonstrate through the acquisition of patents business competence to investors or venture capitalists is an ingredient of success. Understanding the nature of an invention, i.e., whether it is a basic pioneering discovery that can be developed and licensed broadly or a small advance in a crowded field), also is of great value when considering whether to pursue intellectual property protection.

Several additional factors become relevant in the differential selection of intellectual property protection once a decision is made to pursue it. In every such decision-making calculus, the first choice is between trade secrecy and the other modes of intellectual property rights in which public disclosure of an invention or discovery is a prerequisite to obtaining protection. Trade secrecy should be considered when it can be effect as a practical matter without inviting reverse engineering or "me-too products." For example, the parent strains of seed crops might be maintained as trade secrets when hybrid seed, which does not breed true, will be the marketed product. Alternatively, a microorganism can be maintained in secret when its metabolic products, such as antibiotics, will be marketed. As a general rule, internal production processes lend themselves more readily to trade secret protection, but even this is problematic given the large numbers of people involved in industrial production efforts. Thus, the practicalities of such an approach may be daunting. In limited circumstances, trade secrecy will be more useful, albeit risky, in maximizing revenues by delaying inevitable competition. Note that competition develops even for soft drink cola products.

Once trade secret protection has been eliminated as a possibility, the next basic question is whether the Plant Patent Provisions or PVPA apply to the invention or discovery. Note that the former protects plants that are reproduced by asexual means and the latter applies to plants that reproduce by seed. As discussed above, the strongest protection is accorded to plants patented under the

Patent Act. However, the criteria are more stringent for this kind of intellectual property protection than for Section 161 plant patents or PVPA certificates. Note, further, that the costs of acquiring such intellectual property rights increase as the amount of protection against infringers increases.

If patent protection is considered, the costs of pursuing patent protection must be balanced against the value of holding pending patent applications or issued patents. The wisest decisions reflect a hard look at the market for products that will be sold under the scope of a patent. Alternatives to patent protection, such as orphan drug status, should also be considered, although this is beyond the scope of this chapter. As a practical matter, many patents are sought not to enable the marketing of a particular product but to stake out a claim to an area of research with patents that will discourage competitors from venturing too close to one's own products.

For research scientists working in industry, the decision whether to patent is made by corporate patent counsel and business development professionals. In academia, scientists can exercise greater control over the patenting of their inventions subject to their employing institution's reporting requirements to granting agencies and subject to agreements with industrial sponsors. As a general proposition, a scientist is a vital resource for a patent attorney and for the business development or technology transfer professionals who are involved with intellectual property management. Open discussions and a willingness on both sides to ask and answer questions are critical to successful utilization of the intellectual property system.

The timing for seeking patent protection also is a vital concern regardless of the place or nature of a scientist's employment. For example, in most of the world, the first inventor to file a patent application has priority over others who may have made an invention first but filed second. Although a 1-year grace period exists in the United States for filing after publication, patent protection in the most of the rest of the industrialized world is lost if a patent is not already pending once an invention is described in publication. Given the importance of worldwide patent rights to most companies, a patent application should be filed before and invention or discovery is disclosed publicly through publication or otherwise.

As Section II of this chapter described, the patent law continually evolves, with inventions being submitted for patent protection that raise issues of patentability. This has been abundantly demonstrated with respect to Section 101 for living microorganisms, plants, and animals. A similar exercise in testing the system may be reflected by patent applications claiming gene sequences such as have been filed by the National Institutes of Health and the British Medical Research Council. In these cases, even though the biological function of most of its newly discovered genes is unknown, failure to file may have precluded seeking patent protection in most of the industrialized world. Another pertinent issue is whether the publication of such sequences will impair the patentability of future gene-related developments based on those sequences.

V. CONCLUSIONS

The bottom line is that patent protection generally is the most practical mode of intellectual property protection to the extent that it encompasses a particular biotechnological invention. Note, however, the primary exception would be plants protectable under the PVPA that do not satisfy the statutory requirements of the Patent Act. The various intellectual property laws represent an organic entity that changes and evolves over time just as does biotechnology itself. Accordingly, intellectual property practitioners as well as institutions that seek or utilize such rights need to anticipate trends in the law and the expected directions of their own future research and development efforts.

NOTES

1. *Biotechnology in a Global Economy* OTA-BA-495 (U.S. Congress, Office of Technology Assessment, Washington, DC), October 1991 at page 203.
2. See, e.g., R. Adler: Genome research: Fulfilling the public's expectations for knowledge and commercialization. *Science* 257:908 (1992).
3. Demain, Industrial microbiology. *Science* 214:987 (1981).
4. Mendel's work remained obscure until after his death, when it was initially rediscovered about 1900. See generally K. Arms and P. Camp, *Biology*, pp. 192–195 (1979).
5. This discovery was made by Miescher. See. A. Lehninger: *Biochemistry: The Molecular Basis of Cell Structure and Function*, p. 859 (1975).
6. Avery, Macleod, and McCarty: Studies on the chemical nature of the substance inducing transformation of pneumococcal types. *J. Exp. Med.*, 79:37 (1944).
7. J. D. Watson and F. H. L. Crick: A structure for deoxyribonucleic acid. *Nature* 171:737 (1953).
8. Convention of Paris for the Protection of Industrial Property, March 20, 1883, hereinafter "Paris Convention."
9. *Industrial property* is the counterpart term outside of the United States for "intellectual property," which includes the property rights in patents, trademarks, copyrights, and semiconductor chip registrations, among others.
10. Paris Convention, at para. 2.
11. See, e.g., H.R. No. 18851, 59th Cong., 1st Sess. (1906) entitled "A bill to amend the laws of the United States relating to patents in the interest of the originators of horticultural products." Similar legislation was introduced in 1907, 1908, and 1910. Other legislation to protect both plants and animals was proposed but not enacted.
12. This sentiment was based upon an 1889 decision by the Commissioner of Patents in *Ex Parte* Latimer, 1889 Dec. Comm. Pat. 123 (1889).
13. Presently codified at 35 USC §§ 161–164 (1982).

14. H.R. REP. NO. 1129, 71st Cong., 2d Sess. 2 (1930).
15. Idem. at 1–2; S.REP. NO. 315, 71st Cong. 2d Sess. 1–2 (1930).
16. Federal Technology Transfer Act of 1986, 15 USC §§ 3701–3714 (Supp. IV 1986).
17. 35 USC §161 (1982).
18. Idem. § 163.
19. Idem. § 162.
20. S. REP. NO. 315 at 5; H.R. REP NO. 1129 at 6. Some commentators have suggested that an additional reason for excluding sexually reproduced plants may have been the fears of scientists and farmers that such inclusion may have inhibited the free exchange of germplasm (i.e., genetic material).
21. *In re* Arzberger, 112 F.2d 834 (CCPA 1940).
22. A plasmid is a small circular nonchromosomal DNA found in bacteria that often carries the genes for traits such as antibiotic resistance and that has been used industrially to introduce recombinant DNA into an organism for replication and expression of desired cellular products.
23. See *In re* Mancy, 499 F.2d 1289 (CCPA 1974).
24. *Chakrabarty*, 447 U.S. at 316.
25. Bonito Boats, Inc. v. Thunder Craft Boats, Inc., 489 U.S. 41 (1989).
26. 227 USPQ 443 (PTOBAI 1985).
27. Claim 249 from the *Hibberd* patent application.
28. See, e.g., Office of Technology Assessment, ''Patenting Life,'' (OTA-BA-370 U.S. Government Printing Office, Washington, D.C., 1989).
29. 7 USC §2483.

4

Preparing the Patent Application

Bradford J. Duft

Amylin Pharmaceuticals, Inc. and Lyon & Lyon, San Diego, California

I. INTRODUCTION AND OVERVIEW[1]

> There is a popular but mistaken belief that the subject of patents is difficult. So far as the essentials are concerned nothing could be farther from the truth.[26]

This quotation is from the Preface of a remarkable book authored in 1946 by William R. Ballard titled, *There Is No Mystery About Patents.* Although there are intricacies and necessary complications associated with drafting and prosecuting patent applications, as well as sometimes complex strategies, this statement holds true today.

There are two fundamental disciplines embodied in our patent system. The first is the art of preparing and procuring patents, knowing all the laws of patentability and the rules of the U.S. Patent and Trademark Office that govern the patenting process. Most patent attorneys can write a patent application, and many patent lawyers are able to see that those applications will mature into a patent.

The value of those patents may not become clear, however, until they are tested in the marketplace by their ability to generate capital investment or license revenues, or until they withstand ultimate appraisal in a court of law. There is more to patents than getting the science on paper, constructing a few claims, and having an application issue as a patent.

This chapter is devoted to the preparation of patent applications with a mind on the more critical discipline: the art of securing patents that will withstand the challenge of litigation. This amounts to knowing not just how to gain an issued patent but also knowing how and, just as importantly, why patents are litigated. Understanding some of these things, businessmen and inventors alike can play a crucial role in the preparation of patent applications that have real and significant value.

This chapter provides an overview of practical strategies, tactics, and techniques useful in the preparation of a patent application designed to result in an issued patent having maximum strength.

It opens with a discussion of the significance of patents and some of their commonly misunderstood features, and the differences between the various parts that make up a patent application, put in the context of the application process within the U.S. Patent and Trademark Office. The actual drafting of the application is discussed next, with special emphasis upon the important role of the inventor, prior art searches, and documents useful in drafting the application. The obligation of the inventor and his attorney to provide the Patent and Trademark Office with information they know of that is material to patentability, typically scientific publications and the like, is described as an integral part of the patent preparation process. The timing of patent application filing is also discussed, followed by a more detailed discussion on the preparation of the patent application and the importance of format and content in view of the legal requirements for patentability of inventions and the technical ingredients required in a patent application. In conjunction with consideration of the elements that need to be disclosed in the patent application, the deposit of biological materials is addressed, as are certain special considerations regarding inventions having pharmaceutical utility. Finally, aspects of drafting claims to the applicant's invention are reviewed.

Throughout this chapter, reference will be made to one of the early-issued biotechnology patents, U.S. Patent No. 4,376,110, issued on March 8, 1983, to David and Greene for "Immunometric Assays Using Monoclonal Antibodies," and assigned to Hybritech Incorporated. It is directed to monoclonal antibody–based sandwich-type immunoassays for determining the presence or amount of an antigenic substance in a test sample. The patent was also the first biotechnology patent litigated.

The Hybritech patent is attached as Appendix 1. The opinion of the U.S. Court of Appeals for the Federal Circuit in *Hybritech Incorporated v. Monoclonal Antibodies, Inc.*, upholding the validity of the patent, is included as Appendix 2.[2]

The Hybritech patent is relatively short and, having been written well and crisply, is a useful exemplar. It contains all the typical parts of a patent—a title, abstract, field of the invention, background, summary of the invention, brief description of the drawings, detailed description of the invention, an example of the invention, claims, and drawings—and it survived challenges to validity on virtually all grounds on which a patent might be attacked in court.

A. Introduction to Patents: What a Patent Is and Is Not

Ballard's book, *There Is No Mystery About Patents*, begins with the observation that, "A good shortcut to the understanding of any subject is to stake out some of its high spots, so that in what follows one may know what to look for, and may better sense the relative importance, relation and meaning of the statements made."[26] There are several basics—principles that reveal in large measure what patents are and are not—that help in understanding the nature of patents and how they can affect businesses.

Seven things must be understood:

1. *There is always an invention.* But to be the subject of an issued patent, the invention must be *patentable*. Thus, one must always distinguish between "inventions" and "patentable inventions." The point not to be forgotten is that being an *inventor* and having made an *invention* is not changed by the fact that one or more or all of the conditions for *patentability—e.g.*, novelty and nonobviousness—cannot be met.
2. *A patent is not an invention.* A patent is a piece of paper signifying a grant to the inventor of certain rights.
3. *A patent does not grant an inventor the right to do anything with his invention.* Specifically it does not grant him the right to make, use, or sell the patented invention. A patent represents a negative right only—nothing more nor less than the right to *exclude others,* for a limited time, from making, using, or selling the patented invention. This is *all* that one obtains by a patent. The patent grant has never had anything to do with the patentee's right to make, use, or vend. An inventor's right to practice his invention is an inherent common-law right if the rights of others are not thereby violated. The inventor has this right (if at all) whether a patent issues to him or not.
4. *Patents create value for the companies that own them*, in part by attracting capital investment, the cost of which may be cheaper because the investment risk is lower.
5. *Patents do not prevent others from infringing*. That is, patents do not, by virtue of their issuance alone, necessarily stop others from making, using, or selling the patented invention; although a patent may have some deterrent value or provide licensing opportunities, only a lawsuit can end infringement.
6. *Patents are not necessarily worth anything*. The value of a patent often rests entirely upon the perceived value of the underlying invention to those who

may want to make, use, or sell it or upon the extent to which the patent legitimately dominates other embodiments of the invention not yet made.

7. *Even though a patent is issued for a patentable invention, the patent can be invalid.* Reasons for this include improperly drafted claims, failure to comply with disclosure requirements in preparing the application for patent, and dishonesty during prosecution of the application.

Several of the above-styled principles reflect the fact that the patent right represents only a right to exclude. In law, a patent embodies nothing more than a license to sue others. Assuming a willingness to allow others to use the patented invention, a patent can be used to negotiate a license, which is, by definition, merely an agreement not to sue the licensee for conduct which, absent the license, would be actionable. Such agreements not to sue are made in exchange for some consideration from the licensee, typically a royalty stream. In other words, because a patent is an instrument that conveys nothing but the right to exclude others for a time, it may be said, in a strict sense, that if you use a patent at all you can use it only for that purpose. Absent any excluding that can be done by persuasion, such uses virtually always amount to exclusion by a suit in court.[26]

Properly done, the quest for patent protection depends importantly on, if you will, and, in many cases, a way of thinking. It must be understood that the protection of inventions does not begin and end with the preparation of a strategy for obtaining a patent from the Patent and Trademark Office. With respect to inventions that are of value, there will inevitably be a disagreement, which may, if successful from the patent owner's point of view, end in the exclusion of an infringer by a lawsuit or with an agreement to license. As many patent owners that have been disappointed over the years will report, however, the end is not always so cheerful.

B. Introduction to the Real Significance of the Patenting Process: Obtaining a Patent that Will Withstand the Challenge of Litigation

From the outset, it must be understood how patents can be destroyed and measures taken in drafting the patent application to prevent that destruction. It is also of vital importance to understand that devastating evidence that might affect the validity of a patent can often be of the inventor's or the patent owner's own albeit inadvertent, making. Such evidence can take many forms, as discussed below, and is often disclosed to others, published, or found kept in file drawers. Documents and disclosures regarding an invention should always be carefully reviewed before release to third parties and before publication or storage.

There are at least two reasons why patents typically end up in litigation. The first is validity. All issued U.S. patents are presumed to be valid. Nevertheless, aside from any deterrence of others, the right to exclude embodied in that piece of paper may be useless and of no value unless it can be and is enforced. Thus, patents are often litigated because the patent is being ignored by an alleged infringer and it must be asserted in court.

Patents also are litigated based upon what may be termed speculation. In planning a patent strategy, it must be borne in mind that in the real world the determination of whether a patent will be challenged may often not be a function of whether the patent is strong or weak, valid or invalid. Does having a "good" patent keep you out of litigation? The answer is no; it will more likely get you into litigation. As one author put it:

> Perhaps the greatest fallacy in the patent field, widely held among inventors, businessmen, patent practitioners and judges as well, is that strong patents are never litigated and that only weak patents of doubtful validity are subjected to judicial scrutiny. That proposition is simply, empirically, and completely incorrect.[30]

Kayton noted that the realities of business life together with simple mathematics reveal that whether a patent is litigated may be determined by a number of business and legal factors that he organized under an equation termed the *law of expected return*.[30] The equation proposes that the expected return of any given situation is equal to the probability of success multiplied by the total amount that would be produced if the venture is successful.[30] Kayton provided this example:

> Suppose the total potential value of infringement over the life of a patent (that is, the infringer's profits) is $60 million, and the probability of the infringer's winning the infringement suit with the particular facts in the case is only ten percent. This means that even with such a poor probability of success it is well worthwhile for the infringer to infringe and to risk litigation if the litigation expenses are substantially below $6 million.[30]

Although this proposition would not appear to take into account other business and legal matters, such as potential damages or the impact of an injunction on investments and opportunity cost, in the world of biotechnology, we have already seen both types of suits mentioned above—"validity" suits and "speculation" suits based upon the "law of expected return."

A "pure" biotechnology case, one in which the validity of the patent in suit had to be tested because of the infringement of another, is *Hybritech Incorporated* v. *Monoclonal Antibodies, Inc.*, which involved the Appendix 1 patent. It was a suit driven by the need to establish validity. The case is called a "pure" one because it involved a patent on new technology versus the use of that new technology by another. The defendant, Monoclonal Antibodies, Inc., had taken Hybritech's patented technology and based nearly 85% of its business on it. Monoclonal also refused the offer of a license and Hybritech had little choice but to assert its patent and test the presumption of validity. The patent withstood an onslaught of defenses by the infringer, which put forward virtually every defense to the validity of a patent that might be raised. The case took over 3 years to conclude and was expensive, as most litigation can be; that expense being measured in terms of the time required of key scientists and business personnel, as well as in dollars.

The case presented some difficulties that inventors and patent owners can protect against. Unauthorized disclosures about an invention, in the form of "security breaches," for example, can result in questions of inventorship—who is the original inventor—and may well end up putting a competitor in business. That happened in the *Hybritech* case, in one instance through the unauthorized disclosure of the assay invention to a third party by a Hybritech employee while riding in a carpool. The individual to whom the invention was disclosed subsequently went to work at Monoclonal Antibodies, Inc. and, using the knowledge gained during his carpool conversations, was able to put together a prototype of the Hybritech assay in just 1 month. Casual discussions regarding the Hybritech invention also led two individuals outside of Hybritech to file patent applications, alleging that the invention they had been told about by Hybritech employees was in fact their own. All of this complicated the lawsuit and made the case significantly more expensive to discover and try in court.

Publications that arguably cast the patented invention in a poor light must be scrupulously avoided. In the *Hybritech* case, one of its employees published an article allegedly describing the patented invention in terms of "obviousness." Although it was an unwitting characterization, and its true meaning readily explained, it was the focus of intense scrutiny and investigation by the defendant. Nevertheless, it was a slip-up of the most dangerous variety, and it also added to the cost of suit.

The March 1990 issue of *Scientific American* magazine discussed litigation between DuPont and Cetus over patents to Cetus's polymerase chain reaction (PCR) technology. It may be called a "speculation" case. DuPont had apparently begun working on PCR and subsequently tried to license the PCR technology from Cetus. Cetus refused and DuPont brought suit, claiming Cetus's PCR patents to be invalid. The magazine reported that the litigation could easily cost each company on the order of $2 million a year and that it might last more than 2.5 years. The magazine specifically reported the following justification for the litigation, utilizing statements from the interviews of a Cetus and a DuPont employee:

> "So say we make up the numbers" [the Cetus employee] says. "Say the market is worth $500 million." If DuPont thinks it has better than a one-in-a-hundred chance of winning a lawsuit, "it's an even bet," [the Cetus employee] says. "We've done the calculations," [the DuPont employee] replies.[36]

A jury subsequently returned a verdict holding the Cetus PCR patents valid and infringed by DuPont. DuPont then agreed to remove its PCR products from the market.

It is generally understood that enormous amounts of capital are required to sustain a company's research and business efforts during the 6–8 years needed to run the development, preclinical work, and clinical trials required of biotechnology products. Published estimates of the cost to bring a new biotechnology product through the U.S. Food and Drug Administration to the marketplace are typically well over $200 million.

With respect especially to smaller or start-up companies, if important inventions that are coveted by others are not made the subject of effective patent protection, it is not an exaggeration to suggest that those companies will face annihilation. Where large multinationals will sue biotechnology enterprises on patents that are arguably strong, for example, such as the PCR patents, it is plain that where there is no impediment at all to the acquisition of important technology, the start-ups and other less wealthy entities can quickly be outstripped by companies with greater financial and human resources.

In preparing a patent application, then, one must consider both the possibility and the realities of patent litigation. It is equally important to understand that the decision to litigate is not always one's own. In the area of biotechnology, it is imperative to ensure that important patents are being drafted, prosecuted, and continuously thought about with a view toward litigation.

A brief overview of the patenting process, the requirements for patentability, and the significance of the different parts of a patent are provided as a preface to consideration of the fundamentals of preparing a patent application for a biotechnology invention.

C. Introduction to the Patenting Process: Overview from Invention to Issuance

The term *invention* has different meanings, but it can safely be said, chronologically speaking, that an invention first comes into existence when it conceived by the inventor. *Conception* is thus a term having great significance in the patent law. It is defined as the "formation in the mind of the inventor, of a definite and permanent idea of a complete and operative invention, as it is hereafter to be applied in practice."

The idea must be of a specific means, not just a desirable end or result. Thus, the mental formation of the invention is deemed "complete and operative" if the inventor has conceived the means of putting his idea into practice where no more than routine skill would be required to do so. After the complete performance of the mental part of the inventive act, it is said, all that remains to be accomplished belongs to the department of construction, not invention. A subsequent demonstration that an invention will work as conceived is referred to as an "actual reduction to practice."

A patent application is not proper until a workable invention has been conceived. At a time when the complete invention is mentally pictured, that is, fully conceived in an operating form, the inventor may properly pursue patent protection. It is not necessary that the invention had been actually reduced to practice before an application is prepared and filed. Often, however, experimental work is conducted before filing in order to confirm or perfect the invention. This is usually especially important in the area of biotechnology, where preparation of a patent application can be both time consuming and expensive. Furthermore, although not legally required, experimental work will normally undergird most biotechnology patent applications, which are usually directed to discoveries made as a result of laboratory research.

The steps involved in obtaining a patent typically include: preparing a disclosure of the invention; conducting a patentability search for the invention, if needed or appropriate under the circumstances; preparing the patent application; filing the application in the U.S. Patent and Trademark Office (PTO); prosecuting the application before the PTO, and, finally, receiving an issued patent.

Once a patent application is prepared and filed in the PTO, it is given a serial number and filing date. The filing of a proper patent application is referred to as a "constructive reduction to practice." Within the PTO, patent applications are assigned to examiners with specialized technological background or training. Thus, patent applications directed to an invention involving recombinant DNA will be assigned to an examiner with experience in molecular biology.

Examination of applications normally occurs in the order in which they are received by the examiner. In a typical case, the patent examiner first reviews the patent application for compliance with formal requirements. The invention as claimed is then reviewed. With an understanding of the claimed invention, the examiner researches earlier patents and publications, known as the "prior art," in order to determine whether the claims of the patent application define subject matter that is patentable over that art. After this review, and after reaching his conclusions, the examiner prepares an Examiner's Action, stating his position in writing for communication to the patent applicant. Usually, few if any claims are allowed in a first Action.

Most applicant's have a patent lawyer, and the Examiner's Action (also referred to as an Office Action) is mailed directly to the applicant's lawyer. The lawyer will then advise the inventor of the Office Action and request the inventor's assistance in responding to the examiner's position. For example, the inventor may be asked to review the prior art cited by the examiner in order to respond to the examiner's arguments and conclusions on patentability. The applicant is normally given a time period of 3 months (with 3 one-month extensions available, giving a total of up to 6 months) for responding to the Office

Action. The response submitted by the inventor will normally include arguments designed to have the examiner change his position, with or without an amendment modifying the claims of the patent application.

After receiving a response from the applicant, the examiner again considers the case and reviews his position. Although the matter may not be concluded at this stage, there are often no more than two or three actions and responses before the question of patentability of the claims as written is concluded. Of course, the examiner may determine that the application simply is not directed to a patentable invention. Alternatively, he may recognize the existence of a patentable invention and agree with the form and scope of the claims. On the other hand, the examiner may concede that the invention is patentable but disagree with the applicant's lawyer on the scope of the claims that should be granted. Failing an agreement on any aspect, an appeal to the PTO Board of Patent Appeals and Interferences can be pursued. However, frequently agreement is reached and the patent application is ready to issue as a patent. A Notice of Allowance is prepared by the PTO and sent to the applicant's attorney.

After the Notice of Allowance is mailed, a final fee must be paid. There is a multimonth delay after payment of the final fee before a patent will issue. From filing to issue, the average time that a biotechnology patent application is pending before the PTO is about 2.5 years. Applications that are complicated, or where agreement as to patentability or the scope of allowable claims cannot be reached and an appeal is necessary, can take much longer to issue. Patents are issued in numerical order and, by long tradition, are issued every Tuesday. The original of the patent is sent to the applicant's lawyer, who will then forward it to the inventor or his assignee.

In addition to knowing some foundational concepts about patents, patents and litigation, and the patenting process, it will also be useful to have in mind some further patent basics before analyzing the preparation of a patent application in detail. Specifically, one involved in the application process should have at least a basic understanding of the differences between a patent specification and its claims, and should also know about the legal requirements for patentability of inventions—eligible subject matter, utility, novelty, and nonobviousness. These matters are introduced in the following two sections.

D. Introduction to the Difference Between Patent Specifications and Patent Claims: Description Versus Definition

The U.S. patent statutes are set forth in Title 35 of the U.S. Code (usually referenced as "35 U.S.C."), and those laws are identified by code "sections," currently numbered between 1 and 376. The statutes that are of primary importance in this chapter are 35 U.S.C. §§ 101-103, which define the conditions of patentability, and 35 U.S.C. § 112 which mandates, broadly, the information that must be disclosed in a patent application.

In addition to required filing fees, an application for patent, according to the U.S. patent law, must include:

> (1) A *specification* as prescribed by section 112 of this title; (2) a *drawing* as prescribed by section 113 of this title; and (3) an *oath* by the applicant as prescribed by section 115 of this title [35 U.S.C. § 111].

The patent specification is meant to constitute a disclosure of the invention to be claimed. In addition to the technical details of the invention itself, the specification will normally include an introduction to the invention—a discussion of the technology as it existed—that provides a background against which the invention may be contrasted.

The relevant legal requirements regarding the specification are found in the first paragraph of 35 U.S.C. § 112:

> The specification shall contain a written description of the invention, and of the manner and process of making and using it, in such full, clear, concise, and exact terms as to enable any person skilled in the art to which it pertains, or with which it is most nearly connected, to make and use the same, and shall set forth the best mode contemplated by the inventor of carrying out his invention.

The first paragraph of section 112 of 35 U.S.C. thus sets out four requirements against which the adequacy of disclosure in a patent specification will be judged: the "written description" requirement, the "enablement" requirements ("how to make" and "how to use"), and the requirement for a description of the inventor's "best mode". A more detailed discussion of these requirements is provided in Section II.D. of this chapter. Suffice it to say here—and this point should never be lost sight of—that it is the invention *as claimed* that must be adequately described, which must be enabled, and for which a best mode must be specified.

Although the patent specification and the patent claims are typically referred to as separate portions of a patent application, the claims are actually a part of the specification. The second paragraph of section 112 of 35 U.S.C. instructs that:

> The specification shall conclude with one or more claims particularly pointing out and distinctly claiming the subject matter which the applicant regards as his invention.

Although technically a part of the specification, the claims function to define—not to describe—the invention. The sole function of the claims is to delineate the scope of protection sought by the applicant for patent, not to describe what the patentee has invented.

A claim is thus analogous to a land deed. The deed identifies with particularity the metes and bounds of a plot of land. It does not describe the land itself, what resides inside the defined boundaries, but merely tells people where they may not go lest they be trespassers. A patent claim sets up the boundaries of the protection afforded by the patent so that others can know what they may not make, use, or sell without the permission of the patent owner.

In addition to advising others what constitutes the "patented invention," the patent claims also define the invention for purposes of determining whether the invention meets the requirements for patentability (the subject matter, utility, novelty, and nonobviousness requirements) as well as whether the specification complies with the disclosure requirements of section 112 of 35 U.S.C., which require that claim language and claim scope be supported by the specification.

In distinguishing between the specification and claims, it is also useful to bear in mind that the mere fact that a claim may be broader than the specific embodiment or embodiments of the invention disclosed in the specification is in itself of no moment. Some years ago, the Court of Custom and Patent Appeals (which merged in 1982 with the Court of Claims to form the Federal Circuit Court of Appeals) addressed this point:

> A hypothetical situation may make our point clear. If the original specification of a patent application on the scales of justice disclosed only a 1-pound "*lead* weight" as a counterbalance to determine the weight of a pound of flesh, we do not believe the applicant should be prevented, by the so-called "description requirement" of the first paragraph of §112 . . . from later claiming the counterbalance as a "metal weight" or simply as a 1-pound "weight" although both "metal weight" and "weight" would indeed be progressively broader than "lead weight," including even such an undisclosed, but obviously art-recognized equivalent, "weight" as a pound of feathers. The broader claim language would be permitted because the *description of the use and function* of the lead weight as a scale counterbalance in the whole disclosure would immediately convey to any person skilled in the scale art the knowledge that the applicant invented a scale with a 1-pound counterbalance weight, regardless of its composition.[3]

An applicant for patent is entitled to claims as broad as the prior art and his disclosure will allow.

Accordingly, the importance of proper claim drafting can be understood: If a claim is drafted too broadly it may be held invalid as lacking novelty, as obvious in view of the prior art, or as insufficiently supported by the specification; but if a claim is drafted too narrowly, it can undermine the value of a patent by allowing others to make use of the described invention without renumeration to the inventor.

The contrast between the specification and the claims that conclude it is clear. In essence, the specification is drawn to tell the story of the invention. It will normally *describe* the scientific background against which the invention was made, how to make the invention, and provide one or more examples to show its usefulness. The claims, on the other hand, not the specification, measure the invention. The claims *define* the invention that the applicant wishes to prevent others from exploiting without his permission.

Whereas the specification is detailed and specific, the claims are usually written to outline the invention more broadly in order to protect the inventive concept. Thus, as noted above, the claims will often be broader than the particular embodiment of the invention that may have been originally conceived or made and tested prior to the preparation and filing of the application. In the Appendix 1 Hybritech patent, a single embodiment of the assay invention, immunometric assays for the antigen immunoglobulin E, are described in columns 6–8, whereas the broadest claims (claims 1, 10, and 19) are not limited to any particular antigenic substances.

The breadth of claims to which an applicant may be entitled is inextricably intertwined with an inquiry into ''patentability'' of the invention.

E. Introduction to the Requirements for Patentability

The requirements for patentability of inventions are set forth in sections 101, 102, and 103 of Title 35 of the U.S. Code, which make it clear that patents are not granted for all new and useful inventions and discoveries.

The four conditions for patentability are: (1) eligible subject matter, (2) utility, (3) novelty, and (4) nonobviousness.

1. Categories of Inventions Patentable

The subject matter of an invention or discovery must come within the boundaries set forth by 35 U.S.C. § 101 (''Inventions patentable''), which permits patents to be granted only for ''any new and useful *process*, *machine*, *manufacture*, or *composition of matter*, or any new and useful *improvement* thereof.'' This statute has been interpreted broadly by the United States Supreme Court to include ''anything under the sun that is made by man.'' *Diamond v. Chakrabarty*, 447 U.S. 303 (1980). As applied to biotechnology, this has led to the issuance of patents for inventions that can be viewed as ''products of nature,'' such as isolated proteins and deoxyribonucleic acid (DNA) sequences, as well as engineered products such as hybridomas and genetically altered microorganisms. Under the patent law, these inventions are ''compositions of matter'' or ''manufactures.'' Whether such inventions may also be viewed as embracing living matter is irrelevant to the issue of patentability. Indeed, the PTO has now ruled that plant subject matter is properly included as patentable subject matter under section 101 of 35 U.S.C. The PTO also recently announced that it considers nonnaturally occurring, nonhuman multicellular living organisms, including animals, to fall within the scope of inventions that may be patented.

Although the three remaining substantive requirements for patentability of an invention falling within one of the above categories of subject matter are detailed in Chapter 3 of this book, each is briefly reviewed here to reinforce the purpose that underlies the drafting of an application—to put together a document that enhances the likelihood of convincing the PTO, potential business partners, possible business competitors, and, ultimately, the courts that these requirements for patentability are satisfied.

2. Utility

In addition to describing the subject matter eligible for patent protection, 35 U.S.C. § 101 also commands that patents may issue only for inventions claimed that are useful. According to the *Manual of Patent Examining Procedure* (also referred to as the MPEP), a complete specification disclosure should also include a statement of utility. Thus, by way of example with regard to biotechnology patent applications, DNA sequences will be described in the specification as useful for production of the encoded protein, newly discovered proteins may be disclosed to be useful for therapeutic purposes, antibodies may be described for use in various immunological or chromatographic methods, and assays are normally specified as useful for detecting or measuring various antigenic substances.

The Hybritech patent contains no specific statement of utility other than describing the advantages of the assay invention. This is acceptable because the utility of such assay techniques is understood and is without dispute. Furthermore, examples demonstrating that the Hybritech assay works for its intended purpose are provided.

On the other hand, where examples are lacking or an applicant's assertion of utility in an application for patent is so incredible as to create a strong presumption of inoperativeness, such as in a recent case where compounds were asserted to be useful in retarding the aging process,[4] the PTO may properly require the submission of objective evidence sufficient to establish utility. An important point to remember in crafting a patent strategy is that proof of utility may be made later—after patent application filing—by providing evidence of operability that would be convincing to those skilled in the art. Although a use for the invention must nevertheless be disclosed by the specification, it need only be provable, not necessarily proved, and in this way early patent application filing is encouraged.

3. Novelty

The law also requires, in 35 U.S.C. § 102 (''Conditions for patentability; novelty and the loss of right to patent''), that patents issue only for inventions that are new. An invention that is fully described in a prior art reference, and is thus not new, is said to be "anticipated" by that reference. The most common prior art references are published journal articles and previously issued patents.

Anticipation is determined by whether a patent claim "reads on" the reference; that is, by whether each feature of the claims is found in it. These principles underlying the section 102 novelty requirement have been summarized as follows by the Federal Circuit:

> A party asserting that a patent claim is anticipated under 35 USC 102 must demonstrate, among other things, identity of invention . . . [O]ne who seeks such a finding must show that each element of the claim in issue is found . . . in a single prior art reference, or that the claimed invention was previously known or embodied in a single prior art device or practice.[5]

As noted, it is to be borne in mind that it is the invention *as claimed* (not as described in the specification) that must meet this novelty requirement. Thus, the courts have also emphasized the distinction between an invention that is described or taught in a patent and the invention claimed in it:

> The law of anticipation does not require that the reference "teach" what the subject patent teaches. Assuming that a reference is properly "prior art," it is only necessary that the claims under attack . . . "read on" something disclosed in the reference, *i.e.*, all limitations of the claim are found in the reference, or "fully met" by it.[6]

Where only one element of a claimed invention is not disclosed by a prior art reference asserted as an anticipation, the reference will not be sufficient to defeat the novelty of the invention as claimed. See Appendix 2 *Hybritech* opinion.

Instead, such a reference might be used in combination with one or more additional references that do provide the missing element to argue that the claimed invention would have been obvious in light of all the references when viewed together.

4. Nonobviousness

All claim limitations must be found (expressly or inherently) in a single prior art reference to be an anticipation under section 102 of 35 U.S.C., and it is not proper to combine prior art references in determining novelty. Combining prior art references is proper and appropriate, however, to determine obviousness. The nonobviousness requirement extends the field of unpatentable material beyond that which is known to the public under section 102 to include that which could readily be deduced from publicly available material by a person of ordinary skill in the pertinent art. Thus, even though an invention meets the utility and novelty requirements, pursuant to 35 U.S.C. § 103 ("Conditions for patentability; nonobvious subject matter") a patent may not be obtained

> if the differences between the subject matter sought to be patented and the prior art are such that the subject matter as a whole would have been obvious at the time the invention was made to a person having ordinary skill in the art to which said subject matter pertains.

The language of this statute substantiates the first of the seven patent basics ("There is always an invention") introduced in Section I.A. this chapter. Careful reading of section 103 of 35 U.S.C shows that it is not a "standard of invention" and that it is not called a "requirement of invention." The existence of an invention is presumed (by reference to "the time when *the invention* was made"), and the only question to be answered is whether that invention would have been obvious.

Thus, section 103 of 35 U.S.C. speaks of a condition of patentability instead of a requirement for "invention," and the condition is unobviousness. It is important to bear in mind, furthermore, that the unobviousness is *as of a particular time* and *to a particular* legally fictitious, technical *person*. To protect the inventor from hindsight reasoning by others, the time is specified to be *the time when the invention was made*. To prevent the use of too high a standard—which would exclude inventors as a class and defeat the whole patent system—the invention must have been obvious at that time to "a person having *ordinary* skill in the art" to which the subject matter (i.e., the invention) pertains.[34]

But that is not all; what must have been obvious is "the subject matter as a whole." That, of course, is the invention as defined by each patent claim, including all of its characteristics and advantages whether or not specified in the claims. If, for example, a combination of elements is claimed (such as a cloning vector containing various functional nucleic acid sequences), section 103 of 35 U.S.C. requires that to invalidate the claim, it must be shown that *the combination* would have been obvious, not merely its individual components.[34] One might also seek to patent a biopharmaceutical compound that differs only slightly in chemical structure from earlier, known compounds. The law of section 103 requires that all of its biological, pharmaceutical, and other properties be considered in evaluating the nonobviousness of the *compound*, even though the *structure* of the compound would otherwise appear to have been obvious absent consideration of those properties. For example, it might be shown that the compound is easier to make or formulate, that its bioactivity or lack of side effects is preferred over previous compounds, or that its half-life in the body (or shelf-life in the pharmacy) is superior. Such properties might well not be evident from its structure. From the standpoint of patent law, a compound and its structure, all of its properties, and the problems it solves are inseparable. Inventions must be considered in their entirety.

In the end, obviousness of a claimed invention is a question of law that is to be resolved based on the following factual inquiries: (1) the scope and content of the prior art; (2) the differences between the prior art and the claims at issue; (3) the level of ordinary skill in the art; and (4) objective evidence of nonobviousness. With the involved facts determined, the one evaluating nonobviousness is said to confront "a ghost, *i.e.*, 'a person having ordinary skill in the art,' not unlike the 'reasonable man' and other ghosts in the law."[7] To reach a proper conclusion under section 103 of 35 U.S.C., the decision maker is obliged to step backward in time and into the shoes worn by that "person" when the invention was unknown and just before it was made. In light of *all* the evidence, the decision maker will then determine whether the claimed invention *as a whole*, would have been obvious at *that* time to *that* hypothetical person.

Much can be done in drafting the patent application to advance the patentability, i.e., the nonobviousness, of the invention to be claimed.

II. DRAFTING THE APPLICATION

A. Preliminary Considerations

In turning to the actual drafting of the patent application, certain assumptions have been made. First, it is assumed that from a business standpoint, the inventor or his company has determined that the preparation of a patent application for the invention is fitting and proper. This means that the monetary resources necessary to draft and prosecute the application, and to *enforce* the issued patent, have been fully considered. This includes consideration of whether international filing will be undertaken, which is a considerable expense. Additionally, the business purpose of the patent application has been fully analyzed and taken into account and it has been determined that the patent will be of value and will match the business needs of the inventor or his company. The amounts of additional research and development necessary to bring the invention to market, even if patented, have been considered in determining both the timing of preparation of the application and the scientific support that may be required for it. The nature and extent of the competition in the field of the invention have also been factored in.

All in all, prior to drafting the application, the *business* strategy has been considered and established, and both the technical and economic rationale for "taking patent action" is understood.[25] At this point, the inventor will often have a fairly good idea of what to protect, and the patent attorney—together with the inventor or his representatives—are prepared to determine how best to protect it. Other strategic matters relating to patent application preparation, such as conducting prior searches and the timing of application filing, are discussed in more detail below.

1. The Role of the Inventor

After helping to determine whether and why to seek a patent, the inventor plays a critical role in the preparation of every part of a patent application. That role, for each section, will be apparent from the more detailed discussions of those parts below.

Generally, of course, it is the inventor himself who must initially provide a detailed written description of the invention that has been made or conceived. He should also normally try to provide the patent attorney with reviews, journal articles, or other prior art references that help paint the technological background and define the state of the art as it existed before the invention was made. In this regard, inventors are also of particular assistance in defining the invention to be protected and in drafting claims of proper scope. Oftentimes it is inventors who are aware of the most relevant prior art, and they can help analyze that art against the invention with a view toward preparing claims that will distinguish it over earlier inventions.

Additionally, the inventor will contribute to preparation of a specification that will enable others in the art to reproduce the invention as claimed, and he will provide materials useful in preparing examples for the application that reveal work that has already been accomplished. The inventor being required to disclose the best mode contemplated by him for carrying out the claimed invention, he will also assist in ensuring that a specification is prepared that satisfies this obligation as well. These materials can often be provided in preliminary form through an ''invention disclosure,'' as described in Section II.A.3. below.

Once the inventor provides these materials to his attorney and the application is drafted, the inventor will review that draft (and any future drafts) to ensure that the invention is described completely and in a manner that is both accurate and sufficiently detailed. The inventor should also continue to review the specification to ensure that the prior art is accurately described and to ensure that his attorney is aware of any relevant prior art that he may have recently recalled or become aware of. In view of the inventor's duty of disclosure to the PTO, discussed in Section II.A.D. below, the inventor should review the specification a final time just before filing in order to ensure that the document is complete and accurate, that the best mode is fully and properly disclosed and has not changed, and that all known pertinent prior art has been taken into account in crafting the application.

Depending upon the circumstances, the inventor may also wish to conduct a formal prior art search in conjunction with preparation of the application.

2. Searching the Invention

Inventors and businessmen alike give careful study to the investment required to develop and market a new product and, in determining whether to go forward, will balance that investment against the potential value of that product to their company. Similarly, although the value of a patent for a product depends primarily on the commercial importance of the product, there are other important considerations as well. These include the scope and content of the prior art, knowledge of which will be of value in drafting valid claims, in determining the permissible scope of those claims, and will, of course, allow one to evaluate whether a patent may be obtained for the product at all.

Accordingly, if warranted under the circumstances and if sufficient time is available, most patent practitioners will recommend that a prior art search be conducted before the patent application is prepared. On the one hand, a search may reveal that the invention is not novel or that claims of useful scope will likely not be obtained. If a search of the prior art does not uncover any such obstacles, the results of that search may nevertheless be useful in assisting the patent attorney preparing the application first to understand and then to describe and claim the invention and distinguish it from the known prior art. The results can also be used to prepare an Information Disclosure Statement for the PTO. Information Disclosure Statements, in which relevant prior art known to the inventor is typically listed, are now required to be submitted within 3 months of patent application filing pursuant to PTO Rule 97 (further details about Information Disclosure Statements are provided in Section II.A.4. of this chapter).

Many factors will be evaluated in determining whether to conduct a prior art search and, if so, how comprehensive that search should be. In some circumstances, especially in the rapidly advancing fields that compromise "biotechnology," an inventor will already be aware of most, if not all, previous publications that may be relevant in evaluating the breadth and the protectability of the invention to be claimed. In that case, the inventor may wish only to conduct a brief "patentability" search of issued U.S. patents to determine whether the invention or something fairly like it has already been described or patented.

There are other types of searches that may be conducted as well. In addition to "patentability" searches, a "state of the art" search can be carried out that is usually designed to give an overview of the prior patents in a particular area. An "infringement" search examines the claims of unexpired U.S. patents to determine if any unexpired patent would be infringed by practicing the subject invention. Additionally, a review of published foreign patent applications can be undertaken that, although not allowing a clear determination of rights since the claims of such foreign applications have not yet been granted, may give early warning of another's potential patent rights. A "validity" search is a more extensive search that will normally include a review of the scientific literature, both here and abroad, as well as foreign and U.S. patents.

Searches can be carried out in a number of different places. For example, the collection of issued U.S. patents available at the PTO may be reviewed by the public. Foreign patents and some scientific literature are available in the PTO's Scientific Library. Foreign search facilities available to the public are found at various national patent offices, most notably the British Patent Office, and searches can actually be ordered from certain patent office facilities in Europe. The European Patent Office Branch at the Hague in the Netherlands accepts search requests and will report the results in English. The Swedish Patent Office will also conduct a confidential prior art search and report the results in English. The Austrian Patent Office conducts confidential prior art searches and will also render an opinion on patentability. However, although the search request may be submitted in English, the Austrian Patent Office will only report its results in German.

There are also now a number of major data bases that include full-text copies of issued U.S. patents, abstracts of foreign patents and published applications, abstracts of technical publications, and so on. These include Derwent's "World Patent Index," Mead Data's "Lexpat," the "Chemical Abstracts," and Pergamon/Infoline's "Inpadoc." The attorney, the inventor, or another company employee may do some or all of the searching desired in a particular situation. However, a search firm that specializes in these tasks also can be employed.

Although searching can be expensive, so can preparation of a patent application. Thus, it is worthwhile for the inventor and his attorney to carefully consider, in view of those expenses, whether a search is desirable prior to preparation of a patent application. There are a number of factors that might counsel against a search. First, the inventor may be well able to conclude, based upon his knowledge of the field, that the invention is a clearly patentable advance, no matter what the state of the published art. Second, the area may be one with which the inventor is deeply familiar, and he can be relatively sure that no prior art exists that would be of concern. Third, the inventor or his company may view the invention as so important that a patent application should be prepared and filed notwithstanding the prior art. Finally, a patent application might have to be prepared swiftly, with no time to conduct a prior art search. Of course, none of these are absolutes, and some form of search, limited to issued U.S. patents, for example, might desirably be conducted in advance. The question of whether a search is to be conducted, and if so, how extensive that search will be should always be considered by the inventor and his attorney.

In addition to the results of any search, there are a number of documents—those typically prepared during the course of conceiving, developing, making, and testing an invention—that are useful in preparing a patent application.

3. Other Documents Useful in Drafting the Application

The major cost of obtaining a patent is attorney fees, which are ordinarily based upon the amount of attorney time involved in preparing, and subsequently prosecuting, the application. Thus, an inventor or his company can save both time and money by making the attorney's job easier. There are a number of materials that can be prepared or organized by the inventor to help his patent lawyer. These typically include invention disclosures, internal reports, grant applications, draft publications, and earlier (prior art) articles published by the inventor or others.

An invention disclosure is a technical report describing various aspects of the invention. Forms of invention disclosure are available from most patent attorneys. They are designed both to help establish dates of invention and to provide the patent attorney with information needed to begin drafting the patent application. An invention disclosure will usually include a thorough description of the invention, as well as all known information concerning its construction and intended use. The disclosure should also include a description of known related inventions or products and a detailed description of the best ways of practicing the invention. Alternate uses of the invention should also be set out. If there are likely alternative embodiments or aspects of the invention, that is, other ways of carrying out the invention, those should also be described. In the biotechnology invention, such existing and alternative embodiments or aspects may involve various different compounds, protein and nucleic acid sequences, plasmids and other vectors, microorganisms, cell lines, and so on. Although the inventor does not have to detail each and every remotely possible attribute or embodiment of the invention, describing alternatives or modifications of various parts of the invention will assist the lawyer in preparing a specification and drafting claims adequate to provide proper protection for the invention.

With regard to all of the above, laboratory work actually accomplished should be identified and described. Relevant laboratory notebook entries or technical reports can be used to supplement this description. If the inventor has or knows of relevant prior art documents, copies of those documents should be provided to the lawyer together with an indication of their technical relevance to the disclosed invention. In the event that the invention is an improvement over something that went before, an attorney will generally have to analyze the prior art to assess patentability of the invention and to determine how to properly prepare the claims. If the inventor can provide relevant prior art materials, and help to guide such an analysis beforehand, this process can be expedited.

In the company setting, monthly or other periodic reports are also helpful in providing information and in evaluating work done with respect to an invention and can facilitate preparation of an application. Grant applications, even preliminary drafts, usually will contain significant amounts of information useful to the drafter of the patent application.

Another, and often the best way to assist in the preparation of a patent application is to prepare a draft of a full scientific paper as if it were to be submitted for publication. Such a paper will usually include an introduction or background to the subject matter of the experiments, a methods and materials section, a description of the results, and a discussion section. Each of these can be used beneficially as described below. Normally, the more written materials that can be provided by the inventor, the more economically and efficiently the patent application can be prepared by the attorney.

4. Duty of Disclosure and Petitioner Misconduct

It has long been the law that an applicant for patent, and those individuals substantially involved in the preparation and prosecution of his application, such as his attorney, have a duty of candor and good faith in dealing with the PTO, which includes a duty to disclose information known to be material to the patentability of a pending claim. The grant of patent rights is dependent, in large part, on the examiner making a decision on patentability after a thorough consideration of the scope and content of the prior art. Any withholding from the Office of information that directly affects the determination of patentability of an invention may undermine the reliability of the resulting patent. Patent and Trademark Office rules are thus designed to ensure that the examiner considers all pertinent prior art before making a final decision on patentability.

There have been many attempts over the years to define a breach of the duty to disclose information to the PTO. Generally, the nondisclosure or misrepresentation to the PTO of information that is material to the examination of a patent application with an intent to deceive has been held to constitute a breach of the duty of disclosure and is termed *inequitable conduct* by the courts. Inequitable conduct is a complete defense against enforcement of a patent and can result in the patent being held invalid. Accordingly, it is important that the inventor and all others associated with the preparation and prosecution of the patent application understand their obligations of disclosure and that they meet those obligations. Typically, this means discussing with the attorney preparing the application all prior art of which the inventor is aware. Relevant art can then be reviewed for disclosure to the PTO. In this regard, an inventor's own publications should not be overlooked. The first materials that a potential infringer will search for in attempting to prepare a defense against an issued patent is scientific journal articles authored by the named inventor.

To assist inventors and their attorneys in satisfying their disclosure obligations, the PTO has issued a set of rules for the preparation and submission of Information Disclosure Statements. An applicant should carefully review these rules with his attorney at the time of preparation of the patent application. This will ensure that both the inventor and his attorney are reviewing and considering, at the earliest possible moment, not only that prior art that may have to be disclosed to the PTO but art that will be useful in preparing the application itself.

It should be understood that this obligation of disclosure is a continuing one. Accordingly, whenever material information is discovered, through prior art searches or otherwise, it must be disclosed to the PTO in a timely fashion so that it may be considered by the examiner prior to issuance of a patent.

5. Timing of Filing: "Priority" and "Prosecution" Applications

Consideration of timing in the preparation and filing of patent applications can be crucial. This results primarily from the fact that the laws of the United States, as opposed to those of virtually all other economically significant countries in the world, differ in the definition of what is novel. Outside the United States, most countries apply an "absolute novelty" test. In these countries, "an invention is new if it is not part of the 'state of the art'; the state of the art being defined as everything that was available by written or oral publication, use or any other way, in any country in the world," before the date on which the patent application was filed.[28] Thus, in most countries an invention is not novel, and therefore not patentable, if it was described in a publication or talk, for example, by the inventor, or somebody else, prior to the application filing date. Likewise, the invention is not novel in such countries if it was sold, or even if it was offered for sale, prior to the date of filing for the patent application.

There is an exception to that rule in the United States where the inventor is given a 1-year grace period to file his patent application after publication or sale. Most companies and institutions, however, are concerned about preserving patent rights in countries important to the international market, and it is therefore often important that an inventor file a patent application before any public disclosure of the invention.

Another important difference between the U.S. patent system and those of the rest of the world is that, in the United States, a patent is granted to the first to invent and not to the first to file. The "first to invent" and "first to file" rules can be further explained as follows. In the United States, if two or more patent applications claim the same patentable invention, the PTO conducts what is known as an "interference" proceeding to determine who is the first inventor. The rule in the case of such a priority contest is that the patent goes to the first to invent. In the rest of the world, if two patent applications claim the same invention, the patent goes to the person with the earliest filing date. Clearly, under this rule, simply filing your patent application before you publish may not be sufficient to protect your invention. You must file first, before anyone else.

These questions of timing give rise to two forms of patent applications that have been usefully designated by one author as "priority" and "prosecution" applications.[25] Because a patent application filing must be effected before publication to avoid the irrevocable loss of foreign patent rights, it is sometimes necessary to prepare a patent application quickly to establish a filing, or "priority," date. Thus, where time is of the essence because the inventor intends to publish the invention or describe it in a speech, or because the invention has been made in a "hot" area where many researchers are racing to achieve results, a patent specification to establish priority must be quickly prepared.

On the other hand, where time is not of the essence, a patent specification can be more thoughtfully and comprehensively pursued. The strategy with regard to the filing of such a "prosecution" application might even include waiting a period of time while additional experiments are carried out and further data are generated in the laboratory.

In both situations, it is important to bear in mind that early filing is often desirable in and of itself because it avoids the possibility of the publication of new prior art against which the patentability of the invention may ultimately be measured. In either situation, the following drafting objectives and rules will also play some role.

B. The Disclosure: Introduction to Preparing the Specification

1. The Story of the Invention

A patent is much more than simply a legal document. It is also a technical publication and, oftentimes, an effective sales brochure. It is a legal document in that, through the claims, it delineates the boundaries of the patent property. It is a technical document in that it contains a written description of the patented invention so that everyone interested in the field to which the patent pertains may read it and learn how to make and use the invention. A patent is analogous to the sales brochure because it describes what was known before (the prior art) and typically will set forth the characteristics and advantages of the claimed invention and how it is an improvement over the known art.

It has been noted, quite correctly, that virtually every invention has a story and that a patent application should be written to effectively tell that story.[29] Although some inventions represent inadvertent discoveries, most are the result of thoughtful and time-consuming efforts to comprehend and solve a particular problem. The courts understand this as well. Many years ago the Court of Customs and Patent Appeals (predecessor to the Federal Circuit Court of Appeals) perceived that rejections of claimed inventions for want of nonobviousness were being made by the PTO on the basis that it would have been ''obvious to try'' to make that invention. The court swiftly denounced that standard, stating that such an approach merely:

> begs the question, which is obviousness under Section 103 of *compositions* and *methods* not of the direction to be taken in making *efforts* or *attempts* [italics added by the court]. Slight reflection suggests, we think, that there is usually an element of ''obviousness to try'' in any research endeavor, that it is not undertaken with complete blindness but rather with some semblance of a chance of success, and that patentability determinations based on that as a test would not only be contrary to statute but result in a marked deterioration of the entire patent system as an incentive to invest in those efforts and attempts which go by the name of ''research.''[8]

It is important for the inventor to understand that serendipity is not a prerequisite to patentability.

Learning the story behind an invention and how it was made will assist in revealing the important features and utilities of the invention, the problems solved, and so forth. The story will also help determine how to claim the invention so that it avoids the prior art and yet provides adequate commercial protection. The story of the invention also sheds light on the content of the literature, the level of skill in the art, the problems facing the inventor and others working in the field, and the real-world difficulties that were overcome by the inventor.[29] Various portions of the story of an invention can be told in each of the sections that make up a complete patent application.

2. The Parts of a Patent Application

The rules of the U.S. Patent and Trademark Office include rules relating to the parts of an application and their arrangement. Patent Office Rule 77 and Section 601 of the *Manual of Patent Examining Procedure* set forth a preferred layout of the application, stating that the following order of elements is desirable in framing the specification:

1. Title of the invention
2. Cross-reference to related applications (if any)
3. Statement as to rights to inventions made under federally sponsored research and development (if any)
4. Background of the invention, including
   a. Field of the invention
   b. Description of related art (including material disclosed in an Information Disclosure Statement
5. Summary of the invention
6. Brief description of the drawing
7. Detailed description
8. Claim(s)
9. Abstract of the disclosure

The parts of a patent application are briefly summarized here, prefatory to more detailed treatment in the sections that follow.[9]

Following the title of the patent application, any related patent applications are identified and, if appropriate, a statement concerning the extent of government rights in the invention is provided if the invention was made using federal funds under a grant or contract awarded by the government. The following section of the patent application can be titled "Background of the Invention" or "Background and Introduction to the Invention." In it the state of the art that existed before the invention was made is described, with a view also toward setting forth the then extant problems in the art that the invention solves. Information can also be included here that may not represent the state of the art but will serve to introduce the invention; for example, by reference to as yet unpublished work related to the invention.

In the summary of the invention, the invention is described in general terms, including its major and/or important aspects. This section generally also includes a summary, or even a recitation, of the broadest or most important claims.

Some patent applications also include a definitions section, although this is optional or is sometimes included as a part of the Summary of Invention. Such a section can include terms that may have specialized meanings in the context of the application. It can also be used to define terms that may be ambiguous and have varying meanings or connotations in the art.

After the summary or definitions section follows a brief description of the drawings. Because the PTO rules prohibit the inclusion of text on drawings, where drawings are included they are normally briefly described in this section.

A detailed description of the invention follows. In this section, the invention is described in greater detail than the summary, including all its important aspects as well as those aspects of the invention that can be varied in making or using the invention and how. Here, in the case of biopharmaceutical compounds, for example, one will include descriptions of the compounds, including "preferred" (and even "especially preferred") compounds; how the compounds are prepared; and the utility of the compounds, including advantages that may be illustrated with descriptions of results and/or reference to examples. Especially for the purpose of satisfying foreign patent disclosure requirements in the case of a biopharmaceutic intended for use in humans, a detailed description of in vitro or in vivo biological test results, preferred therapeutic routes of administration, and dosages of compounds to be administered should also be included.

The "examples" are a part of and normally conclude the detailed description of the invention, and include either descriptions of results actually accomplished ("working examples") or protocols that are to be accomplished ("prophetic examples").

Following the examples are, of course, the claims.

Finally, a brief abstract of the technical disclosure is provided. The purpose of the abstract, to be no more than 50–250 words, is to enable the PTO and the public generally to determine from a cursory inspection the nature of the technical disclosure.

Each of these sections of the patent application is discussed in more detail below, with appropriate references to the Appendix 1 Hybritech patent and court decision found in Appendix 2. The introductory portions of the specification are considered first; namely, the title, abstract, and background of the invention. Those parts that make up the remainder of the specification and that serve to disclose the invention—the summary, drawings, and detailed description of the invention—are then discussed. Claim drafting is examined last.

C. The Disclosure: Preparing the Introductory Portions of the Specification

1. Title and Abstract of the Disclosure

The considerations in constructing the title of the invention are few. U.S. Patent and Trademark Office Rule 72(a) requires only that the title be "as short and specific as possible," and that it appear as a heading on the first page of the specification. Section 601 of the *Manual of Patent Examining Procedure* further asks that titles of applications be limited to from two to seven words. The title of the Hybritech patent—"Immunometric Assays Using Monoclonal Antibodies"—plainly meets all the above requirements. The title of the application should always be at least commensurate in scope with the broadest claim.

Although placed last in the application, the abstract of the invention is printed on the first page of an issued patent, closely following the title. Pursuant to PTO Rule 72(b), "A brief abstract of the technical disclosure in the specification must be set forth on a separate sheet, preferably following the claims under the heading 'Abstract of the Disclosure'." Although intended to be brief, the content of the Abstract, according to MPEP §608.01(b), should enable its reader, regardless of his degree of familiarity with patent documents, to ascertain quickly the character of the subject matter covered by the technical disclosure and "should include that which is new in the art to which the invention pertains."

By way of example, the abstract for the Hybritech patent reads as follows:

> "Two-site" or "sandwich" immunometric assay techniques for determination of the presence and/or concentration of antigenic substances in fluids using monoclonal antibodies. One monoclonal antibody is presented in a soluble labeled form and a second monoclonal antibody is presented bound to a solid carrier; the soluble and bound monoclonal antibodies may be the products of either the same or different cell lines. Each monoclonal antibody has an affinity for the antigenic substances of at least about 10^8 liters/mole.

As with other parts of the patent application, care should be taken in drafting the abstract, bearing in mind the potential impact of the abstract in later litigation over the patent. Although PTO Rule 72(b) provides that the abstract "shall not be used for interpreting the scope of the claims," the courts have held that the abstract of the disclosure is considered to be a part of the specification and may be used in determining whether an application meets the disclosure requirements of section 112 of 35 U.S.C.

2. Background of the Invention

A significant portion of the work of drafting a specification begins with the description of the background of the invention. The background section should have two parts: a reference to the field of the invention and a description of related art or other introduction to the invention.

a. Field of the Invention Typically, the field of the invention is described in broad terms. The field in which the Hybritech invention was made was identified in the following way:

> This invention relates to methods for detecting and/or determining the concentration of antigenic substances in fluids such as serum. In another aspect in relates to immunometric assay techniques. In yet another aspect it relates to monoclonal antibodies.

In essence, this represents a statement of the field or fields of art to which the invention pertains and assists the PTO in identifying the correct examining group to which the case should be assigned. Because, as described in Chapter 7, this and other parts of the specification can be used by a court in interpreting the meaning and scope of patent claims, it is important not to describe so narrowly the field of the invention that it would appear to imply a limitation to the claims.

As noted above, following the field of the invention, the substantive portion of the background of the invention section is usually written to help illuminate the manner in which the claimed invention differs from, and is nonobvious over, the prior art and provides a measure of the advance over the prior art.

b. Related Art An effective means to introduce the story of an invention is to begin with a brief description of the prior art. Section 608.01(c) of the MPEP ("Background of the Invention") calls for:

> A paragraph(s) describing to the extent practical the state of the prior art or other information disclosed known to applicant, including references to specific prior art or other information where appropriate. Where applicable, the problems involved in the prior art or other information discussed which are solved by the applicant's invention should be indicated.

The amount of prior art references in a particular field, however, may be too great to readily permit a brief description of all of it in the background section. In that case, the references can be combined in the Information Disclosure Statement for submission to the PTO. This permits a more readable, concisely focused statement in the patent application itself.[35]

The background includes not only a summary of the most pertinent prior journal articles, patents or other publications but a characterization of that art, including its limitations, disadvantages, and the desired goals that have been left unmet by it. In other words, the background section is an opportunity to introduce the story of the invention by describing the technical state of affairs that preceded it.

It is generally recognized that the determination of the obviousness question is necessarily somewhat subjective, looking as one must for an answer to that question from a hypothetical person of ordinary skill in the art. Accordingly, courts and litigants alike have come to rely upon objective evidence of nonobviousness that, if available, can usefully be made a part of the patent specification. Such objective evidence includes (1) the prior *failures of others* in the field; (2) the existence of an unmet but *long-standing problem or need*; (3) *unexpected results,* including the surprise or disbelief of experts; (4) *proceeding contrary to accepted wisdom* of the art; (5) *skepticism* about the merits of the claimed invention; (6) *commercial success* attributable to the merits of the claimed invention; (7) *copying* of the claimed invention by others; and (8) *industry respect* for the claimed invention, as may be measured by extant licenses.

Of these, it is safe to say that prior failures and long-felt need are the most likely candidates for discussion in the background portion of a patent application. Many of the remaining objective considerations, best exemplified by those of commercial success and extensive licensing, do not typically come into existence until after—often long after—a patent application has been filed or issued. This is especially so in the area of biotechnology, where a lengthy U.S. Food and Drug Administration approval process is mandated before the commercial sale of most diagnostic and therapeutic inventions.

This was the case with the Hybritech invention and patent. There, persuasive, post-application evidence of commercial success was submitted into evidence at trial by Hybritech, evidence that was subsequently heavily relied upon by the Court of Appeals in upholding the patent (Appendix 2). The undisputed evidence showed that Hybritech's diagnostic invention had an enormous market impact.

> The first diagnostic kit sales occurring in mid-1981, sales increased seven million dollars in just over one year, from $6.9 million in 1983 to an estimated $14.5 million in 1984; sales in 1980 were nonexistent. Competiting with products from industry giants such as Abbott Labs, Hoffman LaRoche [*sic*], Becton-Dickinson, and Baxter-Travenol, Hybritech's HCG kit became the market leader with roughly twenty-five percent of the market at the expense of market shares of the other companies. Its PAP kit ranks second only to a product sold by DuPont's New England Nuclear, surpassing products from Baxter-Travenol, Abbott, and others. Hybritech's other kits, indisputably embodying the invention claimed in the '110 patent, obtained similar substantial market positions.

The court emphasized the lack of any evidence that this commercial success was due to anything but the merits of Hybritech's sandwich assay invention, which was purchased and used by medical personnel in order to evaluate patient condition. In other words, the evidence showed that these kits competed successfully with numerous others for the trust of persons who have to make fast, accurate, and safe diagnoses.

In holding the Hybritech invention to be nonobvious within the meaning of Section 103, the court also relied upon the testimony of witnesses skilled in the diagnostic field. These witnesses, based upon tests done in their laboratories as a result of real-world comparisons in the normal course of research, heralded Hybritech's products and concluded that the diagnostic kits embodying the Hybritech invention unexpectedly solved long-standing problems in the art (Appendix 2).

Additional compelling evidence of nonobviousness, furthermore, was solicited during the cross-examination of the opponent's expert witness at trial. At the time the Hybritech invention was contended to have been obvious that expert had been working with sandwich assays and had even filed a patent application on sandwich assays. At the same time, the expert had also been working with monoclonal antibodies. Yet, she never once mentioned monoclonal antibodies in her patent application. That fact was noted by the Federal Circuit in holding the Hybritech patent valid (Appendix 2). The inconsistency is apparent; if the invention was as obvious as the expert had contended at trial, why didn't she put her monoclonal antibodies and her sandwich assay together?

While after-the-fact, objective evidence of nonobviousness—commercial success, unexpected advantages that satisfied long-felt and widespread needs, and the failures of others—was important in supporting the unobviousness of the Hybritech assay invention in court, the patent itself was drafted to provide then-available evidence of nonobviousness. It showed the superiority of invention over the prior art.

In the background of the Hybritech patent is a brief but accurate description of prior immunoassay methods for the determination of the presence or concentration of antigenic substances. In column 1, competition immunoassay techniques and immunometric assay techniques are described. Prior art immunometric assays run in the "forward" assay format are described, as is the use of such a format in a qualitative "yes/no" assay, and a quantitative assay to determine the amount of target antigen by the use of positive or negative controls is recounted. The patent then sets forth the problems with these prior art assays:

> Despite their great utility, the prior art immunometric assays have been recognized to be slow procedures, in part because two washing steps are required and because lengthy incubation periods are required to reach equilibrium, i.e., the point at which the amount of complex formed does not change with increasing time [column 2, lines 18–23].

In recognition of these difficulties, the patent goes on to state, the prior art proposed the use of "simultaneous" and "reverse" assay formats. This is also described in column 2 of the Hybritech patent. Again, the difficulties with these assay formats are set forth:

> Both the simultaneous and reverse assay techniques require a sufficient excess amount of solid phase antibody to bind most or all of the antigen present to avoid a high dose hook effect where artificially negative or low quantification of antigen is observed at extremely high concentrations of antigen. For this reason, the forward assay has been the approach preferred by the prior art [column 2, lines 46–53].

Column 2 of the Hybritech patent also describes difficulties related to the need in such assays for large amounts of highly purified, active antibody specific to the antigen of interest, which are difficult to obtain from the "polyclonal" antibodies used in prior art processes (column 2, lines 53–60). The patent describes the difficulties, including loss of assay sensitivity, when polyclonal antibody mixtures are utilized in the prior art reverse and simultaneous assay formats.

The background portion of the Hybritech patent closes with the following transition to the remainder of the specification:

> In view of these shortcomings, the limitations to the immunometric procedures known to the prior art are readily apparent. The conventional forward assay is time consuming; the simultaneous and reverse assays are accomplished with fewer steps but require large quantities of solid-phase specific antibody and are not well suited to determination of small concentrations of antigen since formation of a sandwich of the antigen with a multiple number of labeled antibody molecules competes with formation of the sandwich comprising bound antibody:antigen:labeled antibody; and all are subject to misinterpretation of false-positives due to the polyclonal nature of the antibody [column 3, line 58–column 4, line 2].

Thus, the background of the invention portion of the patent can be viewed as a mixture of the most pertinent prior art and the technical story against which the invention unfolds.

The Hybritech patent clearly presents a story in the form of an outline of the problems and difficulties faced, but not solved, by the prior art. In view of the fact that individuals without technical training—in the form of a judge or members of a jury—may be called upon to review, understand, and rule on the merits of a patent, it is useful to attempt to draft the background portion of the patent in both an interesting and a persuasive manner.

Care should be taken in two further aspects when preparing the background of an invention. First, it is hazardous to prepare a patent disclosure without a good working knowledge of the most relevant prior art, for factual statements about the prior art must be accurate. Use of the results of a prior art search conducted on the invention can be quite helpful in this regard to the attorney in the preparation of the application.

Second, care should be taken to avoid admissions about the state of the art that may not be accurate. There is case law to the effect that once an admission has been made, it is not possible to retract it. The patent that issues, including the background of the invention section, will be subjected to careful scrutiny if the patent is ever litigated and every statement in it subjected to the most detailed analysis. If well drafted, of course, the section will facilitate both patent prosecution and patent enforcement.

D. The Disclosure: Preparing the Description of the Invention

Although the introductory portions of the specification are used to effectively introduce the invention, the remainder of the specification is used both to complete the story and to satisfy the legal requirements for description.

1. Introduction: The Written Description, Enablement, and Best Mode Requirements

a. The Written Description Requirement As introduced above, 35 U.S.C. § 112 provides that the "specification shall contain a written description of the invention. . . . " This portion of the statute has been held by the court to impose a description requirement that is distinct from the enablement requirement. The function of that portion of Section 112 that provides that the "specification shall contain a written description of the invention" is to ensure that the invention actually conceived by the inventor is memorialized at the time of filing. The description requirement is wholly distinct from the enablement requirement; the former deals with what has been conceived and then described in a patent application, whereas the latter deals with whether what has been described in the application can be made and used.

The description requirement typically comes into play in the PTO when a claim is added at some stage after an application has been filed and the claim differs in scope from the original claims. If the added claim does not find support in the specification in the sense that it is for an invention not sufficiently described therein, the description requirement is not met and one cannot obtain the benefit of the original filing date for that particular claim. Claims can be barred by the description requirement even though the original specification adequately teaches how to make and use the subject matter of the claims.

Two court cases, one old and one new, emphasize the nature of the description requirement. In *In re Smith*,[10] the court noted that the description requirement can arise in any of three contexts:

> Satisfaction of the description requirement ensures that subject matter presented in the form of a claim subsequent to the filing of the application was

> sufficiently disclosed at the time of filing so that the *prima facie* date of invention can fairly be held to be the filing date of the application.

The court also emphasized that:

> Where the claim is an original claim, the underlying concept of ensuring disclosure as of the filing date is satisfied, and the description requirement has likewise been held to be satisfied.

One might receive a written description requirement rejection from the PTO, therefore, where a claim is rewritten during prosecution to recite elements that the examiner determines to be without support in the original disclosure. This emphasizes the importance of considering and addressing all pertinent utilities for, and embodiments of, an invention in preparing the patent application. Where the inventor of an immunoassay diagnostic method, for example, leaves out a description of the use of that method for evaluation of a particular disease condition, he may well be barred from attempting to specify the use of that assay for evaluating that disease state in a claim.

More recently, the description requirement was considered by the Court of Appeals for the Federal Circuit in *In re Wright.*[11] Wright's invention involved a method of using images using photosensitive microcapsules on a support. Wright amended his claim during prosecution, adding a limitation that stated that the microcapsules were distributed on a support "but not permanently fixed thereto." The PTO rejected the claim under section 112, contending that the phrase "not permanently fixed" was not specified in the disclosure. The court reversed this decision. It noted that compliance with the written description requirement does not require that the claimed subject matter not be described *in haec verba* and, upon reading the specification in light of the claims and the prior art, the court concluded that the added phrase was adequately described in the specification.

In reaching its decision, the court discussed the nature and significance of the demand for a "written description" as follows:

> When the scope of a claim has been changed by amendment in such a way as to justify an assertion that it was directed to a *different invention* than was the original claim, it is proper to inquire whether the newly claimed subject matter was *described* in the patent application when filed as the invention of the applicant. That is the essence of the so-called "description requirement" of §112, first paragraph which opens with the words: "The specification shall contain a written description of the invention . . . " The invention is necessarily the subject matter defined in the claim under consideration.

It is also important to understand that original *claims* (as opposed to amended claims) actually form a part of the original *disclosure*, and can be used in support of compliance with the written description requirement.

Although the consequence of a failure to satisfy the written description requirement is harsh, the courts do not apply a formalistic approach to evaluating a specification for compliance. It is generally enough to satisfy the requirement if the "specification contains a statement of appellant's invention which is as broad as appellant's broadest claims. . . ."[12] However, a lack of literal support for a claim does not mean that the written description requirement is not satisfied. Again, according to the courts:

> How the application . . . achieves compliance is immaterial. . . . It is only required, for example, that the specification describe the invention sufficiently for those of ordinary skill in the art to recognize that the applicant invented the subject matter he now claims.[13]

From the standpoint of patent draftsmanship, it is clearly preferable to use language in the specification that tracks closely, if not exactly, the language used in the actual claims to assure that no question arises as to compliance with the written description requirement. Thus, as discussed in further detail below, the additional limitations added to the broadest independent claims in the development of dependent claims or other claims or narrower scope should also be clearly reflected in the specification.

It is not always possible to predict the course of subsequent prosecution of a patent application, and cases may arise in which a later amended or added claim does not find literal support in the specification. In those cases, resort may be had to the entire specification to establish possession of the invention as required by section 112 of 35 U.S.C. It must be understood, however, that there remains a risk that a specification that supports a broad claim may not contain a written description of the invention embraced by a narrower claim. For example, a patent applicant may wish to amend a claim to recite a range of temperature within a broader disclosed range of temperatures, or a range of proportions of materials in a mixture, which has not been specifically disclosed. In such instances, it may not be possible to find support for a narrower claim even though a broad claim has been adequately described. These considerations should be taken into account when drafting the patent application that recites subject matter that the applicant may wish to subsequently narrow in scope.

In the Hybritech patent, for example, it can be seen that broad claim 1 defines the use of monoclonal antibodies having an affinity for the antigen to be assayed of "at least about 10^8 liters/mole." A narrower claim, claim 4, provides that the affinity is "at least about 10^9 liters/mole." Column 5 of the patent, at lines 35–37, contains an express basis for this narrower claim, reciting that, "Preferably, the monoclonal antibody selected will have an affinity of at least about 10^8 liters/mole and, more preferably, an affinity of about 10^9 liters/mole."

Similarly, Hybritech's claim 1 refers to a process for the determination of the presence or concentration of "an antigenic substance" in a fluid. The claim is not restricted to any particular antigenic substance. Narrower claim 7, however, limits the test substance to immunoglobulin E, hepatitis A, heptatitis B, heptitis Non-A/Non-B, alphafetoprotein, carcinoembryonic antigen, insulin, and human thyroid-stimulating hormone. These terms find express written support in the patent at column 4, lines 57–66, and the written description requirement is unambiguously satisfied.

b. The Enablement Requirements The patent application must also contain a written explanation of how to make and use the claimed invention. According to the courts, this description must be in such full, clear, concise, and exact terms as to enable any person skilled in the art to make and use the invention without "undue experimentation." As with the written description requirement, compliance with the enabling disclosure requirement is critical because it cannot be satisfied by later amending the application to modify or add any further description of the invention not already in place.

A number of factors are considered in determining whether a disclosure is insufficient because it would require "undue experimentation" to practice the subject invention. These include:

1. Quantity of experimentation necessary
2. Amount of direction or guidance presented
3. Presence or absence of working examples
4. Nature of the invention
5. State of the prior art
6. Relative skill of those in the art
7. Predictability or unpredictability of the art
8. Breadth of the claims

It has long been the law, on the other hand, that the fact that some investigation or evaluation is necessary does not preclude enablement, especially where such experimentation is due to the nature of the invention and the involved science. The fact that experimentation may be complex does not necessarily make it undue if the art typically engages in such experimentation. For example, a considerable amount of experimentation is permissible, if it is merely routine, or if the specification in question provides a reasonable amount of guidance with respect to the direction in which the experimentation should proceed.

In the same light, it is understood that there is no need for a "manufacturing specification." Indeed, a patent need not teach, and preferably omits, what is well known in the art (see Appendix 2).

It is also the law that every species that may fall within the scope of a patent claim need not be operative, or optimal. A patentee is not required to foresee every technological problem that may be encountered in adapting his idea to a particular use. Some experimentation and exercise of judgment is to be expected. The enablement provision, furthermore, does not require that the inventor correctly set forth, or even know, how or why his invention works. Even if the theory behind the invention is neither correct nor understood, however, the inventor must teach how to achieve the results of the invention as claimed.

In the *Hybritech* case, at issue was enablement for its immunoassay claims, which required monoclonal antibodies having an affinity for an antigenic substance of at least of about 10^8 liters/mole for each of the labeled and bound antibodies. The patent referenced an art-known procedure for making monoclonal antibodies, the Kohler and Milstein process, and described screening for monoclonals having the desired affinity. The District Court, however, invalidated Hybritech's patent by adopting the defendant's arguments that stated that the patent failed to disclose (1) how to make monoclonal antibodies; (2) how to screen for proper monoclonal antibodies; and (3) how to measure monoclonal antibody affinity. In reversing the lower court, the Federal Circuit wrote:

> The District Court itself stated that the "method for producing monoclonal antibodies *in vitro* was well known prior to the alleged invention of the '110 patent" and used the "sudden availability of monoclonal antibodies" produced by the Kohler and Milstein discovery to support, albeit erroneously, its finding of a lack of nexus between the merits of the claimed invention and its commercial success. The Court then about-faced and held the '110 patent deficient because it fails to teach how to make monoclonal antibodies.

With respect to screening, the court ruled that the only permissible view of the evidence was that screening methods used to identify the necessary antibody characteristics, including the affinity of the monoclonal antibodies to be used in the invention as claimed, were well known to those in the art. The court therefore held, as a matter of law, the Hybritech patent disclosure enabling (see Appendix 2). Hybritech's reference to prior art procedures useful in preparing its assays is clearly proper, as acknowledged by the court.

For purposes of preparing the application, compliance with the enabling disclosure requirement is critical. As noted above, that requirement cannot be satisfied by later amendment to the application. A patent application must be enabling as of the date it is filed in the U.S. Patent and Trademark Office. Getting and keeping an early filing date is important for reasons earlier noted; namely, the "first-to-file" principles that determine entitlement to a patent in most countries other than the United States, and because a filing date establishes a prima facie date of invention such that the PTO cannot use publications occurring after that date as prior art in determining patentability of the invention.

Further issues specific to biotechnology-related patent applications, including the necessity for deposit of biological materials or other starting materials in order to satisfy the enablement requirement are discussed below.

c. The Best Mode Requirement Section 112 of 35 U.S.C. also requires that the specification "set forth the best mode contemplated by the inventor of carrying out his invention.'' The purpose of the best mode requirement is to ensure that the public, in exchange for the rights given the inventor under the patent laws, obtains from the inventor a full disclosure of the preferred embodiment. Satisfaction of the best mode requirement, accordingly, is a question separate and distinct from the question of the sufficiency of the disclosure to comply with the enablement provisions.

Failure to comply with the best mode requirement amounts to concealing the preferred mode contemplated by the applicant at the time of filing. A patent applicant is under no obligation to disclose to the PTO post-filing date improvements and modifications. Thus, the question is one of concealment as of the filing date; that is, whether an applicant has withheld what he considers to be the best mode of carrying out his invention. The best mode requirement does not require one to obtain further knowledge but only to disclose what one knows, or, at least, contemplates.

Some difficulty arises in assessing compliance with the best mode requirement because there is no objective standard by which to judge the adequacy of a best mode disclosure. According to the courts, though, only evidence of concealment (accidental or intentional) is to be considered. In order to reflect a failure to meet the best mode requirement, that evidence must tend to show that the quality of an applicant's best mode disclosure is so poor as to effectively result in concealment. Accordingly, a best mode violation is not indicated where it would require only the application of routine skills to produce the best mode that has been identified in the specification. The difficulty usually arises in situations where the best mode has not been identified with sufficient specificity.

For example, in *Dale Electronics, Inc.* v. *R.C.L. Electronics, Inc.*,[14] the court invalidated a patent on a claimed electrical resistor for failure to disclose a material used in making it that was known to work best. The patent only disclosed a general class of material to be used, some of which did not work well. The patent spoke in terms of making the claimed resistors with ''materials in the classification of epoxys, phenolics or silicons,'' but one specific material (Rogers RX 600) was known to work. The court specifically held that whether the nondisclosure was intentional is irrelevant:

> We do not here accept the proposition that where an inventor, as here, clearly knows a specific material that will make possible the successful reproduction of the effects claimed by his patent, but does not disclose it, speaking instead in terms of broad categories, he may nevertheless be considered as having described the best mode contemplated by him. . . .

The court found that the patentee failed to make work many other materials in the classes described in the patent. In that case, as well as in others, such general disclosures have been held not to comply with the best mode requirement.

In the *Hybritech* case, the defendant Monoclonal argued that Hybritech's patent was invalid for failure to disclose the best modes for making and screening hybridomas. Because not complying with the best mode requirement amounts to concealing the preferred mode contemplated by the applicant at the time of filing, in order to find that the best mode requirement is not satisfied, it must be shown that the applicant knew of and concealed a better mode than he disclosed. The defendant Monoclonal never introduced any such evidence. The court specifically rejected Monoclonal's argument that concealment by Hybritech of a best mode for screening or producing monoclonal antibodies was evident simply because the screening methods used to identify monoclonal antibodies having the claimed characteristics are labor intensive, time consuming, and carried out by sophisticated persons (see Appendix 2).

It is primarily in those portions of the patent specification that follow the "Background and Introduction to the Invention" that are typically relied upon and reviewed in determining satisfaction of the section 112 of 35 U.S.C. written description, enablement, and best mode requirements. Each of these sections is discussed in further detail below, both with a view toward satisfaction of an applicant's disclosure obligations and in particular regard to disclosure issues relating to biotechnology inventions.

2. Summary of the Invention

Rule 73 of the U.S. Patent and Trademark Office requires a brief summary or general statement of the invention indicating its nature and substance. According to the rule, the summary may include a statement of the objects of the invention, and should precede the detailed description of the invention. Furthermore, "Such summary should, once set forth, be commensurate with the invention as claimed and any object recited should be that of the invention as claimed."

The *Manual of Patent Examining Procedure* correctly emphasizes that the summary is separate and distinct from the abstract and is directed toward the invention rather than the disclosure as a whole. According to § 601 of the MPEP:

> The summary may point out the advantages of the invention or how it solves problems previously existent in the art (and preferably indicated in the Background of the Invention). In chemical cases the summary should point out in general terms the utility of the invention. If possible, the nature and gist of the invention or the inventive concept should be set forth. Objects of the invention should be treated briefly and only to the extent that they contribute to an understanding of the invention.

These considerations are reflected in the Summary of the Invention portion of the Hybritech patent. It reads as follows:

> According to the present invention, the polyclonal antibody used in an immunometric assay as the unlabeled antibody bound to a solid support and the antibody used as the soluble labelled antibody are replaced by at least one and usually two or more different monoclonal antibodies, *i.e.*, each antibody specific to a single antigenic site and separately produced by clones derived from unique cell lines. In a preferred embodiment of the invention, the monoclonal antibody used as the antibody bound to the solid support is the product of a different cell line than is the monoclonal antibody used for the labelled antibody and the two monoclonal antibodies are selected to bind the antigenic substance at sites remote from each other so as not to interfere with the others binding to the antigen. The advantages of the present invention, particularly in simultaneous and reverse assays, over prior art methods will become clear after consideration of the accompanying drawings and the following detailed description of the invention.

As indicated by the Hybritech patent, descriptive support for the invention can be provided by paraphrasing or identically including the features of the claims in the Summary of the Invention.

Some patent practitioners, it is said, do not include a Summary of the Invention out of concern that a judge or jury might some day narrowly construe the claims during litigation in light of an unduly restrictive summary.[29] Another reason sometimes given for omitting the summary is that the claims can and often do change during prosecution of the application and thus an accurate summary cannot actually be finalized at the time the application is prepared.[29] One way to handle this is by including a Summary of the Invention under the Detailed Description of the Invention, without a separate heading. However, most practitioners agree that the Summary of the Invention for subject matter of a patent that has been carefully evaluated and well thought through can be used to advantage.

The Summary of the Invention portion of a patent application is a good place in which to take a large step toward meeting the written description obligation imposed on an inventor. Written description support for the invention can be provided here in large measure by paraphrasing the claims and by providing an identification of particularly preferred embodiments as well as other important features, uses, and advantages of the invention. If the invention involves a pharmaceutical or biopharmaceutical compound, the summary will typically contain the broadest generic formula that describes it, as well as a recitation of acceptable chemical substituents in the formula. Methods of preparing and using these compounds as well as preferred species of the compounds also are usually broadly disclosed in the Summary of the Invention as a preface to more detailed treatment further along in the application. Where the claimed application is a process, it will usually be described in the summary broadly and will contain the minimum number of steps required to achieve the intended results. Any additional or preferred steps, or critical or preferred process conditions, will also be set forth broadly in the summary. Similar considerations apply to compositions that have been discovered and that contain a number of ingredients. The ingredients—both broad classes of ingredients and preferred ingredients—will be set forth in the summary and defined to include the minimum number of needed ingredients so that the composition is effective for its intended purpose. Optional ingredients and various and preferred methods of using or preparing such compositions are also summarily identified. In any event, the scope of the invention described in this section should typically be at least as broad as the broadest claim. Preparation of the Summary of the Invention in this fashion will provide effective support for the claimed invention.

It will be noted that, at column 4 of the Hybritech patent, certain "objects" of the invention are described. In that patent, the objects were to provide an improved process for the immunometric assay of antigenic substances, including assays that were speedier and more sensitive, and to provide improved simultaneous and reverse immunometric assays. Some practitioners decline to include an identification of such "objects" for fear that the claimed invention might not live up to one or more of them. Although that fear may be appropriate, the Federal Circuit has refused to rely upon a reference in the specification of a patent to an object of the invention to adversely limit the scope of a patent.[15]

3. Drawings

Drawings are often included in the patent application, but an applicant for patent is required to furnish a drawing only where necessary for the understanding of the subject matter. By law, any included drawings form a part of the specification that may be looked to in determining compliance with the disclosure requirements of section 112 of 35 U.S.C. Accordingly, drawings submitted after the filing date of an application may not be used (1) to overcome any insufficiency of the specification due to lack of an enabling or otherwise inadequate disclosure, or (2) to supplement the original disclosure for the purpose of interpretation of the scope of any claim (35 U.S.C. § 113). Drawings may include illustrations that facilitate an understanding of the inveniton; for example, flow sheets in the case of processes. Photographs are also permissible.

In biotechnology cases, the submission of drawings is common. Illustrations useful in providing a supporting disclosure for biotechnology inventions include, for example, restriction maps of sequenced DNA or recombinant DNA vectors, amino acid and DNA sequences for isolated proteins; reverse phase high-performance liquid chromatograph elution profiles of isolated proteins; autoradiograms; histograms; and flow diagrams showing cloning strategies for the isolation or expression of desired functional DNA fragments.

Frequently, biotechnology applications include drawings that present the results of experiments that have been detailed in the application. Usually, the data are presented in the form of a graph. In the Hybritech patent, for example, two graphs were presented. The first illustrated the results obtained using polyclonal antibodies in four types of immunometric assay for human immunoglobulin E. A second graph presented results showing the difference obtained using monoclonal antibodies according to the claimed invention in the same four types of immunometric assay (simultaneous, reverse, fast forward, and forward). The graphs presented the data from the examples found at columns 6–8 of the Hybritech patent, which data were also presented in tabular form. Because these data were clear, and the results apparent from the text of the application, the drawings were not needed. However, such graphical representation very usefully and efficiently supported Hybritech's determination that their invention was superior over prior art assays.

Although the PTO requirements for drawings are quite particular, informal drawings may be submitted with the application and will be accepted provided that they are readable and reproducible. It is generally desirable to submit informal drawings in order to expedite application filing. Formal drawings can be prepared in compliance with PTO rules and submitted in due time before the patent issues, and formal drawings will not be required by the examiner until the application has received a notice of allowance.

4. Detailed Description

a. Summary Description and the Disclosure of Alternatives Although the Summary of the Invention portion of the patent application and the drawings can be sufficient to fulfill the written description requirement of section 112 of 35 U.S.C., the Detailed Description portion of a patent application is used to memorialize any further description in a more detailed summary of pertinent aspects of the invention, including a detailed disclosure of contemplated or available alternatives useful in practicing the invention. It is also the primary means used to provide a disclosure sufficient to meet the enablement and best mode requirements.

A detailed description of the invention and the drawings follows the Summary of the Invention and Brief Description of the Drawings portions of the patent application. This detailed description, required by Rule 71 of the PTO, must be in such particularity as to enable any person skilled in the pertinent art or science to make and use the invention without unduly extensive experimentation. Although uses for the invention and methods of making the invention can be broadly set forth in the Summary of the Invention portion of the patent application, the detailed description of the invention expands upon that, providing the details of making and using the invention, followed by actual or prophetic examples of making and using the invention. In other words, the detailed description of the invention, in addition to providing a summary description of how to make and use the invention, the best mode, and disclosures of pertinent alternative embodiments, is also that part of the application where work actually accomplished or contemplated is presented in "working" or "prophetic" examples, respectively.

An applicant is ordinarily permitted to use his own terminology as long as it can be understood. Any necessary definitions intended by the applicant for terms used in the application, and which have not been set out in the Summary of the Invention, should be provided in this part of the application. Most patent practitioners understand that all words used in the claims should have an objectively determinable meaning. Preferably, the meaning of each term used in any of the claims should be apparent from the descriptive portion of the specification. The description of the invention portion of the application can thus serve as a dictionary for the claims, if you will, and should provide clear support for all terms used in the claims. Of course, some or all of the words used in the claims will be terms of art having generally accepted meanings. With regard to terms that have more than one meaning or that have specialized meaning in the context of the application, it will be desirable to define those terms. On the other hand, one must always make the determination whether it is more beneficial to leave terms as they are, with no rigid meaning to which the patentee will always be held. Normally, a definition will be omitted or left open-ended where there is concern that the state of the art may change or is evolving and it will be better to leave the definition of certain terms open.

In the Hybritech patent, to ensure that there was no confusion, the inventors define the term *antigen* or *antigenic substance* to refer "broadly to substances to which antibodies can be produced" (column 4, lines 54–57).

In the Detailed Description of the Invention portion, all aspects of making and using the invention are treated, as are alternatives that may be useful in these methods of manufacture and use. The invention is described in greater detail than in the Summary of the Invention, and includes a recitation of all important aspects of the invention as well as those aspects that can be varied and how. Here, for example, one will include descriptions of compounds, including preferred compounds; how compounds are prepared; and the utility of compounds, including advantages that may be illustrated with descriptions of results and/or references to the examples.

The Hybritech patent at columns 4 and 5 cites and briefly describes the art-known process used to prepare monoclonal antibodies useful in the claimed assay invention. The patent also describes alternative assay formats. For example, at columns 5–6, the patent defines a "fast forward" assay, a forward assay that can be accomplished without an intermediate wash step. The patent also notes that, in the case of a forward assay, an alternative embodiment contemplates that the same monoclonal antibody can be used for both the label antibody and the solid-phase antibody. Other alternatives described in column 6 of the patent include examples of different solid phase supports that may be used in the claimed assays, techniques for immobilizing the monoclonal antibodies to the solid supports, and compounds useful as labels for the monoclonal antibodies. Finally, the detailed description of the Hybritech invention introduces the "Example" that concludes it and reveals the advantages of the Hybritech invention.

The examples normally follow or conclude the detailed description of the invention and include either descriptions of results actually accomplished or protocols that are to be accomplished.

b. Examples: Working Examples and Prophetic Examples The *Manual of Patent Examining Procedure* states the following with regard to exemplifying the claimed invention:

> Specific operative embodiments or examples of the invention must be set forth. Examples and a description should be of sufficient scope as to justify the scope of the claims.

Typically, the examples in an application represent actual experiments. These are termed *working examples*. However, in some cases, hypothetical or "prophetic" examples are used. These are also referred to as *paper examples*, and are specifically permitted by the PTO.

The *Manual of Patent Examining Procedure* contains certain guidelines for simulated or predicted test results or prophetic examples:

> Simulated or predicted test results and prophetical examples (paper examples) are permitted in patent applications. Working examples correspond to work actually performed and may describe tests which have actually been conducted and results that were achieved. Paper examples describe the manner and process of making an embodiment of the invention which has not actually been conducted. Paper examples should not be represented as work actually done. No results should be presented as actual results unless they have actually been achieved. Paper examples should not be described using the past tense [MPEP § 608.01(p)D].

Thus, care must be taken in the use of prophetic examples to make it clear that they do not represent actual results. This is typically accomplished by writing only working examples in the past tense, whereas prophetic examples are written in the present tense.

The working examples are distinguished from broad statements or lists of suitable alternative compounds or steps or uses that are also found in the detailed description portion of the specification and elsewhere. Working examples are more detailed than the broader descriptions in the specification, and present the most detailed aspects of the invention in order to teach those skilled in the art how to duplicate and to practice the invention.

In essence, patent application examples are intended to demonstrate exactly how to practice one or more embodiments of the invention to be patented. Although analogous to the materials and methods and results sections of a typical paper from the scientific literature, they do differ. Patent application examples should be drafted in more comprehensive detail in order to avoid any suggestion that the description is not enabling. In other words, the examples should preferably contain (1) all manipulative steps, including the time in which such steps should be taken; (2) a full description of all reagents, including contents, sources and purity; (3) all conditions such as temperature, agitation and pH; and (4) all data-treatment calculations. They should be so free of any ambiguity that they can be replicated by any laboratory technician having the relevant skills. It is important to understand that this is a generally higher descriptive standard than is conventional for the scientific literature.

Also unlike technical papers, patent application examples are written in narrative fashion. This means that they generally begin with a brief description of the objective of the experiment, then introduce materials, reagents, and conditions of the processes conducted, and finally conclude with a report of the results achieved. It is always proper to refer to the published literature, however, for a shorthand treatment of items such as reagent preparation, testing procedures, starting materials, and the like so long as the literature otherwise meets the requirements described above.

Although examples listing all significant steps in the making and using of an invention can be used in applications describing any type of invention, they are often critical in biotechnology cases, where the details of how to make and how to use the invention can be both extensive and complex, and certain aspects of the invention are more critical than others. The number and type of examples required depends primarily upon the nature of the invention but also on the level of skill in the art, as well as the state of the art. This concept has been summarized as follows:

> If, in view of the state of a particular art, a specification that contains no examples is sufficient to apprise the skilled artisan of the method of making and using the invention, then no examples whatever should be required. Similarly, if an invention comprises a number of compounds that are useful for a certain purpose, and the method for making and using them would be readily apparent to the skilled artisan from only a single example describing the preparation and use of one of the compounds, then that single example, together with a description of the other compounds in the group, and an indication that they could be made and used in a similar, or otherwise known manner is sufficient to satisfy the statute.[30]

Repetitive examples are not required. It is the *nature*, not the *number*, of the claimed compounds that determines the sufficiency of the supporting disclosure.

A claim that covers compounds closely related in structure may be supported by a more limited disclosure than those compounds that are related less nearly. As a rule, the selection of examples and other material used in the specification to support a claim must be adequately representative of the subject matter that the claim covers.

It is also well established that, even though desirable in complex technologies, actual working examples are not needed to satisfy section 112 of 35 U.S.C. This has been recognized by both the courts as well as the PTO. Even when prophetic examples may be relied upon, the burden is on a challenger (either the PTO examiner or an infringer) to demonstrate that these prophetic examples together with other parts of the specification are not enabling. The Court of Appeals for the Federal Circuit recently reiterated that the burden of proving nonenablement is on the PTO, and that exhaustive exemplification, even in unpredictable areas of technology is not required.

> [The] *Fisher* and *Bowen* [cases] both involved chemical reactions, recognized by our predecessor Court as having a high degree of unpredictability and therefore requiring an increased enablement disclosure. Yet in *Bowen* the board's non-enablement rejection was reversed where the ''claims literally comprehend numerous polymers in addition to the one specifically described in appellant's specification'' because no persuasive reason was given by the Patent Office why the specification does not realistically enable one skilled in the art to practice the invention as broadly as it is claimed.[16]

Obviously, the more information one has available and includes in the specification by way of examples, the less likely one is to face significance questions about the adequacy of the disclosure. However, the courts have made it clear that there is no obligation to provide a large number of examples in order to establish satisfaction of the enablement requirements and, in fact, a specification with no actual examples at all may comply.

It is important to understand, however, that for reasons that seem primarily to involve concerns about the expansive scope of patent protection that is perceived as often sought by biotechnology applicants, the PTO has generally been rather demanding in biotechnology cases in evaluating the sufficiency of a specification. The number and variety of examples included within an application are necessarily matters of both judgment and reality. One must consider the nature of the invention to be claimed, the requirement that the specification enable others to make and use the invention, together with the amount of work that has already actually been accomplished by the inventor, as well as the competitive atmosphere in the field.

Examples in the biotechnological arts are particularly useful to describe the invention because they provide assurance that the assertions of how to make and use the invention are correct. Furthermore, inclusion of such details usually provides fertile ground for identification of specification support for claims in the event that, sometime during prosecution, the applicant wishes to amend one or more claims to add certain limitations. Those limitations can often be found to be supported by the examples. Examples are also useful for demonstrating that the utility requirement of 35 U.S.C. § 101 has been met as well.

There is also a place for experiments that do not work or, if they do work, do not work well. Such examples can be used to illustrate prior art processes to buttress the description of the prior art and its limitations in the background section. In the example set forth in the Hybritech application, prior art immunometric assays were compared with the assays of the claimed invention (columns 6–8). These results were utilized to conclude that the claimed invention provided significantly more rapid and sensitive simultaneous and reverse assays than the prior art processes. On the other hand, including examples of the claimed invention that are not optimal or work less well than others may also help shield the inventor from criticism in later litigation that he unfairly withheld poor results in order to obtain broad claims. The rule of disclosure in this regard, in any event, is that the actual examples set forth should be "representative" of the results obtained if the data utilized for the examples are selected from a larger data base.[33]

c. *Section 112 and Consideration of the Deposit of Biological Materials*

Some inventions that are the subject of patent applications depend upon or involve the use of novel microorganisms or other biological materials that must be described in the specification in accordance with the disclosure requirements of 35 U.S.C. § 112. In some situations, these biological materials and how to make them can be sufficiently described in words in the patent specification, or a source of the materials can be identified in the application that makes the materials available to the public. In other circumstances, however, where the invention depends upon or involves the use of a microorganism or other biological material that is not known, is not readily available, or is not readily made, an inventor must consider the possibility that a deposit of the biological material may have to be made. Primary tactical considerations revolve around whether to deposit the biological material and, if a deposit needs to be accomplished, the timing of the deposit.

Deposit requirements in the United States have developed by court decision and administrative practice within the PTO. Issues relating to whether to deposit a biological material arise with regard to compliance with two distinct section 112 disclosure requirements: the enablement requirements and best mode requirement.

With regard to enablement, it is plain that biological materials need not be deposited when the invention can be practiced without "undue experimentation" from biological material available in the prior art. Thus, although deposit is a convenient way of complying with the enablement requirement, deposit in a recognized culture bank is not the only way of complying; it is permissive rather than mandatory. The Federal Circuit recently discussed this issue as follows, in a case called *In re Wands*:

> Where an invention depends on the use of living materials such as microorganisms or cultured cells, it may be impossible to enable the public to make the invention (*i.e.*, to obtain these living materials) solely by means of a written description. One means that has been developed for complying with the enablement requirement is to deposit the living materials in cell depositories which will distribute samples to the public who wish to practice the invention after the patent issues. . . . A deposit has been held necessary for enablement where the starting materials (*i.e.*, the living cells used to practice the invention, or cells from which the required cells can be produced) are not readily available to the public. Even when starting materials are available, a deposit has been necessary where it would require undue experimentation to make the cells of the invention from the starting materials.[17]

In the *Wands* case, applicants claimed a method for the immunoassay of hepatitis B surface antigen (HBsAg) by use of high-affinity immunoglobulin M antibodies having a "binding affinity constant for . . . HBsAg determinants of at least 10^9 M^{-1}." The prior art had disfavored use of IgM antibodies and favored immunoglobulin G antibodies because of the sensitivity of the former to reducing agents and their tendency to self aggregate and precipitate. Applicants found that immunoglobulin M antibodies with a high affinity for HBsAg showed unexpected sensitivity and specificity. The PTO rejected claims that were generic to the specified antibody for want of an enabling disclosure. The Federal Circuit reversed that decision, noting that those in the art, relying on the state of the art and applicants' written disclosure could produce and screen new hybridomas secreting other monoclonal antibodies falling within the generic class. Therefore, the Court held, applicants' claims need not be limited to the specific antibody exemplified in their application.

The PTO had erroneously contended that data supplied by applicants on their hybridoma fusion and screening experiments supported the conclusion that an enabling disclosure had not been provided. The Court rejected these contentions, emphasizing that:

> The nature of monoclonal antibody technology is that it involves screening hybridomas to determine which ones secrete antibody with desired characteristics. Practitioners of this art are prepared to screen negative hybridomas in order to find one that makes the desired antibody.

Thus, in the area of biotechnology, as in other areas of technology, it is the law that the fact that some investigation or evaluation work is necessary to practice the invention does not preclude enablement, especially where such experimentation is due to the nature of the invention and the involved science. Furthermore, the fact that experimentation may be complex does not necessarily make it undue if the art typically engages in such experiments.

In the biotechnology area, it was initially held that where the preferred embodiment is disclosed, but not specifically set out in the patent application, the best mode requirement may not be satisfied. In *Scripps Clinic and Research Foundation* v. *Genentech*,[18] Scripps sued on a patent containing a generic disclosure for the use of monoclonal antibodies in the purification of Factor VIII:C, a blood clotting protein. In the *Wands* case, the Federal Circuit had suggested that a deposit of biological material, such as a hybridoma cell line that secretes monoclonal antibodies falling within a generic claim to antibodies of the specified type, may satisfy the best mode requirement. The patent holder in *Scripps* had not only not made a deposit of its preferred biological material but had not identified it or even acknowledged that it existed. Genentech asserted that the failure to identify the best mode of carrying out the invention, namely, the specific monoclonal antibody and hybridoma that had been developed but concealed by Scripps, rendered the patent invalid. Although the District Court agreed and summarily held the patent invalid, since not all monoclonal antibodies disclosed within the genus disclosed by Scripps would achieve the claimed result, and one skilled in the art would not be able to deduce from the patent which antibody would work or where to obtain it, the Federal Circuit Court of Appeals reversed that holding.

Instead, it held that a finding of concealment could not be supported because the patent specification described the inventors' preferred method for obtaining their monoclonal antibodies, the specification describes the properties for which the antibodies were to be screened and evaluated, and because of an alleged lack of evidence that the antibodies used by the inventors "differed from those obtainable according to the process described in the specification."[19] The best mode issue for biotechnology inventions was further clarified in another important case decided by the Federal Circuit.

The patent in *Amgen Inc.* v. *Chugai Pharmaceutical Company*[20] related to human erythropoietin (EPO), a 165–amino acid protein that stimulates red blood cell production. Following isolation of the EPO gene, the inventor transformed mammalian host cells with the gene, which enabled those cells to produce EPO for isolation and purification. The inventor's best mode of practicing the claimed invention was a specific strain of Chinese hampster ovary cells, developed with a technique, gene amplification, that multiplied the cells' copies of the gene and, therefore, their ability to produce EPO. The patent specification identified the cell line and described the method of making it in an example. The preferred cell line was not deposited, however, with a public depository. As in the *Scripps* case, the court held that the patent, which claimed the transformed cells, did not violate the best mode requirement. According to the court, the patentee was not required to deposit its preferred cell culture because the patent specification adequately taught those skilled in the art how to prepare such a culture. The court held that there is no failure to comply with the best mode requirement for lack of a deposit of biological materials

> when the *best mode* of preparing the cells has been disclosed and the best mode cells have been enabled, *i.e.*, they can be prepared by one skilled in the art from known materials using the description in the specification.

The court also held that such a disclosure may be adequate even though it does not allow skilled workers to *exactly* duplicate the inventor's best mode. In light of evidence that no scientist could ever duplicate exactly the best mode used by Amgen, but that those of ordinary skill in the art could produce mammalian host cell strains or lines with similar levels of EPO production, the court held that:

> What is required is an adequate disclosure of the best mode, not a guarantee that every aspect of the specification be precisely and universally reproducible.

The lesson of the *Amgen* and the *Scripps* cases is fairly clear. The inventor should always strive to clearly disclose and to unambiguously identify the best mode contemplated by him for carrying out his invention. Whether or not best mode biological materials need to be placed in a depository should be carefully considered by the inventor with his attorney in light of the enablement requirement and in light of the risks associated with deposit. In the United States, no deposit need be made until the patent is to issue, and if a deposit has been made, it can be held in secret just as the application itself. With respect to applications that will be filed outside the United States, however, access to materials deposited in accordance with foreign laws is more freely permitted. Additionally, one must take into account whether a deposit of biological material will be viewed as an admission that a deposit is necessary; i.e., that the invention would otherwise not be enabled by virtue of undue experimentation that would be required to duplicate or practice it.

d. Aspects of pharmaceutical utility Many biotechnological patent applications include claims directed to the use of various compositions of matter, such as recombinant proteins, antibodies, and the like, for therapeutic purposes. As noted above, especially for the purpose of satisfying foreign patent disclosure requirements, any results from testing such therapeutics in laboratory animals or humans that are available to the inventor should normally be provided in the application. In foreign jurisdictions, disclosures regarding therapeutic routes of administration and dosages of therapeutic compounds to be administered also are often required.

The U.S. Patent and Trademark Office also gives patent applications disclosing drug or pharmaceutical utility enhanced scrutiny, even though the same basic principles of patent law that apply to other inventions are also applicable to therapeutics. Special considerations have evolved with regard to therapeutic inventions.

For example, the PTO takes the position that mere assertions of utility such as "therapeutic agents," for "pharmaceutical purposes," and "biological activity," are normally regarded as insufficient. As indicated above with respect to 35 U.S.C. § 101, if the asserted utility of a compound is believable on its face to a person skilled in the art in view of contemporary knowledge in the art, then the burden is upon the examiner to give adequate support for a rejection asserting lack of utility. On the other hand, incredible statements or statements of utility deemed unlikely to be correct by those skilled in the art require "adequate proof on the part of applicant for patents." MPEP § 608.01 [p] [A]).

According to the PTO, proof of utility in this regard may be established by clinical or by in vivo or in vitro data or combinations of these; again, so long as they would be convincing to those skilled in the art. In the field of biotechnology inventions, it is rarely, if ever, the case that direct evidence of clinical utility in humans will be available at the time a patent application is being prepared. In view of the long efforts required for research and development of therapeutics, as well as the amounts of time required for clinical trials, clinical human data are often not even available by the time a patent is ready to issue. Thus, although human clinical evidence would be desirable, as well as dispositive, animal tests are adequate to demonstrate therapeutic efficacy where those skilled in the art would accept such tests as appropriately correlated with human utility. If there is no assertion of human utility, of course, or if there is an assertion of animal utility, a demonstration of operativeness for use in standard test animals is adequate for patent purposes.

In general, safety is rarely an issue in the PTO. The courts have held that absolute safety is not required to meet the utility requirement of the patent law. Furthermore, the courts have instructed, and the PTO has recognized, that it is the function of the U.S. Food and Drug Administration, not patent examiners, to determine whether a therapeutic is safe and effective for the uses described.

It should also be kept in mind that, in general, it is advantageous to attempt to provide information in a patent application for a therapeutic that describes various formulations of the therapeutic, routes of administration to the patient, and dosages. The PTO has specifically declared that it is not necessary to specify the dosage or method of use of a therapeutic if it is clear to one skilled in the art that such information can be obtained without undue experimentation. In many foreign jurisdictions, however, especially Japan and Europe, it is a preference, if not a requirement, that particular effective dosages be identified.

Additionally, routes of administration in which those dosages are to be provided should be set forth. Although this type of disclosure is sometimes difficult at the time a patent application is initially prepared, by the time corresponding foreign applications are prepared and ready for filing (1 year later), the disclosure may be appropriately updated.

E. The Claims: Scope of Protection

There are entire books, as well entire book chapters, that have been written about and devoted entirely to drafting patent claims. This brief treatment of patent claims is meant to provide some insight into claim format, the drafting of claims, and the various ways that claims may be crafted in order to protect an invention. In view of the fact that claims should be drafted to be infringed, an outline of principles of infringement is also provided.

Claims are treated at the end of this chapter simply because claims are found at the end of a patent. However, many patent practitioners will actually draft their claims, or at least their broadest claims, first, before preparing the remainder of the specification. The reason often given for this is that since it is the claims that must be supported by an enabling specification, they must be prepared in advance because the claims will dictate the contents of that specification.

Because it is the claims that define the invention, claim wording is pivital in determining whether competitors who are marketing a similar product infringe those claims, and they must be drafted with great care. There are a number of things that should be understood about patent claims. First, and as a practical matter, the inventor or appropriate members of his company must participate in formulating a strategy for the claims to be sought. In other words, the invention to be protected must be identified from a business point of view. It must be determined what the invention is, what the inventor or his company contemplates as the product it will manufacture and sell, and the form of that product. Identifying competitors and evaluating whether competitors will likely want to make and use the invention and, if so, what form their product is likely to take is also appropriate.

Second, there are a number of considerations that are essentially legal in nature. Perhaps one of the most important concepts to understand is that—just as the omission of a claimed element from a prior art reference precludes that reference from being an anticipation of the invention—an omitted element in an alleged infringing device precludes infringement. Thus, if unnecessary elements are included within the claims, it may be difficult, if not impossible, to successfully assert the infringement of those claims by a competitor who is manufacturing a similar but plainly altered product. Prior to more detailed consideration of claim drafting, a brief introduction to the law of patent infringement follows.

1. Introduction to Patent Infringement

It is provided in 35 U.S.C. § 271(a) that, "Except as otherwise provided in this title, whoever without authority makes, uses or sells any patented invention, within the United States during the term of the patent therefor, infringes the patent." Perhaps the most significant question in evaluating infringement is centered around the "patented invention." When faced with the words of a patent and a thing accused of infringing that patent, the courts do not always have an easy task in identifying the "patented invention." Disputes in the courts have given rise to case law that addresses issues of claim interpretation, literal infringement, the "doctrine of equivalents," and "prosecution history estoppel." Although the ultimate question is always simply put: Has there been a substantive infringement?—i.e., has the accused appropriated the patented invention—the answer is not always as easily forthcoming.

Courts have held that a determination of patent infringement involves a two-step analysis of each asserted claim. First, the language of the claim must be interpreted, and second, it must be determined whether the claim as interpreted "reads on" the accused infringing item; i.e., whether the claimed invention that is being made, used, or sold by the alleged infringer.

a. Claim interpretation Because it is the claims that measure and define an invention, the first step in determining infringement is their interpretation. In determining the fair meaning and scope of a claim, the claim is read as a whole and the invention is construed not only in light of the specification but also with reference to the prosecution history before the PTO, the other claims, and any offered extrinsic evidence such as the testimony of experts as to how those skilled in the art would interpret the claims.

There are, however, some fairly clear rules. It is established that the restrictions of narrower claims cannot be read into the broader claims either to avoid invalidity or to escape infringement. Although a patentee may be his own lexicographer, words defined in the specification must be given the same meaning when used in a claim, the claims will be construed in accordance with the precepts of English grammar, and unless there is a reason not to, the words of a patent are normally construed as they would be by those of ordinary skill in the art. For example, in evaluating infringement, a claim to "a preparation of substantially pure, recombinant tissue plasminogen activator" would first be examined to determine the meanings of the terms *preparation*, *substantially pure*, and *tissue plasminogen activator*. Once construed, the claims are examined together with the accused infringing subject matter to determine whether they are infringed.

b. Literal infringement Literal infringement, or "identity of invention," is made out where each limitation of a claim as construed is found in an alleged infringing article, device, or process. According to the Federal Circuit Court of

Appeals, "It is elementary that the metes and bounds of a patent right are defined by the claims of the patent, and that if the accused matter falls within the claims, literal infringement is made out."[21] This is similar to the test for lack of novelty described in part Section I.E.3. above. Consideration of the relationship between the test for literal infringement and the test for anticipation, both of which require the presence of all elements of a claimed invention, has led to this legal maxim: That which would literally infringe, if later, will anticipate, if earlier. Thus, just as where the failure of a reference to disclose even one element of a claimed invention prevents that reference from being an anticipation, where one claim limitation is not present in an alleged infringing device, there can be no literal infringement.

On the other hand, it is fundamental that one cannot avoid infringement merely by adding elements if each element recited in the claims is found in the accused subject matter. For example,

> a pencil structurally infringing a patent claim would not become noninfringing when incorporated into a complex machine that limits or controls what the pencil can write. Neither would infringement be negated simply because the patentee failed to contemplate use of the pencil in that environment.[22]

In the tissue plasminogen activator claim example noted above, if the term *preparation* means a pharmaceutically acceptable carrier, the term *substantially pure* means a purity of greater than 95% tissue plasminogen activator protein (as measured by a standard activity assay), and the term *tissue plasminogen activator* refers to a protein having a certain amino acid sequence, literal infringement would be established where a competitor made, used, or sold compositions including these features. The other party would not escape infringement by demonstrating that its mixture contained some other element, such as an additional antithrombotic.

c. *Infringement by equivalents* Where there is no literal infringement, the patent holder must necessarily rely on the doctrine of equivalents to establish liability. The patent specification and claims should be crafted in a manner that facilitates application of the doctrine of equivalents. According to this legal doctrine, even though each claim limitation is not literally found upon comparison of an accused infringing product with the claims, infringement can be made out where the "product performs substantially the same function in substantially the same way to obtain the same result." In essence, the doctrine of equivalents is designed to protect a patentee from an infringer who appropriates a patented invention but avoids the literal language of the claims. In the tissue plasminogen activator illustration, for example, an accused infringer may be held not to infringe literally because he has altered, substituted, or deleted one or more of the amino acids that the patent defines as embodying the claimed "tissue plasminogen activator.'' Nevertheless, marketing this modified tissue plasminogen activator may subject its manufacturer to liability if it is determined that products are equivalent. The patent holder will argue that the modified tissue plasminogen activator is functionally equivalent to the claimed tissue plasminogen activator, that any differences are minor and nonsubstantive, and that even though the literal language of the claim may not cover the modified tissue plasminogen activator, the claims should be interpreted broadly and flexibly to achieve justice and to prevent the pirating of his invention. The patent owner will insist that holding him to his precise claim language would turn the patent grant into a hollow and useless thing, and that the patent law must not make it possible for the unscrupulous copiest to make unimportant and insubstantial changes and substitutions in a claimed invention that, although adding nothing, would be enough to take the copied matter outside the claim and hence outside the reach of the law.

Equivalency is determined against the context of the patent specification, the prior art, and the particular circumstances of the case. An important factor is whether persons reasonably skilled in the art would have known of the interchangeability of an ingredient not contained in the patent with one that was. The test of equivalency extends beyond what is literally stated in a patent specification to be equivalent and encompasses any element that one of ordinary skill in the art would perceive as interchangeable with the claimed element. The courts have also consistently held that in applying the doctrine of equivalents, it is necessary to determine a ''range'' of equivalents to be accorded an invention. Thus, a pioneer invention is entitled to a broad range application of doctrine of equivalents, whereas an invention representing only a modest advance over the prior art is usually given a more restricted (narrow) range of equivalents.

d. Prosecution History Estoppel Before considering the technicalities of claim drafting, one more doctrine must be introduced. It is based upon the precept that one of the limitations on the use of the doctrine of equivalents is that the claims cannot be accorded a construction that would encompass the prior art. Analogously, a patentee may be estopped from asserting a particular range of equivalents based on his activity during prosecution before the PTO. In other words, application of the doctrine of equivalents is limited by the doctrine of prosecution history estoppel.

Prosecution history estoppel limits a patentee's reliance on the doctrine of equivalents by preventing him from contending later in an infringement action that his claim should be interpreted as if limitations added by amendment were not present or that claims abandoned are still present. Generally, a patentee receives in litigation what it sought from the PTO with respect to claim scope. A claim may not include within its range anything that would vitiate limitations expressed before the PTO. When a patentee relies upon the doctrine of equivalents in an infringement suit to extend the scope of his claims, he opens the door to rebuttal based upon prior representations made by him to the PTO that appear in the file history. Thus, there is interplay between the doctrine of equivalents and the doctrine of prosecution history estoppel. With regard to the tissue plasminogen activator example, the alleged infringer might, upon reviewing the file of prosecution before the PTO, attempt to prevent any application of the doctrine of equivalents by the patent holder on the assertion that he had made one or more arguments, or amended his claims in some fashion, such that he should be prevented from casting his patent net over the modified tissue plasminogen activator that is admittedly outside the literal coverage of the claim.

An estoppel that would preclude a patent owners from obtaining a claim construction that would resurrect subject matter surrendered during prosecution applies to claim amendments to overcome based on prior art and to arguments submitted to obtain the patent. However, the mere fact that a claim was amended in light of prior art does not preclude any application of the doctrine of equivalents. According to the Federal Circuit:

> Amendment of claims is a common practice in prosecution of patent applications. No reason or warrant exists for limiting application of the doctrine of equivalents to those comparatively few claims allowed exactly as originally filed and never amended. Amendments may be of different types and may serve different functions. Depending on the nature and purpose of the amendment, it may have a limiting effect within a spectrum ranging from great to small to zero. The effect may or may not be fatal to application of a range of equivalents broad enough to encompass an accused product. It is not fatal to application of the doctrine itself.[23]

Accordingly, it is clear that skill is required not only in drafting a patent application but in prosecuting one, and both activities require an eye on future litigation.

2. Limitations on Claim Scope

It will be recalled that in preparing the detailed description of the invention, it is suggested that the inventor include alternative embodiments; i.e., different ways of making and using the invention. This includes a discussion of which elements or parts of an invention may be varied and how. This type of disclosure provides a firm foundation, if necessary, for application of the doctrine of equivalents to an alleged infringer who may be marketing an invention that does not fall within the literal scope of the patent claims.

The manner in which the patent application specification is drafted can provide limitations on claim scope as well. In addition to the rules surrounding the drafting of patent claims, certain limitations are imposed by 35 U.S.C. § 112 (which sets forth the patent disclosure requirements), and by considerations relating to proof of infringement in later litigation.

Proving infringement of the patented biotechnology invention can be problematic. This difficulty consists of two interrelated aspects: (1) the ability to specifically and properly define the invention and (2) the ability to detect the alleged infringement. Both aspects must be kept in mind during claim drafting such that the accomplishment of the first gives rise to the second. Take the case of a potentially infringing diagnostic assay using a monoclonal antibody for the detection of a protein associated with a particular disease state. If the patent claims are limited only to certain structural characteristics of the antibody, to a particular antibody, or to the antibody produced by an individual cell line, then simple proof that the kit has antibodies or antibody fragments reactive to the antigen may not be enough to meet the burden of proof. On the other hand, claims including antibodies characterized by their chemical nature or function, for example, by their particular affinity for the target antigen, can be fully sufficient to support a finding of infringement. See the Appendix 2 *Hybritech* case.

Limitations in claim drafting are also imposed by the section 112 enablement and written description requirements. A patentee's failure to supply an enabling disclosure is always a defense to validity in an action for infringement. The relevant inquiry is whether the scope of enablement provided to one of ordinary skill in the art by the disclosure is commensurate with the scope of protection sought by the claims. If a claim as issued is directed to material that is not enabled by the specification, the patent will be held invalid. Thus, for example, claims to protein preparations that recite only a lower limit in defining their purity must be supported by a specification that demonstrably permits the production—without undue experimentation or further invention—of a completely or nearly pure protein virtually free of any other contaminants.

In *In re* Fisher, 427 F.2d 833 (CCPA 1970), the applicant claimed a preparation of substances containing adrenocorticotrophic hormones (ACTH). The prior art demonstrated that ACTH preparations could be made having a potency of about 50% of an international standard. Fisher's specification set forth a method of producing ACTH preparations having a potency from 111–230% of standard; i.e., 1.1 to 2.3 international units of ACTH activity per milligram, containing no more than certain limited amounts of two undesirable protein contaminants. Fisher, however, claimed all ACTH preparations having "at least" a potency of standard; i.e., 1.0 units per milligram. The court considered the PTO rejection of this claim as not supported by an enabling disclosure:

> The issue . . . is whether an inventor who is the first to achieve a potency of greater than 1.0 for certain types of compositions, which potency was long desired because of its beneficial effect on humans, should be allowed to dominate all such compositions having potencies greater than 1.0, including

future compositions, having potencies for an excess of those obtainable from his teachings plus ordinary skill.

The rejection of Fisher's "open-ended claims," which had a lower limit, but no upper limit, was sustained. The court concluded that Fisher did not enable the preparation of ACTHs having potencies much greater than 2.3, and the claim recitations of at least "1" rendered the claims insufficiently supported under the first paragraph of 35 U.S.C. § 112. According to the court, its decision was not inconsistent with earlier cases where the involved claims did not present the same breadth problem. In those cases, the possible range of further purification "was either small or non-existent" and each involved claims to "substantially pure compositions."

In *Amgen, Inc.* v. *Chugai Pharmaceutical Co.*,[24] at issue was Amgen's generic patent claim covering all possible DNA sequences that would encode any polypeptide having an amino acid sequence "sufficiently duplicative" of erythropoietin to possess the property of increasing production of red blood cells. The Court of Appeals for the Federal Circuit held this claim invalid for lack of enablement. Noting that erythropoietin contains 165 amino acids, the court emphasized that over 3600 different erythropoietin analogues could be made by substituting at only a single amino acid position, and over a million different analogues could be made by substituting three amino acids. In other words, the number of claimed DNA encoding sequences that could produce an erythropoietinlike product was potentially enormous.

In holding the claim invalid for lack of enablement, the Federal Circuit determined that Amgen had failed to disclose "how to make and use enough sequences to justify grant of the claims sought." The court understood that a specification may, within the meaning of section 112 of 35 U.S.C. contain a written description of a broadly claimed invention without describing all species that the claim encompasses, and that representative examples of an invention are not an end in themselves. The court was swayed, however, by the fact that details for preparing only a few erythropoietin analogue genes were disclosed in Amgen's patent. Suggesting that such a disclosure might justify a generic claim encompassing these and similar analogues, the court stated that it nevertheless does not represent sufficient support for a claim to all erythropoietin gene analogues. Quite simply, the court's determination of nonenablement was based upon the facts that (1) there may be many other genetic sequences that code for erythropoietin-type products, and (2) the patent told how to make and use only a few of them. Amgen's claim to all erythropoietin analogues was held invalid.

Because the claims must advise those in the art how to avoid infringement, limitations in claim drafting are also imposed by the requirement of definiteness of claims under 35 U.S.C. § 112, second paragraph. Assertions of claim indefiniteness can thus have a bearing on the determination of infringement. In the

Hybritech case, for example, defendant Monoclonal Antibodies, Inc. argued that the limitations in Hybritech's patent relating to monoclonal antibody affinities of at least about 10^8 or 10^9 liters/mole were indefinite, apparently on the assertion that no one test for determining affinity was used by every person in the field. The Federal Circuit disagreed:

> The basis of of the district court's holding that the claims are indefinite is that "they do not disclose how infringement may be avoided because antibody affinity cannot be estimated with any consistency." . . . Even if the district court's finding in support of this holding—that "there is no standard set of experimental conditions which are used to estimate affinities"—is accurate, under the law pertaining to indefiniteness—"if the claims, read in light of the specification, reasonably apprise those skilled in the art both of the utilization and scope of the invention, and if the language is as precise as the subject matter permits, the courts can demand no more," *Shatterproof Glass Corp. v. Libbey Owens Ford Co.*, 758 F.2d 613, 624, 225 U.S.P.Q. 634, 641 (Fed. Cir. 1985)—the claims clearly are definite. The evidence of record indisputably shows that calculating affinity was known in the art at the time of filing, and not withstanding that those calculations are not precise, or "standard," the claims, read in light of the specification, reasonably apprise those skilled in the art and are as precise as the subject matter permits. As a matter of law, no court can demand more [Appendix 2].

In drafting the specification and the claims, therefore, it is important to keep an eye out for elements or aspects of the invention which an alleged infringer might claim to be unclear. Although such a lack of clarity can, in the extreme, result in the holding of patent invalidity, it can also make the decision regarding enforcement of the patent less certain.

With these concepts regarding infringement, claim interpretation, claim scope, and claim definiteness in mind, it is appropriate to turn to claim drafting.

3. Claim Formalities and Claim Format

a. Claim formalities 35 U.S.C. § 112 and U.S. Patent and Trademark Office Rule 75 require that the patent specification conclude with at least one claim particularly pointing out and distinctly claiming the subject matter which the applicant regards as his invention or discovery. Typically, more than one claim is presented in an application. The PTO does not object to the number of claims submitted provided that the claims "differ substantially from each other and are not unduly multiplied" (37 C.F.R. § 1.75[b]). Although there is no set statutory form for claims, by convention, each patent claim forms a complete sentence, beginning with a capital letter and ending with a period. Periods are not allowed to be used elsewhere in the claims except for abbreviations. In general, the inventor may craft his claims as he sees fit provided that this format does not result in printing difficulties for the PTO. For example, a claim may be typed with various elements or steps subdivided in paragraph form. This format was used in the Hybritech patent.

The PTO prefers that claims be arranged in order of scope such that the first claim presented in an application is the broadest and the last is the narrowest. Additionally, the PTO prefers that different types of claims (e.g., product and process claims) be separately grouped. According to the PTO, such arrangements are for the purposes of facilitating classification and examination (MPEP § 608.01 [m]).

b. Claim format A patent application may contain both independent and dependent claims. An independent claim is one that does not incorporate any limitations from other claims in the application. The independent claims in the attached Hybritech patent, for example, are claims 1, 10, and 19. The other claims in the Hybritech patent are in dependent form. These claims refer back to and further limit another claim or claims in the patent.

Any dependent claim that refers to more than one other claim (known as a multiple dependent claim) refers to the other claims from which it depends in the alternative. Claims 2 and 3 of the Hybritech patent, both of which depend from independent claim 1, are singular dependent claims. Claim 4 of the Hybritech patent is a multiple dependent claim, adding a particular antibody affinity limitation to claim 1, 2, or 3.

A series of dependent claims is permissible in which a dependent claim refers to a preceding claim that, in turn, refers to another preceding claim. An example of such a series of claims is found in the Hybritech patent at column 12. There, claim 21 depends from independent claim 19, claim 22 is dependent on claim 21 and claim 23 is dependent on claim 22.

With regard to sequence, the PTO instructs that a claim that depends from a dependent claim should not be separated by any claim that does not also depend from that "dependent" claim. However, a dependent claim may always refer back to any preceding independent claim. These are the only restrictions with respect to the sequence of claims and, in general, the PTO instructs its examiners not to change an applicants claim sequence.

Each claim contains three parts: (1) a preamble, (2) a transitional phrase, and (3) a body. The preamble is an introduction to the body of the claim. Although it often recites the broad class to which the claimed invention belongs, for example, "a process for the determination of the presence or concentration of an antigenic substance in a fluid" (see claim 1 of the Appendix 1 Hybritech patent), it may also simply reference one of the broad categories of invention set forth in 35 U.S.C. § 101, for example, "a process." There are potential legal ramifications associated with the form in which a preamble is prepared (i.e., whether the preamble will be used as a limitation in evaluating the scope of the claim in later litigation) and this should be taken into account in preparing the claim. In general, it is best if the preamble is written broadly rather than narrowly.

Following the preamble, but before the body of the claim, is a transitional phrase. The phrases most commonly used are the transitions "comprising," "consisting essentially of," and, "consisting of." Each of these transitional phrases has a specific meaning that is well established in the law and must be used appropriately. The "comprising" transition is synonymous with "including" and leaves a claim to a composition consisting of two elements open to the inclusion of unspecified other elements or ingredients. In other words, it means that the invention includes the recited elements that follow, but is not necessarily limited to just those elements. The transitional phrase "consisting essentially of," on the other hand, permits only the inclusion of unspecified additional elements or ingredients that do not materially affect "the basic and novel" characteristics of the invention. The transitional phrase "consisting of" limits the claimed invention only to those elements specifically recited in the body of the claim, no more and no less. Such a claim is therefore avoided if an alleged infringer either merely adds an element or takes one away. For this reason, this transitional phrase is rarely, if ever, used. It is typically never used in a biotechnology case.

In the Appendix 1 Hybritech patent, it will be seen that the claim 1 preamble—"a process for the determination of the presence or concentration of an antigenic substance in a fluid"—is connected to the body of the claimed process (steps (a)–(e)) by the "comprising" transitional phrase. This is interpreted to mean that any individual who uses an assay process which contains at least steps (a)–(e) would be an infringer. Adding steps to the process would not avoid the claim because of the "comprising" transitional phrase.

There are of course, a number of different type of claims that can be drafted for the biotechnology application.

As indicated in Section I.E. above, patentable subject matter, that is, the categories of inventions that can properly be claimed in a patent application, is identified in 35 U.S.C. § 101. Patents can be granted for:

1. Processes
2. Machines
3. Manufactures
4. Compositions of matter
5. Processes involving new uses of processes; machines, manufacturers, or compositions of matter
6. Improvements in any of the above

The inventor can assist by considering and identifying those aspects of the invention that might be claimed in an application; i.e., those areas in which the invention has utility. For each area of biotechnology—such as diagnostics, biopharmaceuticals, monoclonals, and recombinant DNA, for example, there are a number of items to consider protecting by patent.

With diagnostic inventions, such as an immunoassay, thought should be given to claiming (1) the assay itself (both in kit form and as a series of steps in process format), (2) the use of positive and negative control assays in conjunction with the assay, (3) one or more of the assay reagents, and (4) the use of the assay to evaluate one or more physiological or pathophysiological conditions.

Where the invention is a new pharmaceutical compound, a patent application might include claims to (1) the compound itself, (2) derivatives of the compound (such as salts), (3) pharmaceutical formulations containing the compound, (4) pharmaceutical formulations containing various dosages of the compound, (5) methods of treatment using the compound, (6) methods of identifying and/or evaluating other useful compounds (such as screeing assays), and (7) methods for production of the compound.

With respect to monoclonal antibodies, subject matter for patent application claims includes (1) monoclonal antibodies specific for the target antigen, (2) immunologically functional fragments of the monoclonal antibodies, (3) labeled monoclonal antibodies or fragments thereof, (4) cell lines that secrete the monoclonal antibodies, (5) specific cell lines that have already been made which secrete the monoclonal antibodies, (6) methods for use of the monoclonal antibodies in the treatment of disease conditions, (7) methods for use of the monoclonal antibodies in the diagnosis of various disease conditions, (8) methods for use of the monoclonal antibodies for purification of the target antigen, (9) pharmaceutical formulations containing the monoclonal antibodies, and (10) methods for production or isolation of the monoclonal antibodies.

Inventions relating to recombinant DNA, such as a novel genetic sequence encoding a newly discovered peptide, could result in claims being made to (1) DNA encoding the protein, (2) cloning vectors having DNA encoding the protein, (3) expression vectors having DNA encoding the protein, (4) recombinant host cells containing the cloning or expression vectors, (5) the protein at various levels of purity, (6) methods for producing the protein, (7) intermediates involved in the process of making the protein (such as the pro-form of the molecule, or the molecule conjugated to another protein sequence), (8) immunoassays using the protein as antigen, (9) pharmaceutical compositions containing the protein, (10) methods of treatment using the pharmaceutical compositions, (11) antibodies against the protein, and (12) methods for use of these antibodies (in diagnosis and for purification of the protein).

Although all of the above utilities may not be apparent or properly claimed given the circumstances, these are examples of the types of claims directed to compositions of matter and processes which can be claimed in biotechnology-related patent applications, and to which the inventor should give thought in helping to draft the application.

NOTES

1. The use of the terms *he* and *his* in general discussions in this chapter should be understood to be shortened forms of "he or she" and "his or hers," as individuals of both sexes are involved in all aspects of the subjects dealt with here.
2. 802 F.2d 1367 (Fed. Cir. 1986), *cert. denied*, 480 U.S. 947 (1987). The U.S. Court of Appeals for the Federal Circuit was established in 1982 and given exclusive jurisdiction over patent appeals in the United States, pursuant to which it alone is responsible for appellate review of all cases arising under the patent laws, typically patent infringement suits, and of appeals from decisions of the Patent and Trademark Office rejecting claims in pending patent applications.
3. *In re* Smythe, 480 F.2d 1376 (CCPA 1973).
4. *Ex parte* Heicklen, 16 U.S.P.Q.2d 1463 (Bd. Pat. App. & Int. 1990).
5. Kalman v. Kimberly-Clark Corp., 713 F.2d 760 (Fed. Cir. 1983).
6. *Idem.*
7. Panduit Corp. v. Dennison Manufacturing Co., 810 F.2d 1561 (Fed. Cir. 1987).
8. *In re* Tomlinson, 363 F.2d 928, 931 (CCPA 1966).
9. As a matter of law, statutory requirements govern where in conflict with administrative rules, such as rules made by the U. S. Patent and Trademark Office. Thus, although the foregoing elements are fitting and the above order is preferred by the PTO, all that is truly required is that the applicant for patent provide a specification that (1) satisfies the written description, enablement, and best mode disclosure requirements and (2) concludes with one or more claims that define eligible subject matter meeting the utility, novelty, and nonobviousness conditions of patentability.
10. 481 F.2d 910 (C.C.P.A. 1973).
11. 866 F.2d 422 (Fed. Cir. 1989).
12. *In re* Robins, 429 F.2d 452 (C.C.P.A. 1970).
13. *In re* Voss, 557 F.2d 812 (C.C.P.A. 1977).
14. 488 F.2d 382, 389 (1st Cir. 1973).
15. Rolls-Royce Limited v. GTE Valeron Corp., 800 F.2d 1101 (Fed. Cir. 1986).
16. United States v. Teletronics Inc., 857 F.2d 1778 (Fed. Cir. 1988).
17. 858 F.2d 731 (Fed. Cir. 1988).
18. 707 F. Supp. 1547 (N.D. Cal. 1989).
19. Scripps Clinic and Research Foundation v. Genentech, Inc., 927 F.2d 1565 (Fed. Cir. 1991).
20. 927 F.2d 1200 (Fed. Cir. 1991).
21. Smith International, Inc. v. Hughes Tool Co., 718 F.2d 1573 (Fed. Cir. 1983).
22. A. B. Dick Co. v. Burroughs Corp., 713 F.2d 700 (Fed. Cir. 1983).
23. Hughes Aircraft Co. v. United States, 717 F.2d 1351 (Fed. Cir. 1980).
24. 927 F.2d 1200 (Fed. Cir. 1991).
25. R. A. Armitage: Chemical patent Practice—Drafting the patent application. In *Basic Chemical Practice*. Amer. Intell. Prop. L. Assoc., 1986.

26. W. R. Ballard: *There Is No Mystery About Patents*. New York, J. M. Barrett Corp., 1946.
27. P. W. Grubb: *Patents in Chemistry and Biotechnology*. Oxford, Clarendon Press, 1986.
28. A. C. Johnson: Drafting patents with a view to litigation. In *Intellectual Property Counseling and Litigation*. Matthew Bender, 1990.
29. I. R. Kayton: *Patent Practice*. Washington, D.C., Patent Resources Institute, 1989.
30. *Manual of Patent Examining Procedure*, 5th ed., 1989.
31. W. L. Respess: Preparing a chemical patent application. In *Basic Chemical and Biotechnology Practice Seminar*. Amer. Intell. Prop. L. Assn., 1990.
32. G. S. Rich: Laying the ghost of the ''invention'' requirement. 1 *Am. Pat. L. Assoc. Q. J.* 26 (1972).
33. H. J. Schneider: The first page—Patent specification introductions to success or disaster. In *Intellectual Property Law Review*. Clark Boardman, 1983.
34. Scientific American, page 76–77 (March 1990).

APPENDIX I

United States Patent [11] **4,376,110**
David et al. [45] **Mar. 8, 1983**

[54] **IMMUNOMETRIC ASSAYS USING MONOCLONAL ANTIBODIES**

[75] Inventors: Gary S. David, La Jolla; Howard E. Greene, Carlsbad, both of Calif.

[73] Assignee: Hybritech, Incorporated, La Jolla, Calif.

[21] Appl. No.: 175,133

[22] Filed: Aug. 4, 1980

[51] Int. Cl.^{3} GO1N 33/54; GO1N 33/56

[52] U.S. Cl. 436/513; 435/7; 436/548; 436/529; 436/540

[58] Field of Search .... 23/230 B; 424/12, 1, 424/8; 435/7

References Cited
U.S. PATENT DOCUMENTS

3,654,090	4/1972	Schuurs	435/7
3,791,932	2/1974	Schuurs	435/7
3,867,517	2/1975	Ling	23/230 B X
4,016,043	4/1977	Schuurs	435/7
4,098,876	7/1978	Piasio	424/12 X

OTHER PUBLICATIONS

A. C. Cuello et al., Proc. Natl. Acad. Sci. U.S.A., vol. 76(7), 3532–3536 (Jul. 1979).

Primary Examiner—Sidney Marantz
Attorney, Agent, or Firm—Lyon & Lyon

ABSTRACT

"Two-site" or "sandwich" immunometric assay techniques for determination of the presence and/or concentration of antigenic substances in fluids using monoclonal antibodies. One monoclonal antibody is presented in a soluble labeled form and a second monoclonal antibody is presented bound to a solid carrier; the soluble and bound monoclonal antibodies may be the products of either the same or different cell lines. Each monoclonal antibody has an affinity for the antigenic substances of at least about 10^8 liters/mole.

29 Claims, 2 Drawing Figures

IMMUNOMETRIC ASSAYS USING MONOCLONAL ANTIBODIES

Field of the Invention

This invention relates to methods for detecting and/or determining the concentration of antigenic substances in fluids such as serum. In another aspect it relates to immunometric assay techniques. In yet another aspect it relates to monoclonal antibodies.

Background

The determination of the presence or concentration of antigenic substances, for example, those associated with a wide variety of physiological disorders, in serum or other body fluids relies increasingly upon immunoassay techniques. These techniques are based upon formation of a complex between the antigenic substance being assayed and an antibody or antibodies in which one or the other member of the complex may be labelled, for example, by a radioactive element such as I^{125}, which permits its detection and/or quantitative analysis after separation of the complexed labelled antigen or antibody from uncomplexed labelled antigen or antibody.

In the case of a competition immunoassay technique, the antigenic substance in a sample of fluid being tested for its presence competes with a known quantity of labelled antigen for a limited quantity of antibody binding sites. Thus, the amount of labelled antigen bound to the antibody is inversely proportional to the amount of antigen in the sample. By contrast, immunometric assays employ a labelled antibody. In such an assay, the amount of labelled antibody associated with the complex is directly proportional to the amount of antigenic substance in the fluid sample.

Immunometric assays have been found to be particularly well suited for the detection of polyvalent antigens, i.e., antigenic substances that are able to complex with two or more antibodies at the same time. Such assays employ a quantity of unlabelled antibody bound to a solid support that is insoluble in the fluid being tested and a quantity of soluble antibody bearing a label such as a radioactive isotope that permits detection and/or a quantitative estimate of the amount of the ternary complex formed between solid phase antibody, antigen, and labelled antibody.

In immunometric assays known to the prior art, typically a "forward" assay, in which the antibody bound to the solid phase is first contacted with the sample being tested to extract the antigen from the sample by formation of a binary solid phase antibody: antigen complex, is employed. After a suitable incubation period, the solid support is washed to remove the residue of the fluid sample, including unreacted antigen if any, and then contacted with a solution containing a known quantity of labelled antibody.

After a second incubation period to permit the labelled antibody to complex with the antigen bound to the solid support through the unlabelled antibody, the solid support is washed a second time to remove the unreacted labelled antibody. In a simple "yes/no" assay to determine whether the antigen is present in the sample being tested, the washed solid support is tested to detect the presence of labelled antibody, for example, by measuring emitted radiation if the label is a radioactive element. The amount of labelled antibody detected is compared to that for a negative control sample known to be free of the antigen. Detection of labelled antibody in amounts substantially above the background levels indicated by the negative control is interpreted to indicate the presence of the suspect antigen. Quantitative determinations can be made by comparing the measure of labelled antibody with that obtained for standard samples containing known quantities of the antigen.

This kind of assay is frequently referred to as a "two-site" or "sandwich" assay since the antigen has two antibodies bonded to its surface at different locations. This and related techniques are described by Wide at pp. 199–206 of "Radiommunoassay Methods," Edited by Kirkham and Hunter, E. & S. Livingstone, Edinburgh, 1970. An assay based on this technique for the detection of the antigen associated with serum hepatitis using an 125_I labelled antibody is described in U.S. Pat. No. 3,867,517.

Despite their great utility, the prior art immunometric assays have been recognized to be slow procedures, in part because two washing steps are required and because lengthy incubation periods are required to reach equilibrium, i.e., the point at which the amount of complex formed does not change with increasing time.

To eliminate at least one of the washing steps associated with this procedure, so-called "simultaneous" and "reverse" assays have been proposed. A simultaneous assay involves a single incubation step as the antibody bound to the solid support and the labelled antibody are both added to the sample being tested at the same time. After the incubation is completed, the solid support is washed to remove the residue of fluid sample and uncomplexed labelled antibody. The presence of labelled antibody associated with the solid support is then determined as it would be in a conventional "forward" sandwich assay.

A reverse assay involves the stepwise addition first of a solution of labelled antibody to the fluid sample followed by the addition of unlabelled antibody bound to a solid support after a suitable incubation period. After a second incubation, the solid phase is washed in conventional fashion to free it of the residue of the sample being tested and the solution of unreacted labelled antibody. The determination of labelled antibody associated with the solid support is then determined as in the simultaneous and forward assays.

Both the simultaneous and reverse assay techniques require a sufficient excess amount of solid phase antibody to bind most or all of the antigen present to avoid a high dose hook effect where artificially negative or low quantitation of antigen is observed at extremely high concentrations of antigen. For this reason, the forward assay has been the approach preferred by the prior art. That is because large amounts of highly purified, active antibody specific to the antigen of interest for preparing a solid phase with sufficient antigen binding capacity is difficult to obtain from the "polyclonal" antibodies used in prior art processes. Methods for affinity purifying such antibodies have generally been time consuming and resulted in low yields and loss of high affinity antibodies. When an immunogenic substance is introduced into a living body, the body's immune system reacts by generating antibodies to every site on the immunogen it recognizes. A large immunogenic protein molecule may have dozens of sites and a foreign cell may have hundreds. Thus, while each antibody producing cell produces antibody specific for a single antigenic site the immune system has generated a specie of specific antibody producing cells for each immunogenic site recognized. In addition, the body has produced relatively large quantities of antibodies to antigens other than the one of interest such that most of the antibody in the polyclonal mixture is not specific for the antigen of interest. Accordingly, the antibodies used in prior immunometric assays are necessarily "polyclonal" in nature since the antibodies are derived from antisera raised in a conventional manner in animals and their purification is difficult.

When employing conventional polyclonal antibody mixtures in the reverse and simultaneous assays, the formation of a "sandwich" comprising the antigen complexed by two or more labelled antibodies which complex with the antigen at different sites is possible. These complexes could remain soluble in the sample being tested, be removed by subsequent washing steps, and not "counted" when the solid phase is analyzed for solid phase bound labelled antibody. If this happens to a significant extent, sensitivity of the assay is reduced and erroneous results may arise. However, if the unlabelled bound antibody is added to the sample first as in the forward sandwich assay, steric considerations prevent formation of a sandwich comprising the antigen complexed to two or more unlabelled antibodies where labelled antibody is excluded from also binding to the antigen. Accordingly, the antigen is free to react with a labelled antibody molecule. Nevertheless, it has been proposed to use a simultaneous assay for human thyroid stimulating hormone (HTSH) by employing a large excess of the unlabelled antibody bound to a solid phase to minimize formation by soluble labelled antibodies. See Jeong et al., "Comparison to Radioimmunoassay (RIA) with a Unique, Single-Incubation Two-Site Immunoradiometric Assay (IRMA) as Applied to the Determination of Human Thyroid Stimulating Hormone (HTSH)," Bio-Rad Laboratories, 1979.

It has also been proposed to use a reverse assay for HTSH, hepatitis associated antigen (HAA) and carcinoembryonic antigen (CEA) by employing a quantity of labelled antibody sufficient to assure a labelled antibody:antigen complex but insufficient to form a "sandwich" of all the antigen present in a sample. See U.S. Pat. No. 4,098,876.

Since all three of the procedures known to the prior art use a polyclonal mixture of antibodies, the potential for cross-reaction with other materials in serum or other fluid than the antigen for which the test is intended is increased. The occurrence of cross-reactivity with other antigens also reduces the sensitivity of the test for the suspect antigen and increases the prospect of a "false-positive" assay. Furthermore, the use of polyclonal antibodies in a simultaneous or reverse assay requires a careful consideraiton of the amount of labelled antibody used relative to the amount of solid phase antibody and/or antigen present.

In view of these shortcomings, the limitations to the immunometric procedures known to the prior art are readily apparent. The conventional forward assay is time consuming; the simultaneous and reverse assays are accomplished with fewer steps but require large quantities of solid-phase specific antibody and are not well suited to determination of small concentrations of antigen since formation of a sandwich of the antigen with a multiple number of labelled antibody molecules competes with formation of the sandwich comprising bound antibody:antigen:labelled antibody; and all are subject to misinterpretation of false-positives due to the polyclonal nature of the antibody.

Accordingly, one object of the present invention is to provide an improved process for the immunometric assay for antigenic substances.

More specifically, an object of the present invention is to provide more rapid immunometric assay techniques.

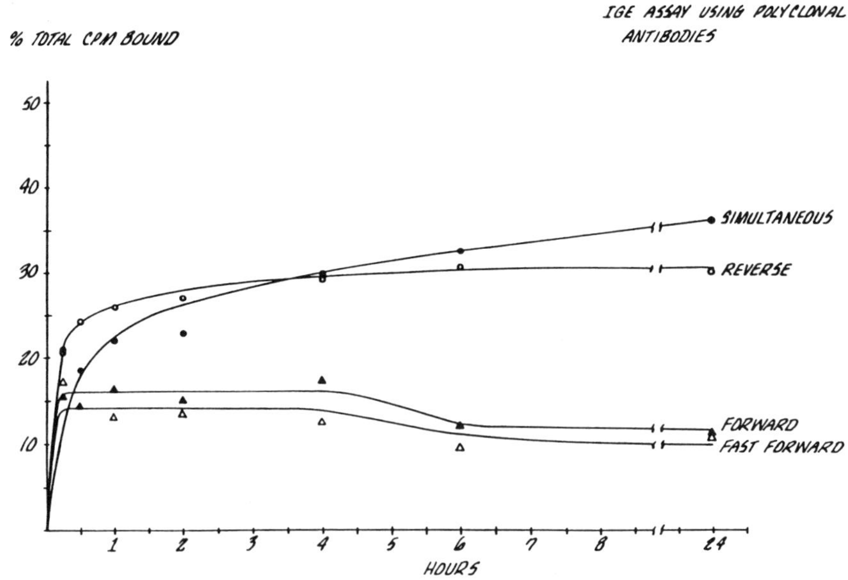

Figure 1

Another object of the present invention is to provide more sensitive immunometric assay techniques.

Yet another object of the present invention is to provide improved "simultaneous" and "reverse" immunometric assays.

The manner in which these and other objects are realized by the present invention will be apparent from the summary and detailed description set forth below.

Summary of the Invention

According to the present invention, the polyclonal antibody used in an immunometric assay as the unlabelled antibody bound to a solid support and the antibody used as the soluble labelled antibody are replaced by at least one and usually two or more different monoclonal antibodies, i.e., each antibody specific to a single antigenic site and separately produced by clones derived from unique cell lines. In a preferred embodiment of the invention, the monoclonal antibody used as the antibody bound to the solid support is the product of a different cell line than is the monoclonal antibody used for the labelled antibody and the two monoclonal antibodies are selected to bind the antigenic substance at sites remote from each other so as to not interfere with the others binding to the antigen. The advantages of the present invention, particularly in simultaneous and reverse assays, over prior art methods will become clear after consideration of the accompanying drawings and the following detailed description of the invention.

Brief Description of the Drawings

FIG. 1 is a graph illustrating the results obtained using polyclonal antibodies in four types of immunometric assay for human IgE.

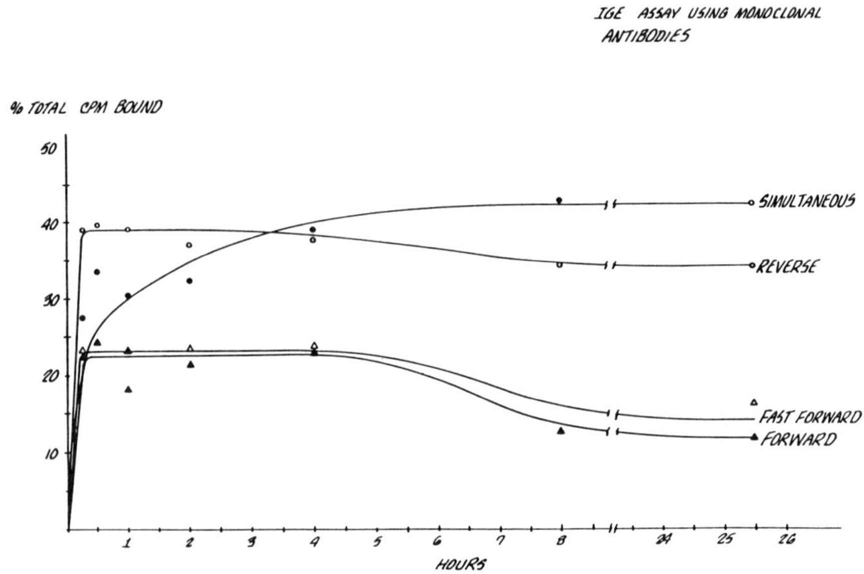

Figure 2

FIG. 2 is a similar graph illustrating the difference in results obtained using monoclonal antibodies in the same four types of immunometric assay for human IgE.

Detailed Description of the Invention

As indicated above, according to the present invention, the polyclonal antibody used in an immunometric assay for an antigenic substance is replaced by a monoclonal antibody. The present invention is useful for the determination of the presence or concentration of a wide variety of polyvalent antigenic substances. Accordingly, as used herein, the term antigen or antigenic substance refers broadly to substances to which antibodies can be produced. Among such substances may be mentioned haptens, hormones such as insulin and human thyroid stimulating hormone (HTSH), gamma globulins, allergens, viruses, virus subunits, bacteria, toxins such as those associated with tetanus and with animal venoms, and even some drugs. Among the specific antigens which may be assayed by the process of the present invention may be mentioned carcinoembryonic antigen (CEA), hepatitis A and B, hepatitis Non A/Non B, IgE and alphafetoprotein.

The monoclonal antibodies useful in the present invention are obtained by the process discussed by Milstein and Kohler and reported in Nature 256 495–497, 1975. The details of this process are well known and will not be repeated here. However, basically it involves injecting a mouse, or other suitable animal, with an immunogen. The mouse is subsequently sacrificed and cells taken from its spleen are fused with myeloma cells. The result is a hybrid cell, referred to as a "hybridoma," that reproduces in vitro. The population of hybridomas are screened to isolate individual clones each of which secrete a single antibody species to the antigen. The individual antibody species obtained in this way are each the product of a single B cell from the immune animal generated in response to a specific antigenic site recognized on the immunogenic substance.

When an immunogenic substance is introduced into a living host, the host's immune system responds by producing antibodies to all the recognizable sites on the substance. This "shotgun" approach to producing antibodies to combat the invader results in the production of antibodies of differing affinities and specificities for the immunogenic substance. Accordingly, after the different hybridoma cell lines are screened to identify those that produce antibody to the desired antigen, the antibodies produced by the individual hybridoma cell lines are preferably screened to identify those having the highest affinity for the immunogenic substance stimulating their original production before selection for use in the present invention. Selection based on this criterion is believed to help provide the increased sensitivity in the immunometric assay of the present invention using monoclonal antibody compared to the polyclonal antibody used in the prior art which, at best, has an affinity for the antigen which is roughly the average of the affinities of all antibodies produced by the immune system. Preferably, the monoclonal antibody selected will have an affinity of at least 10^8 liters/mole and, more preferably, an affinity of at least about 10^9 liters/mole.

Furthermore, those monoclonal antibodies having the highest affinities can be further screened by runing a simulated assay on specimens known to give false positive results with processes employing conventional polyclonal antibodies to identify those monoclonal antibodies which do not cross-react and give false positive results.

Because the two-site immunometric assay relies upon formation of an antibody:antigen:antibody sandwich, usually two different monoclonal antibodies which do not interfere with the binding of each other to the antigen are selected to be the bound antibody and the soluble labelled antibody. Since both are necessary to complete the sandwich, reverse and simultaneous assays can be conducted without concern that a complex of labelled antibody:antigen:labelled antibody will form which will preclude formation of a complex between the antigen and the antibody bound to the solid phase and therein lies a particular advantage of the present invention. Furthermore, a forward assay can be accomplished without the intermediate washing step since the two antibodies bind to different sites. We refer to such a process as a "fast forward" assay.

However, particularly in the case of a forward assay, the same monoclonal antibody can be used for both the labelled antibody and the antibody bound to the solid support when the antigenic substance possesses identical antibody binding sites sufficiently remote from each other to allow more than one antibody molecule to be bound at the same time. In such a system the addition first of the bound antibody to the sample precludes formation of a sandwich because of steric considerations. When the labelled monoclonal antibody is subsequently added, it is also able to complex with the antigen bound to unlabelled antibody on the solid phase.

The unlabelled monoclonal antibody used in the presence of the present invention to extract the antigenic substance from the sample being tested may be immobilized on any of the common supports used in immunometric assays. Among these may be mentioned filter paper, plastic beads or test tubes made from polyethylene, polystyrene, polypropylene or other suitable material. Also useful are particulate materials such as agarose, crosslinked dextran, and other polysaccharides. The techniques for such bonding are well known to those skilled in the art. For example, antibodies may be bound to polysaccharide polymers using the process described in U.S. Pat. No. 3,645,852.

The labelled monoclonal antibody used in the present invention may be provided with the same labels used in prior art immunometric assays. Among these may be mentioned fluorogenic labels for detection by fluorimetry as described in U.S. Pat. No. 3,940,475 and enzymatic markers as described in U.S. Pat. No. 3,645,090. It is presently preferred to label the antibody with a radioisotope such as I^{125} using, for example, the procedure of Hunter and Greenwood, *Nature*, 144 (1962), page 945 or that of David et al., Biochemistry, Vol. 13, pp. 1014–1021, 1974.

In a typical assay, the amount of labelled antibody associated with the insoluble sandwich complex is determined by examination of the insoluble carrier material by suitable means. However, it is also possible to relate the presence or absence of antigen in the fluid sample being assayed to the amount of labelled antibody which does not react during the assay and remains in a soluble form.

The advantages of the present invention in which monoclonal antibodies are used in immunometric assays as compared to polyclonal antibodies are seen by reference to the following example. In this example, four comparative assays, a simultaneous assay, a reverse assay, a forward assay, and a ''fast'' forward assay, were run using both monoclonal antibody and polyclonal antibody using a standard serum containing 100 IU/ml of human IgE as the positive sample. Normal horse serum containing no IgE was used as a negative control.

The polyclonal antibody to IgE used, as the labelled antibody in the example was obtained from Pharmacia Diagnostics of Piscataway, New Jersey. The polyclonal antibody bound to the solid support was obtained from Tago, Inc. of Burlingame, California.

Monoclonal antibody to IgE was obtained using the method of Milstein and Kohler discussed above. The two antibodies selected each exhibited an affinity for IgE of greater than 10^9 liters/mole and did not interfere with the others binding to IgE.

The assays were run using unlabelled antibody bound to agarose by the process of U.S. Pat. No. 3,645,852. Labelling of antibody was by 125, according to the process of David et al. referred to above. Phosphate buffered saline, pH 7.4, was used to wash all samples.

Example

(1) *Simultaneous Assay Method*

Duplicate samples were run in which 100 μl of a suspension of antibody immobilized on agarose particles is mixed with 100 μl of specimen (serum) and 100 μl of soluble ^{125}I-labelled antibody. This mixture as incubated for the specified times shown in Table I (for polyclonal antibody) and Table II (for monoclonal antibody) set forth below, plus 30 minutes. The extra 30 minutes incubation period was added to equalize this assay method with the other assay methods which required an additional 30 minute incubation time for a second added reagent. Following the incubation periods the agarose particles were washed by addition of buffer and centrifuged. After removal of the washing liquid by aspiration, the resulting pellet of agarose particles was then counted for bound ^{125}I-labelled antibody. The counts obtained for each of the complexes after the specified incubation time are set forth in Tables I and II.

(2) *Reverse Assay Method*

Duplicate samples were run in which 100 μl of specimen (serum) is mixed with

TABLE I

	Assay Results Using Polyclonal Antibody							
	Simultaneous Assay		Reverse Assay		Forward Assay		Fast Forward Assay	
Incubation Time (Hrs)	Control Samples	IgE Samples	Control Samples	IgE Samples	Control Samples	IgE Samples	Control Samples	IgE Samples
0.25	372,314	2705,2667	302,243	2568,2581	357,326	2092,2077	396,293	2271,2238
0.50	348,265	2391,2366	284,262	2958,2999	288,233	1905,1817	—,—	—,—
1.00	315,277	2793,2708	305,277	3154,3218	355,424	2157,2255	304,284	1789,1706
2.00	342,356	2897,2887	290,274	3377,3212	302,314	1946,2019	288,312	1728,1867
4.00	421,385	3696,3746	28,280	3413,3651	274,255	2019,2392	283,292	1720,1683
6.00	447,436	4028,4101	296,281	3762,3643	241,267	1750,1452	301,257	1283,1424
24.00	526,577	4564,4628	233,263	3651,3546	320,277	1553,1604	273,256	1450,1470

TABLE II

	Assay Results Using Monoclonal Antibody							
	Simultaneous Assay		Reverse Assay		Forward Assay		Fast Forward Assay	
Incubation Time (Hrs)	Control Samples	IgE Samples	Control Samples	IgE Samples	Control Samples	IgE Samples	Control Samples	IgE Samples
0.25	135,132	5610,5803	388,594	8407,8358	210,205	4618,4894	194,183	4859,4906
0.50	558,459	7472,7115	240,231	8238,8271	223,228	4987,5273	198,197	5024,5152
1.00	268,265	6289,6529	325,265	8010,8377	230,187	3454,4308	215,192	4887,4901
2.00	282,275	6787,6784	255,302	7856,7644	226,197	4834,4268	192,210	4937,4944
4.00	308,272	8150,8155	343,305	8017,7788	231,216	5269,4420	218,187	4929,5071
8.00	549,667	8884,9494	698,850	7870,7358	226,361	2006,3631	405,243	3033,3713
25.55	426,420	8669,9044	497,323	7037,7359	194,201	2465,2586	246,233	3945,2943

100 μl of ^{125}I-labelled soluble antibody and incubated for the specified times shown in Tables I and II. 100 μl of a suspension of antibody immobilized on agarose particles is then added and the mixture was allowed to incubate for another 30 minutes. The agarose particles were then washed and counted as in the simultaneous assay method. The counts are reported in Tables I and II.

(3) *Forward Assay Method*

Duplicate samples were run in which 100 μl of specimen (serum) is mixed with 100 μl of a suspension of antibody immobilized on agarose particles and incubated for the specified times shown in Tables I and II. The agarose particles were then washed once by the addition of 2.5–3.0 ml of buffer which, after mixing, was centrifuged, and the liquid removed by aspiration. 100 μl of ^{125}I-labelled soluble antibody was then added and the mixture incubated an additional 30 minutes. The agarose particles were then washed and counted as in the simultaneous assay method. The counts were reported in Tables I and II.

(4) *Fast Forward Assay Method*

The assay was performed, in duplicate, in a similar manner to the forward assay method except that the wash step between the initial incubation of specimen with antibody immobilized on agarose particles and the addition of soluble ^{125}I-labelled antibody was omitted.

The counts/minute for the duplicate controls and the duplicate assays of the samples containing IgE using polyclonal antibody and monoclonal antibody are shown in Tables I and II, respectively. These data were used to prepare FIGS. I and II in the following way: The average of the counts/minute for a control for a given incubation period was subtracted from the average of the counts for the corresponding IgE assay. The difference was calculated as a percentage of the total counts/minute of labelled antibody added to the sample and is plotted on the Y axis as the percentage of total counts/minute of antibody bound to the solid phase. The incubation time is plotted on the X axis.

A comparison of the plots shown in FIG. 2 displaying the results of assays using monoclonal antibody with those of FIG. 1 of assays using polyclonal antibody shows that in each kind of assay, simultaneous, reverse, forward, and fast forward, the assay using monoclonal antibody was more sensitive as indicated by the higher percentage of total counts bound to the solid phase with 100 IU IgE/ml specimen. Unexpectedly, in the case of the simultaneous and reverse assays, we have found that those run with monoclonal antibody reach equilibrium more rapidly than does the corresponding assay using polyclonal antibody. Therefore, by using a monoclonal antibody in these procedures, the time for the assay can be reduced significantly beyond the time saving achieved by merely eliminating a washing step. In that regard, the reverse assay with monoclonal antibody reached equilibrium in less than one hour. The same assay run with polyclonal assay did not reach equilibrium until after 4 hours. Similarly, in the case of simultaneous assays, the assay using monoclonal antibody reached equilibrium within 8 hours whereas the assay with polyclonal antibody did not reach equilibrium within 24 hours. Accordingly, the present invention substantially provides more rapid and sensitive simultaneous and reverse assays than the prior art processes and eliminates the concern that formation of a soluble "sandwich" complex will compete with formation of the desired insoluble complex.

The examples described above using monoclonal antibody to assay for IgE is merely one exemplar of the use of the present invention. That variations in the actual processes described in the example will be useful will be apparent to those skilled in the art. Therefore, the present invention is to be considered limited only by the appended claims.

We claim:

1. A process for the determination of the presence of concentration of an antigenic substance in a fluid comprising the steps:
   (a) contacting a sample of the fluid with a measured amount of a soluble first monoclonal antibody to the antigenic substance in order to form a soluble complex of the antibody and antigenic substance present in said sample, said first monoclonal antibody being labelled;
   (b) contacting the soluble complex with a second monoclonal antibody to the antigenic substance, said second monoclonal antibody being bound to a solid carrier, said solid carrier being insoluble in said fluid, in order to form an insoluble complex of said first monoclonal antibody, said antigenic substance and said second monoclonal antibody bound to said solid carrier;
   (c) separating said solid carrier from the fluid sample and unreacted labelled antibody;
   (d) measuring either the amount of labelled antibody associated with the solid carrier or the amount of unreacted labelled antibody; and
   (e) relating the amount of labelled antibody measured with the amount of labelled antibody measured for a control sample prepared in accordance with steps (a)–(d), said control sample being known to be free of said antigenic substance, to determine the presence of antigenic substance in said fluid sample, or relating the amount of labelled antibody measured with the amount of labelled antibody measured for samples containing known amounts of antigenic substance prepared in accordance with steps (a)–(d) to determine the concentration of antigenic substance in said fluid sample, the first and second monoclonal antibodies having an affinity for the antigenic substance of at least about 10^8 liters/mole.
2. A process according to claim 1 wherein said first monoclonal antibody is the product of a different cell line than said second monoclonal antibody.
3. A process according to claim 1 wherein said antigen has at least two identical binding sites and said first and second monoclonal antibodies are the product of the same cell line.
4. A process according to claims 1, 2 or 3 wherein the affinity is at least about 10^9 liters/mole.
5. A process according to any of claims 1, 2 or 3 wherein said solid carrier resulting from step (d) is washed to separate the fluid sample from the carrier.
6. A process according to claim 5 wherein the solid carrier is washed with phosphate buffered saline.
7. A process according to claims 1, 2 or 3 wherein the antigenic substance is selected from the group consisting of IgE, hepatitis A, hepatitis B, hepatitis Non A/Non B, alphafetoprotein, carcinoembryonic antigen, insulin and human thyroid stimulating hormone.

8. A process according to claims 1, 2 or 3 wherein the labelled antibody is labelled with a member selected from the group consisting of a radioactive isotope, an enzyme and a fluorogenic material and said examination is by means selected from the group consisting of radiometric means, enzymatic means and fluorometric means.
9. A process according to claim 8 wherein said label is the radioactive isotope ^{125}I.
10. A process for the determination of the presence of an antigenic substance in a fluid comprising the steps:
    (a) simultaneously contacting a sample of the fluid with first and second monoclonal antibodies to said antigenic substance, each monoclonal antibody having an affinity for the antigenic substance of at least about 10^8 liters/mole, said first monoclonal antibody being labelled and soluble in said fluid and being provided for in a measured amount and said second monoclonal antibody being bound to a solid carrier insoluble in said fluid, in order to form an insoluble complex of said first monoclonal antibody, said antigenic substance and said second antibody;
    (b) separating said solid carrier from the fluid sample and unreacted labelled antibody;
    (c) measuring either the amount of labelled antibody associated with the solid carrier or the amount of unreacted labelled antibody; and
    (d) relating the amount of labelled antibody measured with the amount of labelled antibody measured for a control sample prepared in accordance with steps (a)–(c), said control sample being known to be free of said antigenic substance, to determine the presence of antigenic substance in said fluid sample, or relating the amount of labelled antibody measured with the amount of labelled antibody measured for samples containing known amounts of antigenic substance prepared in accordance with steps (a)–(d) to determine the concentration of antigenic substance in said fluid sample.
11. A process according to claim 10 wherein said first monoclonal antibody is the product of a different cell line than said second monoclonal antibody.
12. A process according to claim 10 wherein said antigenic substance has at least two identical binding sites and said first and second monoclonal antibodies are the product of the same cell line.
13. A process according to claims 10, 11 or 12 wherein the affinity is at least about 10^9 liters/mole.
14. A process according to any of claims 10, 11 or 12 wherein said solid carrier resulting from step (b) is washed to separate the fluid sample from the carrier.
15. A process according to claim 14 wherein the solid carrier is washed with phosphate buffered saline.
16. A process according to claims 10, 11 or 12 wherein the antigenic substance is selected from the group consisting of IgE, hepatitis A, hepatitis B, hepatitis Non A/Non B, alphafetoprotein, carcinoembryonic antigen, insulin and human thyroid stimulating hormone.
17. A process according to claims 10, 11 or 12 wherein the labelled antibody is labelled with a member selected from the group consisting of a radioactive isotope, an

enzyme and a fluorogenic material and said examination is by means selected from the group consisting of radiometric means, enzymatic means and fluorometric means.

18. A process according to claim 17 wherein said label is the radioactive isotope ^{125}I.
19. In an immunometric assay to determine the presence or concentration of an antigenic substance in a sample of a fluid comprising forming a ternary complex of a first labelled antibody, said antigenic substance, and a second antibody said second antibody being bound to a solid carrier insoluble in said fluid wherein the presence of the antigenic substance in the samples is determined by measuring either the amount of labelled antibody bound to the solid carrier or the amount of unreacted labelled antibody, the improvement comprising employing monoclonal antibodies having an affinity for the antigenic substance of at least about 10^8 liters/mole for each of said labelled antibody and said antibody bound to a solid carrier.
20 A process according to claim 19 wherein the fluid sample is first contacted with the second monoclonal antibody to form a binary complex of the antigenic substance and said second monoclonal antibody insoluble in the fluid and then contacted with said first labelled monoclonal antibody to form the ternary complex.
21. A process according to claim 19 wherein the fluid sample is first contacted with the second monoclonal antibody to form a binary complex of the antigenic substance and said second monoclonal antibody insoluble in the fluid; the sample separated from the solid carrier and the solid carrier contacted with a solution of said first labelled monoclonal antibody to form said ternary complex.
22. A process according to claim 21 wherein said solid carrier after formation of the ternary complex is washed to separate the fluid sample from the carrier.
23. A process according to claim 22 wherein the solid carrier is washed with phosphate buffered saline.
24. A process according to claims 19, 20 or 21 wherein said first monoclonal antibody is the product of a different cell line than said second monoclonal antibody.
25. A process according to claims 19, 20 or 21 wherein said antigen has at least two identical binding sites and said first and second monoclonal antibodies are the product of the same cell line.
26. A process according to claims 19, 20 or 21 wherein the affinity is at least about 10^9 liters/mole.
27. A process according to claims 19, 20 or 21 wherein the antigen is selected from the group consisting of IgE, hepatitis A, hepatitis B, hepatitis None A/Non B, alphafetoprotein, carcinoembryonic antigen, insulin and human thyroid stimulating hormone.
28. A process according to claims 19, 20 or 21 wherein the labelled antibody is labelled with a member selected from the group consisting of a radioactive isotope, an enzyme and a fluorogenic material.
29. A process according to claim 28 wherein said label is the radioactive isotope ^{125}I.

* * * * *

APPENDIX II

HYBRITECH INCORPORATED, Appellant, v. MONOCLONAL ANTIBODIES, INC., Appellee. Appeal No. 86-531.

United States Court of Appeals,
Federal Circuit.
Sept. 19, 1986.

Suit was brought alleging infringement of patent No. 4,376,110 for immunometric assays using monoclonal antibodies. The United States District Court, Northern District of California, Samuel Conti, J., 623 F.Supp. 1344, held that patent was invalid, and patent holder appealed. The Court of Appeals, Rich, Circuit Judge, held that: (1) laboratory notebooks of patent holder were admissible to show date upon which patent holder first conceived of such assays, though notebooks were not witnessed until a few months to one year after writing; (2) patent was not invalid for obviousness; and (3) patent was not invalid for holder's alleged failure to disclose best mode of carrying out his invention.

Reversed and remanded.

1. Federal Courts—848, 850
District court's findings of fact, which were almost verbatim adoption of findings of fact prepared by prevailing party in advance of trial, were nevertheless findings of court which could not be reversed unless clearly erroneous. Fed.Rules Civ.Proc. Rule 52(a), 28 U.S.C.A.

2. Patents—90(5)
Date on which invention was conceived or reduced to practice is a legal determination, which is subject to review free of clearly erroneous standard. 35 U.S.C.A. § 102(g).

3. Patents—90(5)
Invention is actually reduced to practice, for purposes of determining its priority under federal patent law, only when it works for its intended purpose. 35 U.S.C.A. § 102(g).

4. Evidence—377
Laboratory notebooks of claimed inventor of immunometric assays using monoclonal antibodies were admissible to show date on which inventor first conceived of such assays, for purpose of inventor's patent infringement claim, even though notebooks were not witnessed until a few months to one year after their writing. 35 U.S.C.A. § 102(g).

5. Patents—312(2)
Largely uncorroborated testimony of inventors was not admissible to show date on which invention was first reduced to practice, for purpose of determining whether invention qualified a prior art. 35 U.S.C.A. § 102(g).

6. Patents—324.5
Whether patent is invalid for obviousness constitutes legal determination, which is reviewed free of clearly erroneous standard. 35 U.S.C.A. § 103.

7. Patents—26(2), 36.1(3, 4), 36.2(1)
In determining whether patent is invalid for obviousness, court must consider objective evidence such as patented invention's commercial success, long-felt need for invention, failure of other inventions, and unexpected results of invention. 35 U.S.C.A. § 103.

8. Patents—16.33
Patent No. 4,376,110 for immunometric assays using monoclonal antibodies was not invalid for obviousness, where patented invention enjoyed immediate commercial success not attributable solely to advertising, led to unexpected advances in area of medical diagnosis, and was in no way suggested by scientists' use of monoclonal antibodies of different affinity. 35 U.S.C.A. § 103.

9. Patents—99
Determination that patent enables one skilled in art to make and use claimed invention is not precluded, so that patent is not necessarily invalid, even though some experimentation is necessary. 35 U.S.C.A. § 112.

10. Patents—99
Whether patent enables one skilled in art to make and use claimed invention should be determined, for purpose of evaluating validity of patent, based on information available to those skilled in art when patent application as filed. 35 U.S.C.A. § 112.

11. Patents—99
Patent No. 4,376,110 for immunometric assays using monoclonal antibodies contained sufficient disclosures to permit one skilled in art to make and use invention, so that patent was not on that basis invalid. 35 U.S.C.A. § 112.

12. Patents—99
To show that patent is invalid for failing to comply with best mode requirement, challenger must show what applicant knew of and concealed better mode of carrying out his invention. 35 U.S.C.A. § 112.

13. Patents—312(6)
Evidence that holder of patent for immunometric assays using monoclonal antibodies employed sophisticated competent people to perform screening, and that screening process was labor-intensive and time-consuming, was not sufficient to show that holder knowingly concealed best mode of screening or producing monoclonal antibodies for use in patent process. 35 U.S.C.A. § 112.

14. Patents—99
Patent is not invalid for indefiniteness, where patent reasonably apprises those skilled in art both of utilization and scope of invention, and language is as precise as subject matter permits. 35 U.S.C.A. § 112.

Douglas E. Olson, Lyon & Lyon, Los Angeles, Cal., for appellant. With him on brief were James W. Geriak and Bradford J. Duft.

David J. Brezner, Flehr, Hohback, Test, Albritton & Herbert, San Francisco, Cal., for appellee. Barry E. Bretschneider and Herbert I. Cantor, Washington, D.C., of counsel.

Before RICH, DAVIS and SMITH, Circuit Judges.

RICH, Circuit Judge.

This appeal is from the August 28, 1985, decision of the United States District Court for the Northern District of California, 623 F.Supp. 1344, 227 USPQ 215, in favor of defendant Monoclonal Antibodies, Inc. (Monoclonal) holding that all 29 claims of plaintiff's patent No. 4,376,110 entitled "Immunometric Assays Using Monoclonal Antibodies" ('110 patent), issued to Dr. Gary S. David and Howard E. Greene and assigned to Hybritech Incorporated (Hybritech), are invalid as anticipated under 35 U.S.C. § 102(g), for obviousness under § 103, and under § 112 first and second paragraphs. We reverse and remand.

Background

Vertebrates defend themselves against invasion by microorganisms by producing antibodies, proteins which can complex with the invading microorganisms and target them for destruction or removal. In fact, any foreign molecule of sufficient size can act as a stimulus for antibody production. Such foreign molecules, or antigens, bear particular sites or epitopes that represent antibody recognition sites. B cell lymphocytes, the cells that actually produce antibodies, recognize and respond to an epitope on an antigen by reproducing or cloning themselves and then producing antibodies specific to that epitope. Even if the antigen is highly purified, the lymphocytes will produce antibodies specific to different epitopes on the antigen and so produce antibodies with different specificities. Furthermore, because the body is exposed to many different antigens, the blood of a vertebrate will contain antibodies to many different antigenic substances.

Scientists and clinicians have long employed the ability of antibodies to recognize and complex with antigens as a tool to identify or label particular cells or molecules and to separate them from a mixture. Their source of antibodies has been primarily the serum separated from the blood of a vertebrate immunized or exposed to the antigen. Serum, however, contains a mixture of antibodies directed to numerous antigens and to any number of epitopes on a particular antigen. Because such a mixture of antibodies arises from any different clones of lymphocytes, it is called "polyclonal."

Recent technological advances have made it possible to isolate and cultivate a single clone of lymphocytes to obtain a virtually unlimited supply of antibodies specific to one particular epitope. These antibodies, known as "monoclonal antibodies" because they arise from a single clone of lymphocytes, are produced by a relatively new technology known as the hybridoma. Hybridomas are produced by fusing a particular cancer cell, the myeloma cell, with spleen cells from a mouse that has been injected or immunized with the antigen. These fusions are isolated by transferring them to a growth fluid that kills off the unfused cancer cells, the unfused spleen cells dying off by themselves. The fused hybrid spleen and myeloma cells, called hybridomas, produce antibodies to the antigen initially injected into the mouse. The growth fluid containing the hybridomas is then diluted and put into individual test tubes or wells so that there is only one hybridoma per tube or well. Each hybridoma then reproduces itself and these identical hybridomas each produce identical monoclonal antibodies having the same affinity and specificity. In this way, a virtually unlimited supply of identical antibodies is created, directed to only one epitope on an antigen rather than, as with polyclonal antibodies, to many different epitopes on many different antigens.

In addition to the specificity of antibodies to particular epitopes discussed above, antibodies also have a characteristic "sensitivity," the ability to detect and react to antigens. Sensitivity is expressed in terms of "affinity:" the greater an antibody's ability to bind with a particular antigen, the greater the antibody's affinity. The strength of that antibody-antigen bond is in part dependent upon the antibody's "affinity constant," expressed in liters per mole, for the antigen.

Immunoassays, the subject matter of the '110 patent, are diagnostic methods for determining the presence or amount of antigen in body fluids such as blood or urine by employing the ability of an antibody to recognize and bind to an antigen. Generally, the extent to which the antibody binds to the antigen to be quantitated is an indication of the amount of antigen present in the fluid. Labelling the antibody or, in some cases, the antigen, with either a radioactive substance, I^{125}, or an enzyme makes possible the detection of the antibody-antigen complex. In an extreme case, where the fluid sample contains a very low level of the antigen, binding might not occur unless the antibodies selected or "screened" for the procedure are highly sensitive.

In the case of a "competitive" immunoassay, a labelled antigen reagent is bound to a limited and known quantity of antibody reagent. After that reaction reaches equilibrium, the antigen to be detected is added to the mixture and competes with the labelled antigen for the limited number of antibody binding sites. The amount of labelled antigen reagent displaced, if any, in this second reaction indicates the quantity of the antigen to be detected present in the fluid sample. All of the antigen attached to the antibody will be labelled antigen if there is no antigen in the test fluid sample. The advantage of this method is that only a small amount of antibody is needed, its drawback, generally, that the system must reach equilibrium, and thus produces results slowly.

In the case of a "sandwich" assay, otherwise known as an immunometric assay, the latter being a term coined by Dr. Lawton Miles in 1971, a quantity of unlabelled antibody reagent is bound to a solid support surface such as the inside wall of test tube containing a complex of the fluid sample containing the antigen to be detected and a labelled *antibody* reagent. The result is an insoluble three part complex referred to as a sandwich having antibody bread and antigen filling. This figure is illustrative of the sandwich concept:

The advantage of the sandwich assay is that it is fast and simple, its drawback that enormous quantities of antibodies are needed.

Hybritech

Hybritech, started in 1978 and joined thereafter by coinventors Green and Dr. David, has, since 1979, been in the business of developing diagnostic kits employing monoclonal antibodies that detect numerous antigens and thus a broad range of conditions such as pregnancy, cancer, growth hormone deficiency, or hepatitis. Examples of antigens include influenza viruses, immunoglobulin E (IgE) which indicates allergic reaction, human

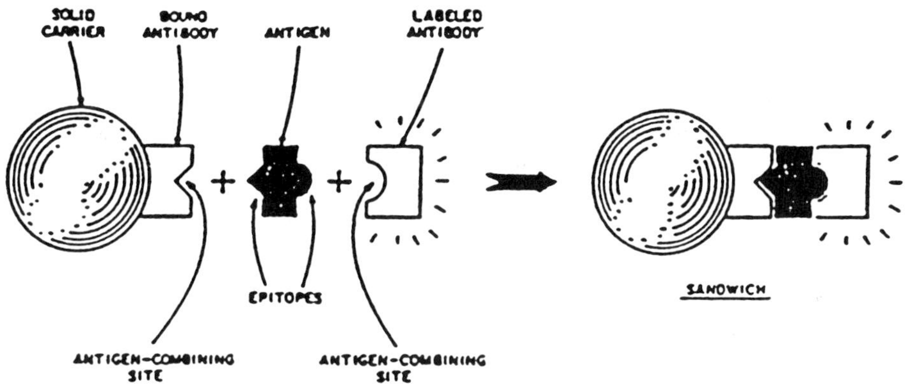

chorionic gonadotropin (HCG) which indicates pregnancy, and prostatic acid phosphatase (PAP) which indicates prostate cancer, to name a few. Dr. Adams, a business-experienced scientist, joined the company in May 1980 as head of research and development. The '110 patent, application for which was filed August 4, 1980, issued March 8, 1983, with claims defining a variety of sandwich assays using monoclonal antibodies. Claim 19, apparently the broadest of the twenty-nine in the patent, is directed generally to a sandwich assay and reads (emphasis ours):

> 19. *In an immunometric assay* to determine the presence or concentration of an antigenic substance in a sample of a fluid comprising forming a ternary complex of a first labelled antibody, said antigenic substance, and a second antibody said second antibody being bound to a solid carrier insoluble in said fluid wherein the presence of the antigenic substance in the samples is determined by measuring either the amount of labelled antibody bound to the solid carrier or the amount of unreacted labelled antibody, *the improvement comprising* employing monoclonal antibodies having an affinity for the antigenic substance of at least about 10^8 liters/mole for each of said labelled antibody and said antibody bound to a solid carrier.

Claim 1, directed particularly to a reverse sandwich assay, explained infra, reads:

> 1. A process for the determination of the presence of [sic, or] concentration of an antigenic substance in a fluid comprising the steps:
>   (a) contacting a sample of the fluid with a measured amount of a soluble first monoclonal antibody to the antigenic substance in order to form a soluble complex of the antibody and antigenic substance present in said sample, said first monoclonal antibody being labelled;
>   (b) contacting the soluble complex with a second monoclonal antibody to the antigenic substance, said second monoclonal antibody being bound to a solid carrier, said solid carrier being insoluble in said fluid, in order to form an insoluble complex of said first monoclonal antibody, said antigenic substance and said second monoclonal antibody bound to said solid carrier;

(c) separating said solid carrier from the fluid sample and unreacted labelled antibody;
(d) measuring either the amount of labelled antibody associated with the solid carrier or the amount of unreacted labelled antibody; and
(e) relating the amount of labelled antibody measured with the amount of labelled antibody measured for a control sample prepared in accordance with steps (a)–(d), said control sample being known to be free of said antigenic substance, to determine the presence of antigenic substance in said fluid sample, or relating the amount of labelled antibody measured with the amount of labelled antibody measured for samples containing known amounts of antigenic substance prepared in accordance with steps (a)–(d) to determine the concentration of antigenic substance in said fluid sample, the first and second monoclonal antibodies having an affinity for the antigenic substance of at least about 10^8 liters/mole.

The District Court Decision

Hybritech sued Monoclonal March 2, 1984, for damages and an injunction alleging that the manufacture and sale of Monoclonal's diagnostic kits infringed the '110 patent. Trial without a jury began on August 5, 1985, and concluded August 23, 1985, thirty witnesses having been heard and over 2,000 pages of transcript generated. The district court produced the reported opinion, findings, and conclusions, which use nearly verbatim Monoclonal's *pre-trial* brief and *pre-trial proposed* findings of fact and conclusions of law, in three days, in support of the judgment now on appeal.

The district court held that the claimed subject matter of the '110 patent was neither conceived nor actually reduced to practice before May 1980, and was anticipated under § 102(g) by the actual reduction to practice of the invention by Drs. Uotila and Ruoslahti at the La Jolla Cancer Research Foundation (LJCRF) as early as November of 1979 and by the actual reduction to practice of the invention by Drs. Oi and Herzenberg (Oi/Herzenberg work) at the Stanford University Laboratory as early as July 1978, later published in December of 1979.

The district court also held the claims of the '110 patent invalid for obviousness from the Oi/Herzenberg work in view of (1) a February 1979 article by M. E. Frankel and W. Gerhard (Frankel article) which discloses high-affinity monoclonal antibodies, and apparently in view of numerous other references including; (2) the work of Nobel Prize winners G. Kohler and C. Milstein disclosing a Nobel Prize-worthy method for producing monoclonal antibodies in vitro (outside the body) published in an August 7, 1975, article; (3) U.S. Patent No. 4,244,940 issued to Jeong et al. disclosing a simultaneous polyclonal assay (Jeong), U.S. Patent No. 4,098,876 to Piasio et al. disclosing a reverse polyclonal sandwich assay (Piasio), U.S. Patent No. 4,016,143 to Schurrs et al. disclosing a forward polyclonal sandwich assay (Schurrs); (4) a July 1979 publication by A. C. Cuello et al. disclosing the use of monoclonal antibodies in competitive assays; and (5) eight articles dated between January 1979 and March 6, 1980, "predicting" that monoclonal antibodies would be used in future immunoassays.[1]

---

[1] With respect to obviousness, one portion of the district court's opinion apparently relies on all of

The district court also invalidated the patent on various grounds based on 35 U.S.C. § 112, first and second paragraphs, as hereinafter discussed.

A. The References

1. *Kohler and Milstein's Nobel Prize-Winning Work: Producing Monoclonal Antibodies In Vitro For the First Time*

In early immunoassay work, polyclonal antibodies produced in vivo (in the body) in mice were used to bind with the antigen to be detected in the body fluid sample. Mice were immunized by injection with antigen so that the lymphocytes in their bodies produced antibodies that attacked the injected antigen. Those polyclonal antibodies were withdrawn from the animal's blood and used in immunoassays. The major problem was that when the mice's immune systems changed or the mice died, the antibodies changed or died too; supply was limited and uncertain.

As the examiner was aware, Kohler and Milstein developed a technique not only for producing antibodies in vitro, independent of a living body, thus eliminating dependence on a particular animal, but for in vitro production of monoclonal antibodies by hybridomas, discussed in the Background section, supra.

Given that sandwich assays require enormous amounts of antibodies, companies like appellant and appellee, which utilize monoclonal antibodies for sandwich assays, would not be in business were it not for the work of Kohler and Milstein.

2. *The Work of Drs. Ruoslahti, Uotila, and Engvall at the La Jolla Cancer Research Foundation (LJCRF) in 1979 and 1980*

Dr. Ruoslahti performed mostly competitive immunoassays using polyclonal antibodies to alphafetoprotein (AFP) antigens at the City of Hope since 1970. Dr. Uotila joined him in late 1978 to perform immunoassays using monoclonal antibodies to AFP. After producing monoclonal antibodies to AFP and performing competitive radio immunoassays (RIA—a competitive assay that uses a radioactive label) with monoclonal antibodies at the City of Hope in mid–1979, Drs. Ruoslahti, Uotila and Engvall left LJCRF.

In the fall of 1979, September or October according to Dr. Uotila, discussion and work began on using monoclonal antibodies to AFP in a sandwich assay. Dr. Uotila, the principal researcher in this particular endeavor, generated six notebooks while at the City of Hope and LJCRF. The next-to-last page of notebook four contained a note to Dr. Uotila from Dr. Ruoslahti reading:

> Sometime you should enzyme label a good monoclonal antibody so that you can set up a sandwich assay. If you use two monoclonal antibodies, you may be able to do the assay with a single incubation, since the monoclonal antibodies are likely to be directed against different determinants and not compete with one another.

---

the above listed references, (1)—(5), for the obviousness holding while a later portion entitled "CONCLUSIONS OF LAW" relies on only the Oi/Herzenberg and Frankel articles. Furthermore, the district court did not state that the LJCRF work was considered for purposes of § 103, although we recognize that § 102(g) prior art can be used for § 103.

Although Dr. Uotila's notebook pages were, for the most part, unsigned, undated, and uncorroborated, Dr. Ruoslahti's testimony, placed the date of this note at about October 1979 by referring to the first pages of notebook five which were dated in early November 1979. Dr. Ruoslahti testified that one curve on one graph on page 43D of notebook five showed a successful simultaneous sandwich assay using monoclonal antibodies about November 5, 1979, although no data supporting that graph could be found elsewhere in the notebook. He further testified that the affinity of the monoclonal antibodies used for that test was not calculated until 1980 but that the raw data necessary for that calculation was generated in 1979.

Dr. Uotila stated in her deposition (she did not testify at trial) that she started work on a sandwich assay using monoclonal antibodies between October 4 and the end of that month, 1979, and that she could not remember the procedure used nor was there enough information in her notebook, including page 43D, to refresh her memory. She did remember, although she continued work on this assay because the tests did not yield repeatedly good curves without which she would not publish her work, that the assay on page 43D was successful. Dr. Engvall testified about a discussion of Dr. Uotila's monoclonal antibody work with her while at the City of Hope and about first performing a sandwich assay after arriving at LJCRF in 1979.

3. *The Work of Drs. Oi and Herzenberg at the Stanford University Laboratory in 1978 Published in December 1979*

Drs. Oi and Herzenberg used monoclonal antibodies to ''map'' epitopes or determine the number and location of different antibody binding sites on a known quantity of IgE antigen by attaching to it an antibody bound to a carrier and exposing that antigen to other monoclonal antibodies. The antibodies either attached to epitopes on the antigen or were blocked from doing so by the other monoclonal antibodies, depending on the location and number of epitopes; if the epitopes on the antigen were too close together and the number of antibodies too great, few antibodies would bind to the antigen. Hybritech points out that both Dr. Herzenberg and Dr. Oi testified that *their work did not involve determining the presence or quantity of antigen*, that they had no idea what the affinities of the monoclonal antibodies used were, and that those values were never calculated.

One unsigned, unwitnessed page from three large laboratory notebooks, which Hybritech argues is insufficient because it does not identify the chemical reagents or protocol used, was relied on by Monoclonal to establish actual reduction to practice of the Oi/Herzenberg work in 1978 to establish a case of § 102(g) prior invention by another. The district court agreed with Monoclonal that the Oi/Herzenberg work anticipated the claimed invention and, in addition, combined this work with the Frankel publication to hold that the claimed subject matter was obvious under § 103.

4. *The Frankel Article: Monoclonal Antibodies Having Affinities of 10^9 liters/mole*

Frankel describes an RIA (radioimmunoassay) method for the rapid determination of affinity constants for monoclonal antibodies produced from hybridomas. The article states that the assay used is applicable only to antibodies with binding constants of about 10^{10} liters/mole and discloses the binding constants for antibodies to several closely related strains of influenza virus.

The district court found that Frankel disclosed monoclonal antibodies having the affinity constants claimed in the '110 patent, 10^8 to over 10^9 liters/mole.

5. *The Cuello Article and the Jeong, Piasio, and Schurr Patents Considered by the Examiner*

Cuello, dated July 1979, states that it describes the usefulness of monoclonal antibodies in the characterization and localization of neurotransmitters such as Substance P, a peptide clearly associated with the transmission of primary sensory information in the spinal cord. The article discloses producing monoclonal antibodies from hybrid myelomas (hybridomas), their use in conventional radioimmunoassay techniques, and the benefits from doing so which flow from the ability to derive permanent cell lines capable of continuous production of highly specific antibodies.

The district court found that the examiner twice rejected all of the claims of the '110 patent based on Cuello alone or in combination with the Jeong, Piasio, and Schurr references which disclose various sandwich assays using polyclonal antibodies. The court also found that the examiner allowed the claims after they were amended to include the 10^8 affinity limitation and after Richard Bartholomew, a Hybritech employee, submitted an affidavit alleging the advantages of using monoclonal rather than polyclonal antibodies in sandwich assays.

Apparently based on the testimony of Monoclonal's expert witness Judith Blakemore, a named inventor of the Jeong patent, manager of antibody programs at BioRad Laboratories from 1975 to 1982, and currently manager of monoclonal antibody therapeutics at Cetus Corporation, a Hybritech competitor in immunoassay diagnostics, the district court stated that the "reasons for allowance were not well-founded because (1) the alleged advantages were expected as naturally flowing from the well-known natural characteristics of monoclonal antibodies . . . ; (2) . . . were not significant . . . ; or (3) were at best minor," although they were "argued to the examiner as if they were" important. These were Monoclonal's words from its pretrial submission adopted by the court.

6. *The References That "Predicted" the Use of Monoclonal Antibodies in Immunoassays*

The district court stated, again in Monoclonal's words, that "it is of the utmost importance" that the advantages of monoclonal antibodies were "predicted by a number of authorities," eight to be exact, not important enough to list here, after the Kohler and Milstein discovery and after monoclonal antibodies became available.

B. The Claimed Subject Matter of the '110 Patent

Hybritech argues that the district court's determination that there is no credible evidence of conception or reduction to practice of the '110 invention before May 1980 is error because Dr. David's laboratory notebooks, Nos. 21 and 24, clearly show successful sandwich assays using monoclonal antibodies in August, September, and October of 1979. At the least, argues Hybritech, the invention was conceived in January of 1979, long before Drs. Ruoslahti, Engvall, and Uotila began work on a sandwich assay using monoclonal antibodies, and diligence was thereafter exercised until constructive reduction to practice occurred by the filing of the '110 patent application on August 4, 1980.

Dr. David and Greene testified that pages 2118 and 2122 of Dr. David's notebook, dated January 4, 1979, and witnessed January 30, 1979, disclose the generic conception of the invention in the context of the physical support structure used to carry out a sandwich assay, and Dr. David testified on redirect that (1) Page 1128 of notebook 21, dated May 27, 1979, recorded an early attempt at a sandwich assay that failed, (2) on August 3, 1979, as recorded at page 1166, a sandwich assay using monoclonal antibody 068 attached to a solid carrier, a radio-labelled 068 antibody, and a hepatitis antigen from an Abbott Labs polyclonal competitive assay kit was successfully performed, and (3) a sandwich assay using a bound 259 antibody, a radio-labelled 068 antibody, and a hepatitis antigen was successfully performed on September 21, 1979. Hybritech also urges that work in October 1979 directed to determining whether certain monoclonal antibodies were recognizing the same or different determinants, was a reduction to practice.

Monoclonal points out that these notebook pages do not expressly state that monoclonal antibodies of 10^8 liters/mole affinity were used in a sandwich assay and that the May, August, and September notebook entries were not witnessed until about the time Dr. Adams, experienced in patent matters, joined Hybritech and advised its researchers on properly recording laboratory work. They therefore claim that actual reduction to practice was not shown before May 1980.

OPINION

I. Review Under Rule 52(a)
Fed.R.Civ.P.

[1] Rule 52(a) "ensures care in the preparation of an opinion . . . and provides appellate courts with the benefit of the District Court's insights into a case," *Pentec, Inc. v. Graphic Controls Corp.*, 776 F.2d 309, 318, 227 USPQ 766, 772 (Fed.Cir. 1985) (Harvey, Senior District Judge, concurring) by requiring a district court to "find the facts specially and state separately its conclusions of law thereon." With the exception of the first eight paragraphs, the first half of the district court's opinion here is Monoclonal's *pretrial* brief and the last three pages of the opinion are Monoclonal's *pretrial* findings of fact and conclusions of law. The district court adopted the above documents virtually verbatim, with the exception of portions of each concerning inequitable conduct and non-infringement, apparently without inviting a response from Hybritech, resulting in a repetitious (as the district court admitted in the opinion), sometimes internally inconsistent, and hard to follow opinion that presents us with a difficult task in gleaning the basis for many of the conclusions. For some of the findings, submitted before trial, no supporting evidence was introduced at trial.

The Supreme Court, in *Anderson v. City of Bessemer City, N.C.*, 470 U.S. 564, 105 S.Ct. 1504, 84 L.Ed.2d 518 (1985), strongly criticized the practice of "verbatim adoption of findings of fact prepared by prevailing parties, particularly when those findings have taken the form of conclusory statements unsupported by citation to the record." *Anderson*, supra, 105 S.Ct. at 1511. This court also has cautioned against the adoption of findings, especially when proposed by a party before trial, as here, and stated that the likelihood of clear error in those findings increases in such a situation. *Lindemann Maschinenfabrik v. American Hoist and Derrick*, 730 F.2d 1452, 1457, 221 USPQ 481, 485 (Fed.Cir. 1984). Notwithstanding our misgivings about whether the findings in this case, prepared before any evidence was introduced, satisfy the objectives of Rule 52(a)—a carefully prepared opinion providing the reviewing court with the benefit of the district court's *reasoned insights* into the case—those findings are the district court's and may be reversed only if clearly erroneous. *See Anderson*, supra, 105 S.Ct. at 1511; *Lindemann*, 730 F.2d at 1457, 221 USPQ at 485.

"A finding is clearly erroneous when, although there is evidence to support it, the reviewing court on the entire evidence is left with the definite and firm conviction that a mistake has been committed." *United States v. United States Gypsum Co.*, 333 U.S. 364, 395, 68 S.Ct. 525, 542, 92 L.Ed. 746 (1948). "This standard plainly does not entitle a reviewing court to reverse the finding of the trier of fact simply because it is convinced that it would have decided the case differently." *Anderson*, supra, 105 S.Ct. at 1511. In other words, "if the district court's account of the evidence is plausible in light of the record viewed in its entirety" or "where there are two permissible views of the evidence," the factfinder cannot be clearly erroneous. *Anderson*, supra, at 1511 (quoting *United States v. Yellow Cab Co.*, 338 U.S. 338, 342, 70 S.Ct. 177, 179, 94 L.Ed. 150 (1949)). This is so, stated the Court in dictum, *see Anderson*, supra, 105 S.Ct. at 1516 (Blackmun, J., concurring), even when the district court's findings rest on physical or documentary evidence or inferences from other facts and not on credibility determinations. *See also* Rule 52(a) Fed.R.Civ.P. (as amended Aug. 1, 1985). If the latter are involved, "Rule 52 demands even greater deference to the trial court's findings" but a trial judge may not "insulate his findings from review by denominating them credibility determinations"; if documents or objective evidence contradict the witness's story, clear error may be found even in a finding purportedly based on a credibility determination. *Anderson*, supra, at 1512–13. We proceed in light of all these principles.

II. Presumption of Validity

Under 35 U.S.C. § 282, a patent is presumed valid, and the one attacking validity has the burden of proving invalidity by clear and convincing evidence. *See, e.g., American Hoist & Derrick Co. v. Sowa & Sons, Inc.*, 725 F.2d 1350, 1360, 220 USPQ 763, 770 (Fed.Cir. 1984). Notwithstanding that the introduction of prior art not before the examiner may facilitate the challenger's meeting the burden of proof on invalidity, the presumption remains intact and on the challenger throughout the litigation, and the clear and convincing standard does not change. *See, e.g., Jervis B. Webb Co. v. Southern Systems, Inc.*, 742 F.2d 1388, 1392 & n. 4, 222 USPQ 943, 945 & n. 4 (Fed.Cir. 1984). The only indication that the district court recognized the presumption of validity and its proper application was its statement that "[t]he key issue in this case is whether the defendant has overcome the presumption of nonobviousness." That statement, however, speaks only part of the truth; the presumption of validity goes to validity of the patent in relation to the patent statute *as a whole*, not just to nonobviousness under section 103.

III. Prior Invention of Another, 35 U.S.C. § 102(g)

Section 102(g) states that a person shall be entitled to a patent unless "before the applicant's invention thereof the invention was made in this country by another who had not abandoned, suppressed, or concealed it." Section 102(g) "relates to prior inventorship by another in this country" and "retains the rules governing the determination of priority of invention. . . . " *Kimberly-Clark Corp. v. Johnson & Johnson*, 745 F.2d 1437, 1444, 223 USPQ 603, 606 (Fed.Cir.1984) (quoting P.J. Federico, *Commentary on the New Patent Act*, 35 USCA page 1, at 19 (1954)). Section 102(g) says: "In determining priority of invention there shall be considered not only the respective dates of conception and reduction to practice of the invention, but also the reasonable diligence of one who was first to conceive and last to reduce to practice, from a time prior to conception by the other."

[2] Reduction to practice, and conception as well, is a legal determination subject to review free of the clearly erroneous standard. *Barmag Barmer Maschinenfabrik AG v. Murato Machinery, Ltd.*, 731 F.2d 831, 837, 221 USPQ 561, 565–66 (Fed.Cir.1984); *D.L. Auld Co. v. Chroma Graphics Corp.*, 714 F.2d 1144, 1151, 219 USPQ 13, 18 (Fed.Cir.1983). Findings of fact supporting that legal conclusion are, of course, reviewed under the clearly erroneous standard.

[3] Conception is the "formation in the mind of the inventor, of a definite and permanent idea of the complete and operative invention, as it is hereafter to be applied in practice." 1 *Robinson On Patents* 532 (1890); *Coleman v. Dines*, 754 F.2d 353, 359, 224 USPQ 857, 862 (Fed.Cir.1985). Actual reduction to practice requires that the claimed invention work for its intended purpose, *see, e.g., Great Northern Corp. v. Davis Core & Pad Co.*, 782 F.2d 159, 165, 228 USPQ 356, 358, (Fed.Cir.1986), and, as has long been the law, constructive reduction to practice occurs when a patent application on the claimed invention is filed. *Weil v. Fritz*, 572 F.2d 856, 865 n. 16, 196 USPQ 600, 608 n. 16 (CCPA 1978) (citing with approval *Automatic Weighing Machine Co. v. Pneumatic Scale Corp.*, 166 F. 288 (1st Cir.1909)).

After a review of the record in its entirety, including the numerous corroborating Hybritech laboratory notebooks, internal documents, and pertinent testimony, we hold clearly erroneous the district court's finding that there is no clear or corroborated evidence "with regard to when before May 1980, the idea of actually using monoclonals in sandwich assays" was conceived or, more properly, or when the *claimed invention* was conceived, and therefore reverse the court's holding, as a matter of law, that Hybritech's inventors did not conceive the claimed invention before May 1980.

Hybritech's claim of conception, generally, is evidenced by the sometimes sparsely documented work of a start-up company whose first small advances evolved into the myriad activities of a mature company with efforts directed toward developing the claimed invention by first employing the Kohler and Milstein technology to produce the necessary monoclonal antibodies and using those antibodies in diagnostic sandwich assay kits. There is no doubt that exploiting monoclonal antibodies for use in sandwich assays was one of the major objectives of Hybritech. In a letter to Pharmacia Fine Chemicals dated April 26, 1979, Greene, in responding to Pharmacia's interest in Hybritech's products, outlined the latter's "efforts to bring the exciting new hybridoma technology into routine medical use" and its exploration of "several intriguing concepts for which monoclonals may open up new immunodiagnostic techniques heretofore infeasible with animal serums." Although company minutes in early 1979 contain little about the claimed subject matter and some of the discussions thereon, such as Greene's and Dr. Adams' conversation about monoclonal sandwich assays when the former was trying to woo Dr. Adams to join Hybritech were unrecorded, the Hybritech laboratory notebooks and the nature of

Hybritech's research program fully corroborate the testimonial evidence of conception and thus clearly support our holding that Hybritech conceived the claimed invention before LJCRF.

Dr. David's January 1979 notebook describes, in detail, as explained by Greene and Dr. David at trial, a nylon apparatus that undoubtedly could be used for performing a sandwich assay using monoclonal antibodies, although Dr. David testified on cross-examination that at that time Hybritech had not yet developed any monoclonal antibodies, including attaching one of the reagents to a solid carrier ring, contacting that ring with a fluid sample in a microtiter plate well, adding a labelled reagent to the well after rinsing, and then "counting" or measuring the amount of either the labelled or unlabelled reagent after a prescribed time and second rinsing. The notebook then describes the procedure for detecting an antibody "(a–x)" to an antigen "(x)" complete with diagrams and text, both illuminated by Dr. David at trial. The notebook further states, "Alternatively, if one wished to quantitate an antigen, y, the identical procedure would be followed, except that reagents would be reversed, i.e. the reaction would be:" and there follows a clear illustration of an antibody attached to a solid carrier reacting with an antigen to form a complex, and that complex reacting with a second labelled antibody. The notebook was signed by Dr. David on January 4, 1979, and witnessed and signed on January 30 of the same year by Dr. Curry, the first cell biologist hired at Hybritech to set up the hybridoma production program.

Dr. David testified on direct that monoclonal antibodies were developed in the following months: antigens were purchased from outside sources and purified before being injected into mice; the spleen cells from those mice were fused with myelomas; and the resultant hybridomas were separated into well plates for development, and a radioimmunoassay procedure was carried out to determine the affinity of the antibodies.

The May 1979 failed sandwich assay, witnessed in May 1980, corroborates Dr. David's testimony that a polyclonal antibody bound to a solid carrier and a labelled monoclonal antibody were used in a sandwich assay with an antigen from Abbott Labs' Ausria polyclonal diagnostic kit for hepatitis. No binding was detected.

Dr. David testified about the experiment documented in the August 1979 notebook, a sandwich assay with a hepatitis antigen from an Abbott Labs Ausria kit with two Hybritech 068 monoclonal antibodies, one attached to a solid carrier bead and the other labelled; the purpose of the experiment was to quantitate the antigen. The notebook corroborates Dr. David's testimony that the test was positive and lists the counts per minute of the labelled antibody. Defendant Monoclonal's expert Ciotti testified about this experiment:

> Also, of course, it is limited to—it is limited to hepatitis antigen. And without a generic conception, it would just be merely a—if it did work for its intended purpose—which I would assume for purposes of discussion—*it would be a reduction to practice of one embodiment*. And without a corresponding generic conception, I don't think it would be held to be the making of the invention in terms of, for instance, in claim 19. [Emphasis ours.]

Dr. David further testified that the September 21, 1979, record in David's notebook, witnessed months later, shows a reverse sandwich assay using a bound 259 monoclonal antibody and a labelled 068 monoclonal antibody with a hepatitis antigen with results confirmed by a dose response curve.[2] Hybritech further alleges that a laboratory notebook page dated October 1979 is a reduction to practice of the claimed invention but fails to cite any related testimony or other evidence in support thereof.

Finally, the record shows that the claimed affinity limitation "of at least about 10^8 liters/mole" was determined and appreciated during the course of the development of the claimed subject matter. Dr. David and Dr. Adams separately testified that the screening procedures used by Hybritech ensured that only monoclonal antibodies having at least 10^8 liters/mole affinity would be used in assays. An October 1979 internal memorandum from Greene to the staff states, "To improve comparisons we will express all affinities to the base ten to the eighth which represents the lower end of the useable range."

[4] We are left with the definite and firm conviction that a mistake has been committed because the district court's account of the evidence that "there was no credible evidence of conception before May 1980" is insupportable. There is such evidence. The laboratory notebooks, alone, are enough to show clear error in the findings that underlie the holding that the invention was not conceived before May 1980. That some of the notebooks were not witnessed until a few months to one year after their writing does not make them incredible or necessarily of little corroborative value. Admittedly, Hybritech was a young, growing company in 1979 that failed to have witnesses sign the inventors' notebooks contemporaneously with their writing. Under a reasoned analysis and evaluation of all pertinent evidence, however, we cannot ignore that Hybritech, within a reasonable time thereafter, prudently had researchers other than those who performed the particular experiments witness the notebooks in response to Tom Adams' advice. The notebooks clearly show facts underlying and contemporaneous with conception of the claimed invention and in conjunction with the testimony of Dr. David and Greene, and others, are altogether legally adequate documentary evidence, under the law pertaining to conception, of the formation in the minds of the inventors of a definite and permanent idea of the complete and operative invention as it was thereafter applied in practice. We thus are not moved by Monoclonal's argument that the findings of fact underlying conception are based on credibility determinations and are more sacrosanct than usual. *See Anderson*, supra, 105 S.Ct. at 1512–13.

1. LJCRF Is Not Prior Art

Hybritech laboratory notebooks and the uncontradicted testimony of Dr. David and Mr. Greene show that development of the claimed invention proceeded diligently through the rest of 1979 and 1980, there being absolutely no evidence of record nor even argument by Monoclonal that Hybritech was not diligent in its efforts to reduce to practice the claimed invention during the period January 1979 to the '110 application filing date of August 4, 1980. We therefore hold as a matter of law that Hybritech's conception, which was before LJCRF conceived the claimed invention, coupled by diligence to its constructive reduction to practice by the filing of the '110 application, entitle Hybritech to priority over LJCRF. *See* 35 U.S.C. § 102(g). The work of LJCRF is therefore not prior art.

---

[2] A dose response curve is antigen concentration plotted against the signal produced by labelled antibody in an immunoassay. The signal increases with increasing antigen concentration in a successful assay but at some point decreases when the antigen concentration becomes too high.

[5] We also note that there is inadequate factual basis for the district court's holding that LJCRF reduced the claimed invention to practice as early as November 1979 because the only evidence that corroborates the testimony of Ruoslahti, Uotila, and Engvall is the note from Ruoslahti to Uotila, see section A, 2, supra, which indisputably is not the claimed invention, and the *one* curve from *one* graph from only one page, 43D, of the six Uotila notebooks. After a reasoned examination, analysis, and evaluation of this pertinent evidence we conclude that it falls far short of showing the "formation in the mind of the inventor, of a definite and permanent idea of the complete and operative invention, as it is hereafter to be applied in practice," *see Coleman*, 754 F.2d at 359, 224 USPQ at 862, and therefore is legally inadequate to support even a holding of *conception* of the claimed invention by LJCRF personnel in 1979.

(1) It is undisputed that page 43D was not signed, witnessed, or dated; (2) the deposition testimony of Uotila was that she could not remember the procedure used to arrive at the dose-response curve on page 43D and there was not enough information in her notebook to refresh her memory; (3) the testimony of Ruoslahti was that he could find *no* data in the notebook supporting that graph, none of the *later* graphs shown there represented successful assays and that "especially after this was done, we ran into more severe problems. And it took us a while to do away with the problems;" (4) Ruoslahti also testified that they never determined, in 1979, the affinities of the monoclonal antibodies they used, and that the title of page 43D had been altered at some point—the word "inhibition" had been crossed out and "sandwich" written in; and (5) the testimony of Engvall was that there was nothing about the shape of those curves which indicates that they were sandwich assays. We also note, as evidence bearing upon the credibility of Ruoslahti's testimony (that LJCRF actually reduced the claimed invention to practice in 1979), that when LJCRF attempted to provoke an interference in the PTO with Hybritech based on the U.S. filing of an application that was the counterpart to a Swedish application disclosing similar subject matter, LJCRF could not demonstrate even a *prima facie* reduction to practice prior to Hybritech's August 4, 1980, filing date. During that proceeding, the earliest dates Ruoslahti set down on paper to support conception and reduction to practice were in 1980.

2. The Work of Oi/Herzenberg Is Not the Claimed Invention

It is axiomatic that for prior art to anticipate under § 102 it has to meet every element of the claimed invention, and that such a determination is one of fact. *See, e.g., Lindemann*, supra, 730 F.2d at 1458, 221 USPQ at 485; *Great Northern Corp. v. Davis Core & Pad Co.*, 782 F.2d 159, 165, 228 USPQ 356, 358 (Fed.Cir.1986). Section 102(g) upon which the district court relied is one type of "anticipation," i.e., prior invention by another of the same invention. Drs. Oi and Herzenberg testified that their work did not involve detecting the presence of or quantitating antigen but a determination of the number and location of epitopes on a *known* quantity of antigen. Although this work did involve a sandwich assay to the extent that an antigen was sandwiched between two monoclonal antibodies, it is clear that the similarity between that work and the claimed invention goes no further. Furthermore, both doctors testified that they did not know the affinities of the antibodies that were used in their mapping work and in fact never calculated them. Ciotti, Monoclonal's expert, testified that the 10^8 affinity limitation cannot be found anywhere in the Oi/Herzenberg work. Again we are left with a definite and firm conviction that a mistake was made because that work does not meet every element of the claimed invention. The district court's finding to the contrary is clearly erroneous.

We note that the district court, in also holding the patent invalid under § 103, next considered, combined the Oi/Herzenberg work with the Frankel reference, one justifiable inference therefrom being that the court recognized that Frankel discloses a claim *element* that Oi/Herzenberg does not, namely, at least about 10^8 liters/mole affinity.

IV. Obviousness, 35 U.S.C. § 103

[6, 7] A section 103 obviousness determination—whether the claimed invention *would have been* (not "would be" as the court repeatedly stated because Monoclonal's pretrial papers used that improper language) obvious at the time the invention was made is reviewed free of the clearly erroneous standard although the underlying factual inquiries—scope and content of the prior art, level of ordinary skill in the art,[3] and differences between the prior art and the claimed invention—integral parts of the subjective determination involved in § 103, are reviewed under that standard. Objective evidence such as commercial success, failure of others, long-felt need, and unexpected results must be considered *before* a conclusion on obviousness is reached and is not merely "icing on the cake," as the district court stated at trial. *See Lindemann*, supra, 730 F.2d at 1461, 221 USPQ at 488; *Stratoflex, Inc. v. Aeroquip Corp.*, 713 F.2d 1530, 218 USPQ 871 (Fed.Cir.1983); *Kansas Jack, Inc. v. Kuhn*, 719 F.2d 1144, 219 USPQ 856 (Fed.Cir.1983); *W. L. Gore & Associates v. Garlock, Inc.*, 721 F.2d 1540, 220 USPQ 303, 314 (Fed.Cir.1983).

1. *The Eight Articles "Predicting" Widespread Use of Monoclonal Antibodies*

[8] Before discussing the more pertinent references in this case—the Oi/Herzenberg and Frankel works—we cull the other prior art references relied on by the trial court.

First, the latest four of the eight articles that the court stated were of the "utmost importance" because they "predicted" that the breakthrough in production of monoclonal antibodies by Kohler and Milstein would lead to widespread use of monoclonal antibodies in immunoassays are neither 102(a)/103 nor 102(b)/103 prior art because they are dated between late 1979 and March 6, 1980, well after the date of conception and within one year of the filing date of the '110 patent.

The earliest four of the eight articles, on the other hand, although clearly prior art, discuss *production* of monoclonal antibodies—admittedly old after Kohler and Milstein showed how to produce them—but none discloses sandwich assays. At *most*, these articles are invitations to try monoclonal antibodies in immunoassays but do not suggest how that end might be accomplished. To the extent the district court relied upon these references to establish that it would have been *obvious to try* monoclonal antibodies of 10^8

[3] Although the district court failed expressly to find the level of ordinary skill in the art at the time the invention was made, it did make reference to "[p]eople working in immunology aware of the Kohler and Milstein discovery" which we deem an accurate finding for the purposes of that portion of the *Graham* factual inquiries.

liters/mole affinity in a sandwich immunoassay that detects the presence of or quantitates antigen, the court was in error. *See Jones v. Hardy*, 727 F.2d 1524, 1530, 220 USPQ 1021, 1026 (Fed.Cir.1984) ("Obvious to try" is improper consideration in adjudicating obviousness issue).[4]

2. *The Kohler and Milstein Work, the Cuello Article and the Jeong, Piasio, and Schurr Patents Considered by the Examiner*

The district court's finding that Kohler and Milstein developed a method for producing monoclonal antibodies in vitro is correct, but that finding proves no more; although it made possible all later work in that it paved the way for a supply of monoclonal antibodies, it indisputably does not suggest using monoclonal antibodies in a sandwich assay in accordance with the invention claimed in the '110 patent.

The Cuello reference discloses monoclonal antibodies but not in a sandwich assay. The competitive assay in Cuello, moreover, uses only one monoclonal antibody and thus in no way suggests the claimed invention wherein a ternary complex of two monoclonal antibodies and an antigen form a sandwich. Furthermore, the court did not explain how this art, by itself or in combination with any of the other art, suggests the claimed subject matter and thus why that combination would have been obvious. We are of the opinion that it does not.

The district court correctly found that the use of polyclonal antibodies in sandwich assays was well known. The Jeong patent discloses the use of polyclonal antibodies in a simultaneous sandwich assay, with no suggestion that monoclonal antibodies be so used. It is prior art by virtue of § 102(e), application for the patent having been filed September 5, 1978, its effective date as a reference. The Piasio patent, disclosing a reverse sandwich assay using polyclonal antibodies, and Schurrs, disclosing a forward sandwich assay using the same, both § 102(a) prior art, are likewise devoid of any suggestion that monoclonal antibodies can be used in a similar fashion.

3. *The Oi/Herzenberg Work and the Frankel Article*

Clearly, the most pertinent items of prior art not cited by the examiner are the Oi/Herzenberg work, as described in section A, 3, supra, and the Frankel article. As stated in the discussion of Prior Invention of Another (section III, 2, supra), the Oi/Herzenberg work involved mapping epitopes on a known quantity of antigen. It was not concerned

---

[4] Finding 10, which states that the invention was contemporaneously developed and disclosed in at least five publications and patent applications not listed above *and dated well after the filing date of the '110 patent but before its issuance* is irrelevant for purposes of the hypothesis based on the three factual inquiries required by § 103 as interpreted by *Graham v. John Deere*, 383 U.S. 1, 86 S.Ct. 684, 15 L.Ed.2d 545, 148 USPQ 459 (1966) because obviousness must be determined as of the time the invention was made. Additionally, they are of little probative value in this case because they are dated December 1981 at the earliest, more than a year after the August 4, 1980, filing date here and roughly two years after conception occurred. Furthermore, simultaneous development may or may not be indicative of obviousness, the latter being the case here for the above reasons and because the other evidence of nonobviousness is adequate, such occurrences having been provided for in 35 U.S.C. § 135. *Lindemann*, supra, 730 F.2d at 1460–61, 221 USPQ at 487; *Environmental Designs, Ltd. v. Union Oil Co. of California*, 713 F.2d 693, 698 n. 7, 218 USPQ 865, 869 n. 7 (Fed.Cir.1983).

with and does not disclose using monoclonal antibodies of at least 10^8 liters/mole affinity. Oi and Herzenberg testified that they did not know the affinity of the antibodies used, and Ciotti testified that nowhere in that work is there mention of monoclonal antibody affinity of at least 10^8 liters/mole. On this basis, we conclude that the Oi/Herzenberg work is qualitatively different than the claimed invention; the former is directed to mapping epitopes on a known quantity of antigen and the latter to determining the "presence or concentration of an antigenic substance in a sample of fluid. . . . " We disagree with Monoclonal that these are "essentially the same thing." Furthermore, it is perfectly clear that this work in no way suggests using monoclonal antibodies of the affinity claimed in the '110 patent. It is because of these differences between the Oi/Herzenberg work and the claimed invention that the fact that an antigen was sandwiched between two monoclonal antibodies in the course of Oi's and Herzenberg's work is not sufficient basis to conclude that the claimed invention would have been obvious at the time it was made to a person of ordinary skill in the art.

Likewise, a conclusion that the invention would have been obvious cannot properly be reached when the Oi/Herzenberg work is considered in view of the Frankel article. Frankel teaches a method for rapid determination of affinity constants for monoclonal antibodies, some of which clearly have affinities of the order defined by the claims, but does not in any way suggest using two of those antibodies in a sandwich to assay an antigen by forming a ternary complex of labelled antibody, the antigenic substance, and a bound antibody wherein the presence of the antigenic substance is determined by measuring either the amount of labelled antibody bound to a solid carrier or the amount of unreacted labelled antibody. The mere existence of prior art disclosing how to measure the affinity of high affinity monoclonal antibodies is insufficient to support a holding of obviousness. Hybritech's claims define a *process* that *employs* monoclonal antibodies, and does not merely claim antibodies of high affinity. In view of the fact that the Oi/Herzenberg work is not directed to an assay as claimed and does not disclose antibodies of at least 10^8 liters/mole affinity, and further that Frankel fails to suggest using such antibodies in a sandwich assay, the Frankel article does not compensate for the substantial difference between the Oi/Herzenberg work and the claimed subject matter, and therefore those references in combination cannot support a holding of obviousness.

4. *Objective Evidence of Nonobviousness*

In one part of its opinion the court found that "the commercial success of the kits *may* well be attributed to the business expertise and acumen of the plaintiff's personnel, together with its capital base and marketing abilities" (emphasis ours) and later that "[w]here commercial success is based on the sudden availability of starting materials, in this instance the availability of monoclonal antibodies as a result of the Kohler and Milstein discovery, business acumen, marketing ability, and capital sources, no causal relationship is proven." (Citation omitted.)

i. Commercial Success: Hybritech's Diagnostic Kits Grabbed a Substantial Market Share

The undisputed evidence is that Hybritech's diagnostic kits had a substantial market impact. The first diagnostic kit sales occurring in mid–1981, sales increased seven million dollars in just over one year, from $6.9 million in 1983 to an estimated $14.5 million in 1984; sales in 1980 were nonexistent. Competing with products from industry giants such as Abbott Labs, Hoffman LaRoche, Becton-Dickinson, and Baxter-Travenol, Hybritech's HCG kit became the market leader with roughly twenty-five percent of the market at the expense of market shares of the other companies. Its PAP kit ranks second only to a product sold by Dupont's New England Nuclear, surpassing products from Baxter-Travenol, Abbott, and others. Hybritech's other kits, indisputably embodying the invention claimed in the '110 patent, obtained similar substantial market positions.

Although the district court did not provide its insights into why commercial success was due to business acumen and not to the merits of the claimed invention, Monoclonal urges in support that it was due to Hybritech's spending disproportionate sums on marketing, 25–30% of income. The undisputed evidence was that expenditures of *mature* companies in this field are between 17 and 32%. Furthermore, the record shows that advertising makes those in the industry—hospitals, doctors, and clinical laboratories—aware of the diagnostic kits but does not make these potential users buy them; the products have to work, and there is no evidence that that is not the case here or that the success was not due to the merits of the claimed sandwich assays—clearly contrary to the district court's finding.

The trial court's finding that the "sudden availability of monoclonals" was the reason for the commercial success of Hybritech's diagnostic kits (Finding 11) is unsupported by the record and clearly erroneous. Monoclonal admits that monoclonal antibodies were available in the United States in *1978*, and the evidence clearly reflects that. Thus, at least *three years* passed between the time monoclonal antibodies were available in adequate supply and the time Hybritech began selling its kits. Especially in the fast-moving biotechnology field, as the evidence shows, that is anything but sudden availability.

ii. Unexpected Advantages

Hybritech points to the testimony of three witnesses skilled in the diagnostic field who state that, based on tests done in their laboratories as a result of real-world comparisons in the normal course of research, the diagnostic kits that embody the '110 invention unexpectedly solved longstanding problems. Dr. Hussa, the head of a large referral laboratory and a world-wide consultant, testified that until Hybritech introduced its kits, he and others were very skeptical and had almost exclusively used competitive assays with a radioactive tracer (RIAs).[5] In relation to an HCG Hybritech kit, he testified that he had first thought that the Hybritech HCG kit would not give accurate results for low antigen concentrations because that condition is indicated in the Hybritech kit by a low radioactivity reading, a reading difficult to differentiate from control samples containing no antigen. He also stated that in the past, RIA kits falsely detected HCG in nonpregnant

[5] Monoclonal's expert Blakemore testified that of 425 assays on the market in 1979 less than 1% were sandwich assays. Today, sandwich assays constitute the majority of all assays sold.
The record also shows that Blakemore, who testified extensively for Monoclonal that the claimed invention would have been obvious, never used monoclonal antibodies in sandwich assays at Bio-Rad before 1980. Additionally, she did not even mention them in the Jeong patent, of which she was a coinventor, which issued January 13, 1981, long after the beginning of Hybritech's work in this area in 1979.

women, a condition which would indicate cancer and surgery. He stated that when he employed the Hybritech HCG kit in such instances it demonstrated, correctly and absent any difficulty interpreting the data, that no HCG was present.

Dr. Blethen, an M.D. holding a Ph.D. in biochemistry, testified that she did not think that the Hybritech HGH kit, for detecting growth hormone in children, would offer any advantage, but she determined that it detected HGH deficiencies in children where conventional RIAs failed to do so. She also stated that the kit does not give false positive readings as do conventional RIA kits, an opinion shared by Dr. Hussa. A third witness, Dr. Herschman, who holds a master's degree in chemistry, testified that he spent years working on the development of an assay that would determine the presence of TSH (thyroid stimulating hormone) with greater sensitivity. He succeeded but discovered that the Hybritech TSH kit had the same sensitivity, the test being performed in four hours rather than the three days his kit required.

Having considered the evidence of nonobviousness required by § 103 and *Graham*, supra, we hold, as a matter of law, that the claimed subject matter of the '110 patent would not have been obvious to one of ordinary skill in the art at the time the invention was made and therefore reverse the court's judgment to the contrary. The large number of references, as a whole, relied upon by the district court to show obviousness, about twenty in number, skirt all around but do not as a whole suggest the claimed invention, which they must, to overcome the presumed validity, *Lindemann*, 730 F.2d at 1462, 221 USPQ at 488, *as a whole. See* 35 U.S.C. § 103; *Jones v. Hardy*, 727 F.2d 1524, 1529, 220 USPQ 1021, 1024 (Fed.Cir.1984). Focusing on the obviousness of substitutions and differences instead of on the invention as a whole, as the district court did in frequently describing the claimed invention as the mere substitution of monoclonal for polyclonal antibodies in a sandwich assay, was a legally improper way to simplify the difficult determination of obviousness. *See generally Hodosh v. Block Drug Co.*, 786 F.2d 1136, 229 USPQ 182 (Fed.Cir.1986).[6]

With respect to the objective indicia of nonobviousness, while there is evidence that marketing and financing played a role in the success of Hybritech's kits, as they do with any product, it is clear to us on the entire record that the commercial success here was due to the merits of the claimed invention. It cannot be argued on this record that Hybritech's success would have been as great and as prolonged as admittedly it has been if that success were not due to the merits of the invention. The evidence is that these kits

---

[6] It bears repeating that it is crucial that counsel set forth the law accurately. More particularly, it is the duty of counsel to impart to the judge that the obviousness question properly is whether the *claimed* invention *as a whole would have been* obvious to one of *ordinary* skill in the art *at the time the invention was made*, and that the district court must *expressly* make the three factual determinations required by *Graham* and consider objective evidence of obviousness *before* the legal conclusion of obviousness vel non is made. Submitting to the court language like "any differences . . . would have been obvious," as was done here, violates the axiom that the question is not whether the differences would have been obvious but the claimed invention *as a whole*. Furthermore, arguing that "it would be obvious" rather than that it would *have been* obvious shifts the court's focus to the wrong period of time, namely to a time long after the invention was made, in which, more likely than not, the prior art and the level of ordinary skill in the art are more advanced. *See* 35 U.S.C. § 103.

compete successfully with numerous others for the trust of persons who have to make fast, accurate, and safe diagnoses. This is not the kind of merchandise that can be sold by advertising hyperbole.

V. Enablement, Best Mode, and Definiteness Under § 112

The section 112 defense appears to have been an afterthought of both Monoclonal, who briefly but unsuccessfully attempts to defend this utterly baseless determination, and of the district court which adopted the defense from Monoclonal's pretrial papers apparently without knowledge of the applicable law, to highlight, as it stated at trial, that it was part of its job to see that "whoever wins wins all the way or whoever loses loses all the way." Taken as a whole, the court's comments on § 112—split into two parts, one from Monoclonal's pretrial brief and the other from the adopted pretrial findings and conclusions—are internally inconsistent. The opinion states that the patent fails to disclose how (1) to make monoclonal antibodies; (2) to screen for proper monoclonal antibodies; and (3) to measure monoclonal antibody affinity and therefore the specification is nonenabling and does not satisfy the best mode requirement, and the claims are indefinite. We discuss each of these in turn.

1. *Enablement*

[9, 10] Enablement is a legal determination of whether a patent enables one skilled in the art to make and use the claimed invention, *Raytheon Co. v. Roper Corp.*, 724 F.2d 951, 960, 220 USPQ 592, 599 (Fed.Cir.1983), is not precluded even if some experimentation is necessary, although the amount of experimentation needed must not be unduly extensive, *Atlas Powder Co. v. E.I. Du Pont De Nemours & Co.*, 750 F.2d 1569, 1576, 224 USPQ 409, 413 (Fed.Cir.1984), and is determined as of the filing date of the patent application, which was August 4, 1980. *See W.L. Gore and Associates v. Garlock, Inc.*, 721 F.2d 1540, 1556, 220 USPQ 303, 315 (Fed.Cir.1983). Furthermore, a patent need not teach, and preferably omits, what is well known in the art. *Lindemann*, 730 F.2d at 1463, 221 USPQ at 489.

[11] The record fully supports the '110 patent's statement that

> The monoclonal antibodies used for the present invention are obtained by the [hybridoma] process discussed by Milstein and Kohler. . . . The details of this process are well known and not repeated here.

The district court itself stated that the "method for producing monoclonal antibodies in vitro was well known prior to the alleged invention of the '110 patent," and used the "sudden availability of monoclonal antibodies" produced by the Kohler and Milstein discovery to support, albeit erroneously, its finding of a lack of nexus between the merits of the claimed invention and its commercial success. The court then about-faced and held the '110 patent deficient because it fails to teach how to make monoclonal antibodies.

With respect to screening, the only permissible view of the evidence is that screening methods used to identify the necessary characteristics, including affinity, of the monoclonal antibodies used in the invention were known in the art and that the '110 patent contemplated one of those. At trial, Monoclonal's counsel stated "it is a procedure that was known in '78." In similar fashion, the district court held that the claimed subject matter would have been obvious in part because the "existence of monoclonal antibodies *having the affinity constants claimed in the patent was well known* prior to the alleged invention. . . . " [Emphasis ours.] Furthermore, there was not a shred of evidence that undue experimentation was required by those skilled in the art to practice the invention. We hold as a matter of law that the '110 patent disclosure is enabling.

2. *Best Mode*

[12, 13] "The specification . . . shall set forth the best mode contemplated by the inventor of carrying out his invention." 35 U.S.C. § 112. Because not complying with the best mode requirement amounts to concealing the preferred mode contemplated by the applicant at the time of filing, in order to find that the best mode requirement is not satisfied, it must be shown that the applicant knew of and concealed a better mode than he disclosed. *DeGeorge v. Bernier*, 768 F.2d 1318, 1324, 226 USPQ 758, 763 (Fed.Cir.1985) (quoting with approval *In re Sherwood*, 613 F.2d 809, 204 USPQ 537 (CCPA 1980)). The only evidence even colorably relating to concealment is testimony by various Hybritech employees that sophisticated, competent people perform the screening and that the screening process is labor-intensive and time-consuming. It is not plausible that this evidence amounts to proof of concealment of a best mode for screening or producing monoclonal antibodies for use in the claimed '110 process, and therefore we are of the firm conviction that the district court's finding that the best mode requirement was not satisfied is clearly erroneous.

3. *Indefiniteness*

[14] The basis of the district court's holding that the claims are indefinite is that "they do not disclose how infringement may be avoided because antibody affinity cannot be estimated with any consistency." (Conclusion 6.) Even if the district court's finding in support of this holding—that "there is no standard set of experimental conditions which are used to estimate affinities"—is accurate, under the law pertaining to indefiniteness—"if the claims, read in light of the specification, reasonably apprise those skilled in the art both of the utilization and scope of the invention, and if the language is as precise as the subject matter permits, the courts can demand no more," *Shatterproof Glass Corp. v. Libbey-Owens Ford Co.*, 758 F.2d 613, 624, 225 USPQ 634, 641 (Fed.Cir.1985)—the claims clearly are definite. The evidence of record indisputably shows that calculating affinity was known in the art at the time of filing, and notwithstanding the fact that those calculations are not precise, or "standard," the claims, read in light of the specification, reasonably apprise those skilled in the art and are as precise as the subject matter permits. As a matter of law, no court can demand more.

VI. Motions

Monoclonal's motion to strike Appendices A and B of Hybritech's reply brief as being beyond the page limit applicable to reply briefs is granted as to Appendix A but denied as to Appendix B, the latter having been helpful in culling the often non-supportive citations to the record by Monoclonal.

Hybritech's motion to supplement the record with a Monoclonal advertisement not considered at trial is denied. Any adverse impact that the disposition of these two motions has upon either party is more than outweighed by this court's patience with the seemingly endless flow of post-argument argumentative papers.

VII. Conclusion

The judgment of the district court holding the patent in suit invalid is *reversed* in all respects, and the case is *remanded* for a determination of the issue of infringement which the court held was moot.

REVERSED AND REMANDED.

5

Inventorship in the Research Laboratory

Edward T. Lentz

SmithKline Beecham Corp., Philadelphia, Pennsylvania

I. INTRODUCTION

This chapter attempts to explain why inventorship is important and how inventorship is determined.

The answer to why inventorship is important is simple: United States law requires that a patent be applied for by, and only by, the inventor or inventors of the invention being claimed.[1] A consequence of failing to correctly designate inventors can be invalidation of the patent, if the error in designation was made with deceptive intent.[2] Correction of an incorrect inventorship determination is only possible if the initial determination was made in good faith.

The inventorship determination is important for other reasons as well. These include:

(i) the inventorship determination can affect the date of invention

(ii) the inventorship determination can affect a patent applicant's legal ability to claim the benefit of the filing date of an earlier filed patent application; and

(iii) the inventorship determination can affect the ownership of a patent.

These matters are discussed further below.[3]

II. WHAT IS AN INVENTOR

American concepts of inventorship are rooted in the days of the "attic inventors," the creative geniuses who tinkered away in virtual isolation. In this context, inventorship, obviously, is usually simple to determine.

However, whether we like it or not, such days are largely gone. Most research today is carried out in large institutional research laboratories and involves large teams of researchers. In this context, the conception, first reduction to practice, and optimization of inventions usually result from the combined efforts of teams of researchers, sometimes working collaboratively and sometimes not. Thus, selecting the inventor or joint inventors from among the many persons involved in such efforts can be, and often is, difficult.

Patent lawyers and courts have struggled with inventorship determinations countless times. Out of these struggles, certain principles have emerged. Nevertheless, any attempt at a black letter definition of "inventor" must come with a warning that inventorship determinations (1) are fact intensive and, therefore, can only be made on a case-by-case basis and (2) are things about which reasonable persons can differ. Having stated this warning, the following definition is offered as a guide: An inventor is the first person to have a complete conception of an invention and who does not thereafter abandon the invention.

A complete conception is, "the formation in the mind of the inventor of a complete and operative invention as it is thereafter to be reduced to practice."[4] In other words, a complete conception occurs when an

> inventive idea is crystallized in all of its essential attributes and becomes so clearly defined in the mind of the inventor as to be capable of being converted into reality and reduced to practice by the inventor or by one skilled in the art.[5]

In cases of unpredictable arts, because of the requirement that a complete conception requires knowledge of how to reduce an invention to practice, it may not be possible for a complete conception to be made in the absence of a reduction to practice.[6] This principle was called upon in the recent patent war between Amgen and Genetics Institute/Chugai regarding recombinant human erythropoietin (EPO).[7] The issue arose because one of the parties asserted that its researcher was the first to conceive of a deoxyribonucleic acid (DNA) sequence encoding EPO. The alleged conception resided in an idea to use two sets of degenerate probes from two different regions of the EPO gene to screen a genomic DNA library, which strategy was eventually successfully applied.

The court ruled that the idea was not a complete conception because there was no way to know what the DNA sequence was at the time the idea was conceived. The idea was only "an objective to make an invention which he could not then adequately describe or define."[8] In explaining what would have constituted a complete conception, the Court stated

> Conception does not occur unless one has a mental picture of the structure of the chemical, or is able to define it by its method of preparation, its physical or chemical properties, or whatever characteristics sufficiently distin-

guish it. It is not sufficient to define it solely by its principal biological property, *e.g.*, encoding human erythropoietin, because an alleged conception having no more specificity than that is simply a wish to know the identity of any material with that biological property. We hold that when an inventor is unable to envision the detailed constitution of a gene so as to distinguish it from other materials, as well as a method for obtaining it, conception has not been achieved until reduction to practice has occurred, *i.e.*, until after the gene has been isolated.''[9]

Does this mean that a complete conception of a DNA sequence can never occur until after it has been cloned? The answer is probably not. One of the facts relied upon by the court was that at the time the cloning strategy was conceived, the amino acid sequence of EPO was not known and, therefore, it was not possible at that time to design the necessary probes. Had the entire amino acid sequence of EPO, or at least some significant part thereof, been known, the result may have been different because, in that case, there would have been sufficient information such that one of ordinary skill in the art could reduce the invention to practice without undue experimentation and such that the DNA sequence encoding EPO could have been distinguished from other DNA sequences.[10]

When more than one person is the first to have such complete conception, as is usually the case in modern times, then the several persons are joint inventors (or co-inventors). Joint inventorship was once aptly described as ''one of the muddiest concepts in the muddy metaphysics of the patent law.''[11]

In 1984, the law regarding joint inventorship was clarified (some would say liberalized). The law now states:

> When an invention is made by two or more persons jointly, they shall apply for a patent jointly and each make the required oath, except as otherwise provided in this title. Inventors may apply for a patent jointly even though (1) they did not physically work together or at the same time, (2) each did not make the same type or amount of contribution, or (3) each did not make a contribution to the subject matter of every claim of the patent.[12]

Although the law now makes clear that joint inventors need not have *physically* worked together, there continues to be some debate over whether or not joint inventors must have worked together, i.e., in collaboration, in any sense. The view that seems to be correct is that joint inventors must have collaborated with each other; inventors cannot be joint inventors if each operated independently of and isolated from the other.[13]

Persons can become joint inventors by making contemporaneous contributions to a complete conception or by making sequential contributions. Contemporaneous contributions come about typically from discussions or other interactions in the course of which each of several persons stimulates the thoughts of the others, thereby progressing their combined ideas toward the conception, notwithstanding that one of the several must have actually been the first to have the conception.

Sequential contributions occur in cases in which one or more persons has an incomplete conception that is subsequently completed as a result of a contribution by another. Thus, a person who conceives of an idea is a sole inventor of the idea if he or she knows how to reduce the idea to practice, or if a means for doing so is within the ordinary skill in the relevant art notwithstanding that he or she does not himself or herself know of the means. If, however, the reduction to practice requires further exercise of inventive skill, i.e., the conception is incomplete, then, assuming collaboration, the original conceiver and the subsequent contributor are joint inventors.

In the same vein, an element of reduction to practice is using an invention.[14] Therefore, it follows that without a conception of a utility, conception is incomplete.[15] Thus, a person who conceives of or actually makes something, e.g., a new chemical entity, but who does not know of a use for the thing probably has not made a patentable invention and is not, therefore, an inventor unless and until a utility is discovered. At that time, a patentable invention will have been made and the person who conceived of the utility is probably a joint inventor with the person who conceived of the thing.

In some cases, the novelty of an invention is not appreciated by the person who conceives of the invention but rather is discovered subsequently. In these cases, conception of the invention is not complete until the novelty is appreciated.[16]

Joint inventors can also be named in a patent application based on their contributions to various aspects of the invention. For example, person A may have a complete conception of an invention, e.g., a certain monoclonal antibody. Person B may conceive of a new use for the antibody of which a novel modification of the antibody changes its properties. Person B is not a joint inventor of the antibody, but persons A and B are probably joint inventors of the new use or modification. In this case, certain of the claims in a patent application may be directed only to the antibody, of which person A is the inventor, and other of the claims may be directed to the new use or modification, of which persons A and B are the inventors.

III. THE INVENTORSHIP DETERMINATION PROCESS

In the course of drafting a patent application, the patent attorney or agent will collect information and data from a number of persons who may have been involved in some way in the research effort that produced an invention. He or she will also, at some point before making the inventorship determination, collect information concerning the roles and contributions of persons involved in the research that produced the invention for which a patent is being sought. However, inventorship is determined on the basis of the invention as defined by the claims of a patent application. Significant data bearing on the uses or properties of an invention are disclosed in the specification of a patent application. Nevertheless, the persons who generated such data are not necessarily inventors. For example, persons who discover an unexpected property that confers patentability on an invention are not by virtue of such discovery inventors.[17] Therefore, the inventorship determination is usually not made until after the claims have been settled.

At that point, the attorney or agent must determine inventorship objectively based on the facts before him or her. A real problem often faced by patent attorneys and researchers in research-based entities is that management often places great emphasis on inventorship on patent applications. The attorney, however, cannot be influenced by such extraneous factors, notwithstanding that they may be of vital professional or career importance to persons involved in the research.

A second problem that may be encountered in determining inventorship in modern research laboratories is that an invention may be conceived but not acted upon, only to be later independently conceived and reduced to practice by another. In this case, if the second conception was truly independent, then the first to conceive is probably not an inventor because the conception was abandoned by the first to conceive.[18] This is probably the correct result even if the first to conceive wanted his or her conception to be pursued but was prevented from working on it by his or her management. The result could be different, however, if the person who first conceived was diligent in pursuing the invention in the sense of continuing to press his or her management for resources to develop the invention.

Another problem that frequently occurs is expression of an inventive concept at a group meeting without anyone having the ability (or courage) to claim or attribute the conception. In such case, it may be possible to identify a small group of persons who engaged in a discussion out of which the conception sprang. In this case, as in any case, the small group is probably to be credited with jointly conceiving the invention. If this is not possible, then it may be necessary to designate the person who decided to pursue the conception after the group meeting as an inventor.

Another important point to bear in mind is that authorship on a scientific publication describing the invention is not synonymous with invention. Authors are named for a variety of different reasons, most of which are irrelevant to the inventorship determination.

The peculiarities of the patent law thus oftentimes lead to apparent inequities. Specifically, persons who put a lot of time and effort into reduction to practice and/or practical development of an invention may not be inventors.

Unfortunately, most researched-based institutions do not have a satisfactory mechanism for rewarding noninventor contributors. The problem is particularly acute in the case of institutions which reward their inventors such as with a cash award or with credit that enhances career advancement.

These kind of difficulties often make the inventorship determination difficult and sometimes painful. They can also easily cause errors to be made.

IV. HYPOTHETICAL INVENTORSHIP DETERMINATIONS

As stated by Judge Alan D. Lourie, "A gene is a chemical compound, albeit a complex one."[19] Therefore, biotechnology inventions are governed by the practices established for chemical inventions.

A typical case in chemical patent practice involves a chemist who conceives of a series of compounds, a technician who reduces the invention to practice by synthesizing the compounds and a third person, (for the purposes of this discussion, a biologist) who tests the compounds. In the typical case, the chemist conceives of the compounds with a particular chemical or biological activity in mind, the technician utilizes standard synthetic chemistry techniques to make the compounds, and the third person, e.g., the biologist, utilizes standard assay procedures to confirm the activity.

In such case, the inventor is the chemist, notwithstanding that the technician and the biologist may each have done more physical work than the chemist and that both may well be named as co-authors in a scientific publication describing the invention.

The result does not change if the biologist discovers a second, unforeseen activity, even if the second activity is of far greater importance.[20] Nor does the result change if the biologist discovers that certain of the compounds have greater activity than others or that a few of the compounds in the series are inactive.[21]

The case of a natural product discovered in the course of a collection, screening, and isolation program presents special considerations. Such case typically involves a specimen collector who collects soil samples or other natural specimens, a biologist who screens extracts of the samples for one or more particular activities, and a chemist who isolates an active compound and a second chemist who determines the structure of the active compound.

In this case, it is likely that none of the individuals involved conceived of the invention *a priori*, in the sense of planning to discover a particular compound with a particular activity.

In a usual case, the collector randomly collects a sample. The biologist applies standard protocols to prepare extracts of the specimen and to assay the extracts. When an active fraction is identified, he passes it on to the first chemist for isolation. The first chemist prepares various partially purified samples using different purification techniques. These are returned to the biologist for screening. Active samples are again returned to the chemist for further work-up and the process is repeated until a highly purified, active extract is obtained. The structure of the active compound(s) is then determined by another, or perhaps the same, chemist.

Whether or not any or all of these individuals are co-inventors will depend upon the specific facts of a given case. The specimen collector is probably an inventor, provided he or she exercised some level of independent judgment in selecting the specimen; e.g., based on location, appearance or other attributes of the specimen that distinguish it from materials not collected.

The biologist may or may not be an inventor based on similar considerations; i.e., did he or she act merely as a pair of hands, failing to exercise independent judgment, or did he or she contribute to the selection of assays, the interpretation of data, or the selection of samples for further work-up?

The first chemist is probably an inventor, based on the assumption that he or she selected samples for further work-up, independently or in collaboration with the biologist, or exercised judgment in selecting particular purification procedures or conditions. Once again, however, if the chemist acted merely as a pair of hands by running samples selected by someone else through predetermined protocols, he or she is probably not an inventor.

The chemist who determines the structure of the compound is probably not an inventor because he or she has merely identified the structural properties of an invention already made. However, certain factors could indicate a different result, e.g., whether or not future supplies of the compound can be obtained without knowing the structure (i.e., without relying on synthetic chemistry), and whether or not the compound was believed to be novel before the structure was determined, e.g., based on physical or biological properties.[22]

Similar considerations apply in determining inventorship of the organism, assuming the organism is patentable. For example, the collector is probably an inventor based on his or her selection of the specimen. The biologist may or may not be depending upon, e.g., whether he or she is selecting assays or merely applying predetermined protocols. However, neither of the chemists is likely to be an inventor of the organism because the invention will have been made prior to their involvement.

Biotechnology inventions bear more or fewer similarities to the case of a synthetically produced chemical compound or to the discovery of a natural product, depending upon the nature of the biotechnology invention.

An invention of a complementary DNA (cDNA) for a known protein involves a preconceived endpoint with certain predicted physical and chemical properties. Thus, like the chemist who conceives of a new compound, the person who conceives of cloning and expressing the cDNA is likely to be an inventor. In the case of some patentable cDNA inventions, there is some unusual obstacle that had to be overcome, such that the invention was not capable of being reduced to practice by exercise of ordinary skill. In such cases, the person who discovered a technique for surmounting the obstacle is probably a co-inventor. If the cDNA is patentable because it possesses an unusual property rather than because some obstacle had to be overcome, then the discoverer of the property is probably not an inventor.[23]

If the invention is a particular expression system comprising the cDNA, then inventorship will depend upon whether or not the expression system (without the cDNA) was known, i.e., had been used for other gene products, and whether or not the insertion and expression of the cDNA in the system was accomplished routinely or required that some unexpected obstacle be overcome. If the expression system was known and the expression was accomplished routinely, the inventor of the system with the gene insert is probably the person who first thought of using the system.

In the case of a monoclonal antibody to a known protein, the inventor is probably the person who conceived of making the antibody. As in the case of a cDNA, consideration must also be given to the person who overcame an unusual obstacle, if any. Again, the person who discovers an unusual property that causes the antibody to be patentable is probably not an inventor.

With increasing frequency, biotechnology inventions resemble classic chemical inventions. Specifically, biotechnology inventions now often concern hybrid or variant monoclonal antibodies or other proteins that are designed in advance to have specific properties. In these cases, as in the cases of synthesized chemical compounds, the inventor is usually the person who conceived of the hybrid or variant.

V. EFFECTS OF THE INVENTORSHIP DETERMINATION

As noted at the outset of this chapter, the inventorship determination can have a variety of effects on a patent application, or issued patent, which effects can sometimes have serious consequences.

A. Validity

Perhaps the most dramatic consequence of an incorrect designation is that a patent can be declared invalid by a court during an infringement suit if the court finds that the error was made with deceptive intent.

If the error occurred in spite of a good-faith effort to name the correct inventors, the error can usually be corrected without loss of patent rights. Incorrect inventorship is generally regarded as a technical defense and, therefore, "disfavored" as a reason for a court to hold a patent invalid.[24] Nevertheless, a good-faith effort to identify the correct inventor(s) of an invention being claimed in a patent application is necessary in order to be able to establish later that any error that may have been made was made without deceptive intent.

Courts tend to look more favorably on errors of misjoinder, i.e., inclusion of persons who are not inventors than on errors of nonjoinder, i.e., omission of persons who are inventors. Therefore, it is better to err on the side of misjoinder rather than nonjoinder.

B. Date of Invention

An early date of invention may be important to overcome a rejection based on the prior art by "swearing behind" the prior art.[25] An early date of invention may also be necessary to establish priority of invention as between two or more interfering patent applicants.

1. Swearing Behind the Prior Art

It is commonplace during examination of a patent application for the application to be rejected on the grounds that the invention lacks novelty or is obvious in view of a prior publication or use. If the patent applicant cannot persuade the Patent Examiner that the invention is in fact novel and unobvious, he or she may be able to demonstrate that the prior publication or use is not "prior art" under the patent law because it became known to the public only after the applicant made his or her invention and less than 1 year prior to the date on which the patent application was filed.[26] This practice is commonly referred to as swearing behind the prior art, the rules for which are set out in federal regulations, as noted above. It may also be necessary to swear behind the prior art during litigation in which an infringer alleges that the patented invention is not patentable because it lacks novelty or is obvious over the prior art.

The importance of the inventorship determination to swearing behind the prior art stems from the fact that an invention requires complete conception. By implication, the contributions of all jointly named inventors were required for the complete conception to have been made. Therefore, to establish the earliest date of invention requires proof that all contributions were made as of such date. Thus, the difficulty of establishing the earliest date of conception can be increased by the naming of multiple inventors. Worse yet, the earliest date of complete conception may be pushed forward to a later date by which date all inventive contributions were made.

Of course, it is not always necessary to prove the dates on which all joint inventors made their respective contributions. For example, in many cases, the earliest date of invention is not required; rather all that is required is to demonstrate a date of invention earlier than the date of the prior art. In other cases, it may be that certain joint inventors were named based on their contributions to aspects of the invention that are not part of the basic invention and, therefore, appear in some claims but not in a main claim. In this case, the dates of their contributions may be irrelevant to the date of complete conception of the main invention.

In cases in which the prior art asserted by a Patent Examiner, or an infringer, is a publication by the inventors, there is no need to swear behind the prior art because public disclosure of one's own invention can never be used a prior art against a patent application on the invention unless, of course, the public disclosure occurred more than 1 year earlier than the filing date of the patent application.[27]

Of course, if the public disclosure is by the inventor himself, then the fact that the disclosure is of the inventor's own invention will be apparent. However, it commonly happens that the public disclosure is by someone other than the inventor because of the difference between co-authorship and inventorship. For example, if persons A, B, and C are the named inventors in a patent application and the invention was disclosed in a paper co-authored by persons A, B, C, and D, then the public disclosure is by another because the inventive entity A, B, and C is different from the authorship entity A, B, C, and D.

This problem can be overcome by establishing that the invention disclosed by A, B, C, and D is actually the invention of A, B, and C. This is done by submitting a declaration by the author-noninventor to the U.S. Patent and Trademark Office attesting that the disclosed invention is the invention of A, B, and C.[28] Alternatively, a declaration can be submitted by the authors-inventors attesting to the fact that they are the correct inventors and that D is not an inventor.[29]

In any event, in addition to an earlier complete conception, swearing behind the prior art requires a showing of diligence from the date on which the complete conception was made to the date of filing of the patent application. Diligence means the continued application of effort by the inventor or his or her representative toward reduction to practice, actual or constructive. The representative need not be acting under the control or authority of the inventor. Thus, efforts by colleagues or co-employees of the same company are counted. The representative can also be a legal representative, e.g., a patent lawyer working toward drafting and filing of a patent application. In at least one case, a court has held that the efforts must be either toward an actual or a constructive reduction to practice; efforts toward both cannot be combined to establish overall diligence.[30]

In swearing behind the prior art, it may not be necessary to establish that the full scope of an invention was conceived and/or reduced to practice prior to the effective date of the prior art. It is only necessary to establish priority over that portion of the claimed invention that is disclosed in the prior art reference. Thus, if a claimed invention encompasses a class of mutant proteins and the prior art reference teaches how to make and use only one of such proteins, then it is only necessary to establish that the one prior art protein was conceived and/or reduced to practice prior to the effective date of the invention.[31]

2. Interference

Establishing an early date of invention can be critical in an interference. An interference is an administrative proceeding conducted by the U.S. Patent and Trademark Office to determine who, as between two or more applicants for a patent on the same or substantially the same invention, was actually the first to invent. It is a complicated, often lengthy and expensive process, the rules for which are set out in Federal Regulations.[32] Interferences are discussed elsewhere in this book.

In essence, the interference process is designed to determine, based on documentary evidence and, in some cases, on oral testimony, who among the competing inventive entities was the first to make a complete conception because that inventive entity was the first to invent. However, if that entity, in spite of being the first to conceive, was the last to reduce the invention to practice, then that inventive entity cannot prevail in the interference unless it can demonstrate that it was diligent in progressing the conception toward a reduction to practice. Therefore, interferences are also designed to determine with similar evidence the diligence of one who was first to conceive an invention but last to reduce it to practice.

Reduction to practice can be actual or constructive. An "actual" reduction to practice is just that, an actual making of a machine or composition or an actual carrying out of the steps of a process or method. What is required for reduction to practice depends upon the subject matter of the invention and the predictability of the operability of the invention. For example, if the invention is a class of variant proteins, then reduction to practice will ordinarily require expression or synthesis of at least one member of the class, and, unless the utility of the protein can be foretold with certainty, testing of such protein to establish activity.[33] If the invention is a method for treating a certain disease state, then the reduction to practice probably requires conducting of a predictive assay but can, in the case of a speculative treatment, require an actual treatment of a subject.

A "constructive" reduction to practice is a description of a complete means for making and using an invention in a duly filed patent application. Note that a constructive reduction to practice can only be accomplished by filing of a patent application containing such written description.

As in the case of swearing behind the prior art, the inventorship designation can affect the date of invention by making the date as late as the date of the inventive contribution of the inventor who was last to contribute. It also complicates the proof of early invention by necessitating more evidence. In addition, in an interference, the testimony, written or oral, of an inventor must be corroborated by a noninventor; thus further complicating the proof of date of invention.

The requirement for corroboration affects (or should affect) the record-keeping practices of all researchers. Specifically, notebook records of all researchers should be witnessed, more or less contemporaneously, by another researcher who can confirm that the efforts recorded in the notebook pages were in fact performed as described and on the dates indicated.

It is not uncommon during an interference for a party to attempt to correct inventorship in order to claim benefit of the filing date or an earlier application, as discussed below, or to alter the date of invention or the proof required to establish the date of invention.

C. Continuations-in-Part

Claiming the benefit of the filing date of an earlier filed patent application on related subject matter can be an important means for overcoming a rejection based on obviousness over intervening prior art, i.e., a publication or other prior art that arises between the filing dates of two patent applications. This is accomplished by converting the second application into what is known as "continuation-in-part" (CIP) application because it contains some of the subject matter of an earlier filed application; i.e., it continues certain subject matter but also includes new matter not in the earlier filed application.

Filing of a CIP can also be done to overcome a rejection based on an earlier filed patent application that discloses similar subject matter.[34]

However, benefit of the filing date of an earlier filed patent application can only be claimed if there is at least one common inventor. In other words, if the first application names inventors A, B, and C, then at least one of inventors A, B, and C must also be named as an inventor in the second application.

D. Ownership

In the absence of an agreement to the contrary, the inventors jointly own their invention. An important consequence of this result is that in the U.S. joint owners are free to make, use, and sell the invention without any obligation to the other owners.[35]

In other words, if researchers employed by different entities collaborate without an agreement and jointly make an invention, then all of the joint inventors, or more likely their respective employers, jointly own the invention. In addition, each of the joint inventors (or their employers) have the right to make, have made, use, and sell the invention and to license such rights to third parties without obligation to the other joint inventors.

It is primarily for this reason that corporate patent attorneys insist that collaborative research agreements be entered into prior to initiation of collaborations with researchers from other entities.

In addition to the commercial implications of joint ownership, joint ownership can also affect patentability. Specifically, under U.S. patent law, an invention made by another, even if not publicly disclosed, generally can be used as prior art in determining the novelty and obviousness of a later made invention. However, if the first invention and the later made invention were subject to a common obligation of assignment at the times they were made, then the first invention cannot be used as prior art against the later made invention.[36]

This can be important in the case of an invention that results from a collaboration that is based upon prior research at one or more of collaborating institutions because in these situations it is likely that an invention will be jointly made by researchers from both institutions. If the researchers are both obligated by their employment contracts to assign their inventions to their respective employers, then the invention will not have the same obligation of assignment as the developments made prior to the collaboration. Such developments are, therefore, available as prior art against the joint invention.

VI. CORRECTION OF INVENTORSHIP

Correction of an incorrect inventorship designation can be made at any time; i.e., while the application is pending in the Patent and Trademark Office or after it issues. The rules for correcting inventorship are set out in Federal regulations.[37]

While an application is pending in the U.S. Patent and Trademark Office, inventorship can be corrected if the inventorship designation is incorrect owing to mistake or to amendment, cancellation, or addition of claims.

In the first case, the inventorship can be corrected only if the error in inventorship was made through error without any deceptive intent. The regulations state, in part,

> Such amendment must be diligently made and must be accompanied by (a) a petition including a statement of facts verified by the original named inventor or inventors establishing when the error without deceptive intention was discovered and how it occurred; (b) an oath or declaration by each actual inventor or inventors as required by Section 1.63; (c) the fee set forth in Section 1.17 (h) and (d) the written consent of any assignee.[38]

The same procedure must be followed to add an inventor owing to addition of claims.

To remove an inventor because of amendment or cancellation of claims during prosecution requires diligence, a petition acknowledging that the misjoined inventor's invention is no longer being claimed, and the $120.00 fee.

After a patent is issued, the inventorship can be corrected by the U.S. Patent and Trademark Office by substantially the same procedure as for deleting inventors outlined above or by order of a court. Even a court, however, can only order correction of inventorship if the incorrect inventorship designation was made through error without deceptive intent.

VI. SUMMARY

Inventorship is based on complete conception; i.e., an inventor is the first person who has a complete conception of an invention and who does not thereafter abandon the invention. Inventorship is not the same as authorship.

A complete conception includes conception of a means for reducing the invention to practice, i.e., for making and using the invention (or such means must be within the ordinary skill in the art), and requires that the inventor appreciate that the invention is novel.

If an inventorship designation is incorrect, a patent can be declared invalid if the error was made with deceptive intent. If the error was made without deceptive intent, it can normally be corrected. Nevertheless, the problems associated with an incorrect inventorship determination, including the difficulties of correcting the error, are such that it is far better to get it right the first time even if doing so causes a certain amount of effort, anguish, and even pain.

NOTES

1. The basis for this rule of law is found in the U.S. patent statutes, at 35 U.S.C. 101 and 35 U.S.C. 102 (f). In contrast, the European Patent Convention provides, "The inventor shall have the right, vis-a-vis the applicant for or proprietor of a European patent, to be mentioned as such before the European Patent Office." European Patent Convention, Article 62.
2. 35 U.S.C. 116, 35 U.S.C. 256 and 37 C.F.R. 1.48.
3. One further point, not discussed below, is that an incorrect inventorship determination can subject the owner of a patent to liability for tortious misrepresentation. See, e.g., Misani v. Ortho Pharmaceutical Corporation, 145 U.S.P.Q. 625 (N.J. Supreme Ct. 1965).
4. Deller's Walker on Patents, Section 45, at 191 (1964).
5. Ibid., at 192. See also GAF Corporation v. Amchem Products, 211 U.S.P.Q. 172, 192 (E.D. Pa. 1981), and Technitrol v. United States, 169 U.S.P.Q. 732, 736–7 (Ct. Cl. 1971).
6. See Smith v. Bosquet, 45 U.S.P.Q. 347 (C.C.P.A. 1949), and Alpert v. Slatin, 134 U.S.P.Q. 296 (C.C.P.A. 1962).
7. Amgen v. Chugai Pharmaceutical Co. Ltd., 18 U.S.P.Q.2d 1016 (Fed. Cir. 1991).
8. Ibid., at 1021.
9. idem., at 1021.
10. The question of what constitutes a complete conception in the biotech area was also considered in Hybritech v. Monoclonal Antibodies, 231 U.S.P.Q. 81 (Fed. Cir. 1986). In that case, the court ruled that a conception of using monoclonal antibodies in an immunoassay constituted a complete conception.
11. Mueller Brass v. Reading Industries, 176 U.S.P.Q. 361, 372 (E.D. Pa. 1972).
12. 35 U.S.C. 116.
13. See Carstens, "Joint Inventorship Under 35 U.S.C. 116," J.P.T.O.S., *73*: 616–640 (1991).

14. See, e.g., Kawai v. Metlesics and Sternbach, 178 U.S.P.Q. 158, 163 (C.C.P.A. 1973).
15. This view was taken by the U.S. Patent and Trademark office in D'Amico v. Brown, 155 U.S.P.Q. 534 (Patent Office Bd. Pat. Inter. 1967). In Rey-Bellet v. Englehardt, 181 U.S.P.Q. 453 (C.C.P.A. 1974); however, the court expressly reserved judgment on this issue.
16. Silvestri v. Grant, 181 U.S.P.Q. 706, 708 (C.C.P.A. 1974).
17. Eli Lilly & Co. v. Premo Pharmaceutical Laboratories, 207 U.S.P.Q. 719 (3d. Cir. 1980).
18. The U.S. law states, "A person shall be entitled to a patent unless . . . he has abandoned the invention." 35 U.S.C. 102 (c).
19. Amgen v. Chugai Pharmaceutical Co. Ltd., 18 U.S.P.Q. 2d 1016, 1021, (Fed. Cir. 1991).
20. Howell v. Taborsky, 164 USPQ 58 (USPTO Bd. Pat. Int. 1969).
21. MacMillan v. Moffet, 167 USPQ 550 (C.C.P.A. 1970).
22. Silvestri v. Grant, 181 U.S.P.Q. 706, 708 (C.C.P.A. 1974).
23. Eli Lilly & Co. v. Premo Pharmaceutical Laboratories, 207 USPQ 719 (3d Cir. 1980).
24. In re Certain Steel Rod Treating Apparatus, 215 USPQ 237, 255 (U.S.I.T.C. 1981), citing Chisum, Section 2.03[4].
25. See, 37 C.F.R. 1.131
26. 35 U.S.C. 102 (a), (b). The one year "grace period" is virtually unique to the United States. Japan has a 6 month grace period for disclosures by the inventor. Many patent laws, including the European Patent Convention, have grace periods, usually six months, for disclosures at certain international exhibitions or for misappropriated inventions, i.e., disclosures of inventions without authorization from the proprietor.
27. *In re* Facius, 161 U.S.P.Q. 294, 302 (C.C.P.A. 1969).
28. Declarations of this type can be submitted under 37 C.F.R. 1.132.
29. *In re* Katz, 215 U.S.P.Q. 14 (C.C.P.A. 1982).
30. *In re* Mulder, 219 U.S.P.Q. 189 (Fed. Cir. 1983).
31. *In re* Stempel, 113 U.S.P.Q. 77 (C.C.P.A. 1957).
32. 37 C.F.R. 1.600 *et seq.*
33. De Solms v. Schoenwald, 15 USPQ 2d 1507, 1510–1511 (B.P.A.I. 1990).
34. Such rejection is typically characterized as a "provisional 102 (e) rejection," meaning that it is based on 35 U.S.C. 102 (e) which provides, in effect, that an invention is not patentable if it is obvious over subject matter disclosed in an issued U.S. patent, an application for which was filed before the date on which the rejected invention was made. Until the prior patent application issues, the rejection can be overcome by refiling the prior and the subsequent patent applications as a single CIP. Once the prior patent application is issued, the rejection becomes actual, rather than provisional, and it can no longer be overcome by refiling the applications as a CIP. Interestingly, whether or not the prior patent is prior art under 35 U.S.C. 102 (e) also depends upon inventorship, as it only qualifies as prior art if the inventive entity is different from the inventive entity designated in the alter application.

35. 35 U.S.C. 262. This is not so in all countries.
36. 35 U.S.C. 102 (g). Rejections of this type can also be based on double patenting.
37. 37 C.F.R. 1.48 and 1.324.
38. 37 C.F.R. 1.48. The fee set forth in Section 1.17 (h) is $120.00

Additional Notes

Any question over whether or not joint inventors must have collaborated with each other appears to have been removed by the decision in *Kimberly-Clark v. Proctor & Gamble*, 23 USPQ2d, 1921 (Fed. Cir. 1992), wherein the Court stated,

> What is clear is that the statutory word "jointly" is not mere surplusage. For persons to be joint inventors under Section 116, there must be some element of joint behavior, such as collaboration or working under common direction, one inventor seeing a relevant report and building upon it or hearing another's suggestion at a meeting."

23 USPQ2d, at 1926.

The Court also affirmed the viability of *Monsanto v. Kamp*, 154 USPQ 259 (D.D.C. 1967) in which the Court, in addition to requiring some element of collaboration in order for an invention to have been jointly made, stated,

> To constitute a joint invention, it is necessary that each of the inventors work on the same subject matter and make some contribution to the final result. Each needs to perform but a part of the task if an invention emerges from all of the steps taken together. It is not necessary that the entire inventive concept should occur to each of the joint inventors, or that the two should physically work on the project together. One may take one step at one time, the other an approach at different times. One may do more of the experimental work while the other makes suggestions from time to time. The fact that each of the inventors plays a different role and that the contribution of one may not be as great as that of another does not detract from the fact that the invention is joint if each makes some original contribution, though partial, to the final solution of the problem.

154 USPQ, at 262.

6

Ownership of Laboratory Discoveries and Work Product

William L. LaFuze and Peter E. Mims

Vinson & Elkins, Houston, Texas

I. INTRODUCTION

Once a discovery is made in the laboratory, the question arises: Who owns it? The research scientist who spent many days and late nights may believe he or she should own the discovery. The corporation or university employing the research scientists may believe it should own the discovery. Sometimes the parties agree in advance about ownership of discoveries but sometimes they do not.

Ownership in the discovery is not always the only issue. (By "discovery," we mean an idea that is protectable under the patent or trade secret laws. For example, a research scientist might discover a new and useful cell line for the treatment of a disease.) There may also be laboratory notebooks, documents, production information, software programs, and manuals and other writings associated with the discovery that are valuable but do not rise to the level of patentable invention that can be referred to as "work product." For example, the research scientist who discovered the cell line may have employed a particular method in the laboratory or developed a computer software program to aid in practicing the discovery. Similar questions may arise regarding the ownership of work product.

The ownership of laboratory discoveries and work product involves the law of property, contracts, and torts. Depending on the factual setting, one body of law may be controlling in a particular ownership dispute. In the absence of a written agreement, ownership often turns on the facts and circumstances of the dispute.

Furthermore, because the field of biotechnology sometimes raises unique issues, the courts and legislatures may sometimes treat ownership disputes differently in this field.

Indeed, some ownership disputes may turn on policy issues unique to the field of biotechnology. For example, some courts may be receptive to ownership claims that would not be seriously considered in other areas of innovation because the involvement of human tissue and other life forms. In the absence of legislative pronouncements regarding ownership in such disputes, a court might have to settle a dispute concerning ownership rights between the research scientist and possibly the supplier of the naturally occurring organism; e.g., the person providing his or her tissue to the scientist.

Other ownership disputes in the biotechnology field will be resolved in a more traditional fashion using legal principles applicable in other areas of intellectual property. For example, a dispute regarding ownership between an employee and an employer will likely turn on what are called common law rules; i.e., judge-made rules of law that have been developed by courts to resolve similar disputes concerning other areas of technology.

Biotechnology research is often funded by third parties, such as the federal government or private industry. Determining the ownership interests in discoveries and work product made in federally funded research will turn on traditional legal concepts of statutory interpretation and administrative regulations pertaining to government contracts and not judge-made policy decisions. Funding by private industry will usually turn on the terms of a complex agreement, often drafted by attorneys, before any research was done.

No matter what the context, however, the scientist discovering the laboratory innovation or developing some useful work product in the laboratory needs to know who owns what. This chapter will discuss the ownership interests of intellectual property rights in biotechnology. Our goal is to acquaint the research scientist with the ownership issues that arise when a new and useful discovery in biotechnology is made.

Because of the nature of the biotechnology research, rarely will a discovery in the biotechnology area involve a single inventor without input from other parties, whether they be employers, colleagues, suppliers of naturally occurring materials, or the government. The simplicity of a single discoverer unencumbered with obligations to a university, private corporation, or government is likely the exception in the field of biotechnology. One recent study reported that approximately 80% of patents issued in the United States were assigned at the time of issue.[1] Consequently, most scientists will seek answers to questions of ownership in far more complex situations. The ultimate issue discussed in this chapter then is who owns what and what are some of the important ownership issues that confront the laboratory scientist in the field of biotechnology?

II. GENERAL OVERVIEW OF COMMON LAW OWNERSHIP OF INTELLECTUAL PROPERTY RIGHTS IN LABORATORY DISCOVERIES AND WORK PRODUCT

As an initial matter, the basic legal concepts of property law bear discussion. Once a claim of ownership in property is recognized by the law, it becomes a legal "right." If the owner can prove someone has violated a right, then a court will protect the right by compensating the owner for the infringement, possibly by awarding monetary damages or injunctive relief. Transferability of ownership is also an important right. Patents, trade secrets, and copyrights, which are often referred to as intellectual property rights, may be assigned, licensed, or otherwise transferred by their owners like other forms of property.

There are different types of property. For example, property may be categorized as tangible and intangible. In the case of laboratory discoveries and work product, we are generally concerned with intangible property rights, specifically what are called "intellectual property rights." Like ownership of more familiar tangible property rights, such as real property, ownership of intellectual property rights, such as patents, copyrights, trademarks, and trade secrets, gives the owner certain enforceable rights. For example, the owner of real property has the right to exclude others from his or her property. Similarly, an owner of an intellectual property right, such as a patent or copyright, has the right to exclude others from infringing his or her intellectual property rights. In other cases, such as a trade secret, it may be the right to enjoin one who learned the secret in confidence from using or disclosing it.

This chapter assumes that patent protection is the most likely form of protection in the biotechnology industry. Although trade secret and copyright protection may be available for certain aspects of biotechnology, this chapter focuses primarily on ownership issues in a patent law context. In the course of discussing ownership of patent rights, however, this chapter also discusses trade secret and copyright protection where it appears most applicable. With this background on the meaning of "ownership," we can turn to the inquiry of this chapter: Who owns the discovery and the work product associated with the discovery?

As a general rule, when a single research scientist discovers something new, he or she "owns" it. But what rights does the scientist have in his or her discovery? The research scientist can take a number of different courses of action to protect the ownership in his or her discovery. He or she can disclose the invention to the U.S. Patent Office in an application for letters patent in hopes of obtaining a patent. A patent gives its owner the right to exclude others from practicing his or her invention as claimed. If the government awards the research scientist a patent, he or she has protected the rights in his or her invention as described in the claims in the issued patent. As a result, the inventor has the ''property'' right to exclude others from making, using, or selling his or her invention for 17 years from the date of issuance of the patent.

Seeking patent protection is not the only choice for the research scientist. He or she can choose to keep it secret and exploit it as a trade secret. By maintaining the discovery in secrecy, the inventor protects his or her rights in the discovery. Trade secret protection can also be used in conjunction with patent protection. Often, during the pendency of a patent application, an inventor will maintain the discovery and associated work product in confidence. If the patent does not issue, then the inventor will simply continue to maintain the discovery and work product as a trade secret. However, as discussed above in Chapter 3, once others independently discover and publish the ''secret'' invention or discovery, the property right in the trade secret may be lost. However, it is also possible that the second person independently discovering the trade secret will also maintain the discovery in confidence. In such a case, two people, completely unconnected, ''own'' the same trade secret. The scientist can also sell or transfer the invention or discovery and allow the buyer to determine whether to patent or maintain the discovery in secrecy.

A final option, of course, is to simply publish the discovery and dedicate the discovery to the public. Of course, the inventor would own the copyright to the publication, but he or she would not be able to exclude others from practicing the discovery. In the academic world, perhaps this choice to publish may not seem so unusual.

The law has recognized the ownership interest in the individual who makes a patentable discovery or who authors an original writing. For example, the U.S. Constitution acknowledges an inventor's ownership in his or her discoveries and an author's ownership in his or her writings: ''The Congress shall have power . . . [t]o promote the Progress of Science and useful Arts, by securing for limited Times to Authors and Inventors the exclusive Right to *their* respective Writings and Discoveries.''[2] Pursuant to this constitutional power to legislate regarding the ''Discoveries'' of ''Inventors,'' Congress enacted the federal patent laws. Expounding on the peculiar nature of the act of invention, the U.S. Supreme Court described the act of invention as:

> . . . consisting of neither in finding out the laws of nature, nor in fruitful research as to the operations of natural laws, but in discovering how those laws may be utilized or applied for some beneficial purpose, by a process, a device or a machine. It is the result of an inventive act, the birth of an idea and its reduction to practice; the product of original thought; a concept demonstrated to be true by practical application or embodiment in tangible form.[3]

Similarly, pursuant to this constitutional power to legislate the ''Writings'' of ''Authors,'' Congress enacted the federal copyright laws.

Notwithstanding this apparent clarity of this constitutional recognition of the ''Inventor's'' right to his or her ''Discoveries,'' and ''Authors'' for their ''Writings,'' the determination of ownership of the discovery and work product often involves more than identifying the inventor[4] or the author. These other concerns sometimes lead to disputes and, in turn, to lawsuits. Over time, in the course of resolving disputes regarding ownership rights, the federal and state courts have developed rules for deciding the disputes. These federal and state law rules can be seen as the result of a number of courts addressing similar issues and developing fair and reasonable solutions. At times, however, the state law rules regarding ownership issues vary from jurisdiction to jurisdiction. The courts are not the only source of laws on ownership; other ownership issues turn on the interpretation of federal and state statutes. Identifying the applicable law is not the end of the analysis. Disputes regarding ownership often will also turn heavily on the facts.

Because biotechnology involves technological advances that may involve the human body, certain biotechnology discoveries raise ownership issues that have not been previously addressed or decided by statute or the common law. For example, until recently, state law had never addressed the ownership of biotechnology discoveries that involve the use of human tissues and cells as a raw material. Other areas of technology do not involve an invention or discovery based on the manipulation of tissues removed from a human. In *Moore* v. *Regents of the University of California*,[5] the California Supreme Court refused to recognize a common law property right in the person whose cells were used to develop a new biotechnology product. (In *Moore*, a research scientist developed a new cell line and derivative products using tissue from a patient whose spleen was removed as part of his treatment for leukemia.) This holding does not mean other jurisdictions will agree with the California Supreme Court. Indeed, the lower appellate court of California[6] did recognize a property right in that individual's tissues or cells that ultimately led to a profitable product. Even if the courts refuse to recognize such a right, compensation could be required by future legislation.

Property law is not the only body of law that can complicate exploitation of such a new discovery and associated work product. Even though the California Supreme Court did not recognize a property right, it did hold that Moore, the patient whose removed spleen contained the valuable cells, did have a state law tort claim for a breach of fiduciary duty against the doctor because the doctor did not disclose his potential profit motive in removing the spleen. Although such a claim may not give the patient ownership rights in the invention or discovery, the claim can pose significant risks for the doctor or researching party who fails to disclose the potential profit motive to the patient prior to the medical operation. A common law claim of breach of the fiduciary duty allows the patient to sue for actual and punitive damages for the breach of duty. Such a common law claim may be as effective a negotiating tool as an actual property right in the discovery itself.

The argument that the donor may have a property interest in the new discovery raises a similar issue about the source of all existing materials used in biotechnology experiments. It is possible that a third party has intellectual property rights in the source material used for a discovery. In the event there is an agreement restricting the research scientist's use of the source material, the restrictions on any "progeny" of the source material should control any ownership dispute regarding the discovery. Thus, these restrictions on the use of the source material will ultimately serve as a qualification or limitation on the title to any subsequent discovery.

A research scientist may also have an obligation of confidence regarding the discovery and work product. For example, if the employer treats such discoveries as trade secrets, state law possibly obligates the employee to maintain the discovery in confidence. Similarly, a researcher will be obligated to maintain in secrecy work product information associated with the trade secret. In such a case, the employer's ownership interest in the trade secret provides the owner with a state law clam to preclude the research scientist from disclosing or using the secret discovery.

In other cases, the research scientist, because of a fiduciary relationship, may have an obligation to assign any discoveries or work product. To be a fiduciary, one must hold a special position of trust or confidence. A fiduciary relationship can arise when the research scientist is a partner, officer, director, or controlling shareholder. In the appropriate case, a court may order an assignment by the inventor to a corporation or partnership because of the inventor's fiduciary duty to such entity.[7]

III. THE EMPLOYEE WITH NO WRITTEN AGREEMENT

Many biotechnologists are not self-employed; rather they are employees of corporations, nonprofit research centers, or universities. A common question is to determine ownership of discoveries and work product made by such an employee who has no written agreement. In the absence of a written agreement, state law generally governs disputes regarding ownership of discoveries, although special rules apply to work product that is the subject of the federal copyright laws. This portion of the chapter addresses the general rules of ownership in the absence of a written employment agreement. This discussion assumes, however, that the individual who made the discovery or prepared the work product is an employee and not an independent contractor, which is defined and discussed below.

A. Common Law Rules of Employee Ownership

Many employers would be surprised to learn that not all inventions made by employees that relate to their business are automatically owned by the employer. In fact, the general common law rule is that an employer does not own its employee's inventions[8] even though that discovery was conceived or reduced to practice during their employment. The Restatement of Agency, Second, which is a restatement of the law by eminent judges and lawyers, has restated the common law rule on employee inventions as follows:

> When Agent Has Right to Patents. Unless otherwise agreed, a person employed by another to do non-inventive work is entitled to patents which are the result of his invention although the invention is due to the work for which he is employed.[9]

Consequently, in the absence of an agreement, unless an employee is hired to do "inventive work," the inventions of the employee will not be owned by the employer. This is in keeping with the courts' recognition of the special nature of the act of invention. This rule regarding an invention is equally applicable to all discoveries; i.e., is, both patentable and unpatentable ideas.

Mere employment of an inventor at the time that his discovery or invention was made is not adequate to vest an employer with rights of ownership. Instead, an employer must derive his ownership rights to the invention and to the intellectual property associated with the invention through either an express or implied contract to assign inventions or other circumstances from which the employer, as a matter of equity, will have some or all rights to the invention.

The law acknowledges that people, and not business or nonprofit entities, make discoveries. For example, the U.S. patent laws recognizes only real persons as inventors.[10] Thus, in the absence of an agreement, express or implied, a patent on a discovery belongs to the individual inventor. Indeed, one basic policy of the patent system is to encourage technological invention by awarding patent rights to inventors.

Ownership of trade secrets is generally treated the same as patents.[11] However, one noted commentator has stated that most case law involves the ownership of patented inventions and trade secret ownership "would appear equally applicable to trade secret principles under appropriate circumstances."[12] However, the same commentator then advises that because there is no clear authority applicable to trade secrets, an employer should address the issue by agreement. At common law, an invention made by an employee, whether it is ultimately patented or held in confidence as a trade secret, will be the property of the employee unless by the terms of an employment agreement or as a matter of equity, the employee agreed to transfer ownership of such discovery to the employer.

This does not mean the common law leaves the employer with no rights at all with regard to discoveries. Even if the employee owns the invention or discovery, the employer may have a "shop right"; i.e., the right to practice the discovery in its business on a nontransferable, royalty-free basis. This "shop right" is discussed below.

B. The Employee Hired to Invent

The common law has recognized an exception to the common law rule that an employee owns his or her inventions. It is well established that an employer owns the discoveries or inventions of an employee that is "hired to invent." In the landmark case of *Solomons* v. *United States*, the U.S. Supreme Court wrote as follows:

> If one is employed to devise or perfect an instrument, or a means for accomplishing a prescribed result, he cannot, after successfully accomplishing the work for which he was employed, plead title thereto as against his employer. That which he has been employed and paid to accomplish becomes, when accomplished, the property of his employer. Whatever rights as an individual he may have had in and to his inventive powers, and that which they are able to accomplish, he has sold in advance to his employer.[13]

This rule has been applied to both patentable and unpatentable inventions.[14] In essence, because the purpose of the employment is to invent, the law *implies* an agreement by the employee to assign inventions or discoveries to the employer. Obviously, no express agreement exists; however, the courts imply this agreement presumably out of a sense of fairness and equity, based on the circumstances of the employment.

Similarly, if an employee is assigned the task of solving a specific problem, then any discovery or invention resulting from that work will be owned by the employer. In an often-cited case, one court explained the employer's right to the discovery in terms of the employee's duty to the employer:

> It is contended, however, that the rule which we have discussed has no application here, because it is said that defendant was not employed by the government as an inventor, or to invent the fumigant gas which is the subject of the patent, but merely to do ordinary work as a chemist, such as the analyzing of dust samples. The trouble with this argument is that it gives too narrow a meaning to the word "employed." The right of the employer to the invention or discovery of the employee depends, not upon the terms of the original contract of hiring, but upon the nature of the service in which the employee is engaged at the time he makes the discovery or invention, and arises, not out of the terms of the contract of hiring, but out of the duty which the em-

> ployee owes to his employer with respect to the service in which he is engaged. It matters not in what capacity the employee may originally have been hired, if he be set to experimenting with the view of making an invention, and accepts pay for such work, it is his duty to disclose to his employer what he discovers in making the experiments, and what he accomplishes by the experiments belongs to the employer. During the period that he is so engaged, he is "employed to invent," and the results of his efforts at invention belong to his employer in the same way as would the product of his efforts in any other direction.[15]

This rule should apply equally to associated work product. To the extent work product is produced as a result of such efforts to invent, the employer should own such trade secrets.

In determining ownership to a laboratory innovation and work product, the research scientist should ask himself or herself if he or she was hired to invent or was hired for the task of solving a particular problem that has now resulted in a potentially valuable discovery. The courts look closely at the specificity of the task assigned to the employee. One commentator has identified a number of factors that may lead a court to find an employee is hired to invent:

> previous inventions by the employee have been assigned to the employer;
>
> the company has a customary practice for other similarly situated employees to assign inventions;
>
> the invention was conceived during employment period;
>
> the problem solved by the inventor was originally posed by the employer;
>
> the employee had the authority in the company to choose the individual who would attempt to solve the problem;
>
> the idea or invention is important to the employer's business;
>
> the employer has not taken a previous inconsistent portion regarding inventorship by the employee;
>
> the employer had agreed to pay royalties to the employee;
>
> the employer paid fees for the prosecution of the patent; and
>
> the employer did not fail to express interest in the invention or discovery when it was first presented by the employee.[16]

Thus, when a biotechnologist is assessing whether he or she was hired to invent, this list provides at least a checklist of the evidence a court would consider.

C. Shop Rights of Employer

Although the common law has protected the employee's rights to his or her inventions or discoveries, it has also recognized a limited right of the employer to practice or use the employee's invention. This right or privilege, known as a "shop right," arises as a matter of equity when the employee has made use of the employer's resources in developing his invention or discovery. For example, the employee may have used the time, material, equipment, or facilities of the employer in making his discovery. When a shop right exists, the common law employer has an irrevocable, royalty-free, nontransferable, nonexclusive license to use or practice the discovery or invention in its business.

The equitable principles underlying the shop right doctrine were articulated by the United States Supreme Court in *United States* v. *Dublier Condenser Corp.*:

> [W]here a servant, during his hours of employment, working with his master's materials and appliances, conceives and perfects an invention for which he obtains a patent, he must accord his master a non-exclusive right to practice the invention. . . . Since the servant uses his master's time, facilities, and materials to attain a concrete result, the latter is in equity entitled to use that which embodies his own property and to duplicate it as often as he may find occasion to employ similar appliances in his business.[17]

Although the employee has title or owns the discovery, the shop right allows the employer to use the discovery in its regular business without compensation to the inventor. Thus, when a patent issues on the invention, the employer will still have the right to practice the invention without payment of a royalty to the employee. In the event the employee wishes to assign or license the patent rights to third parties, the third parties will take subject to the employer's irrevocable, nontransferable, royalty-free license. Similarly, with similar facts involving a trade secret, courts have recognized a shop right in the trade secrets.[18] With an important discovery, a shop right may be a very valuable right.

The crucial factual inquiry that determines the existence of shop rights is the extent of the employer's contribution to the discovery. Employment alone at the time of the discovery is clearly not enough to vest any rights in the employer. Generally speaking, if the employee makes "significant" use of the employer's resources, then a shop right exists. Significant use adequate to justify a shop right might include using the time and assistance of other employees, using the employer's facilities, materials or equipment, and working on the discovery partly or completely while on the job.

One commonly litigated case is that of the employee who conceives his invention and reduces it to practice completely on his own time and at his own expense. Subsequently, however, the employer aids the employee in additional development of the invention; e.g., through the manufacture or commercialization of the invention. In such a case, some courts have held the employer is entitled to a shop right.

The shop right is not exclusively associated with the employment relationship. Some courts have found such a right to arise from other business relationships.[19] Generally, in these other cases, the courts can rely on the doctrines of estoppel and implied licenses to find a shop right in other types of relationships. These other contexts will be discussed further below.

D. Work Product of Employees and Copyright Laws

Work product involves some of the same as well as different issues. To the extent work product involves tangible writings, the federal copyright laws will be applicable. The copyright laws provide that a writing or document prepared by an employee in the scope of his or her employment will be considered a "work made for hire," and thus title will vest in the employer as a matter of law.[20] In short, the law considers the employer to be the "author" of the particular document actually written by the employee or group of employees. To the extent work product involves trade secrets, such as a computer program, laboratory books, laboratory methods, or other related information, then the employer will similarly have an argument that it owns the trade secrets and can obligate the employee to maintain them in confidence.

IV. THE EMPLOYEE WITH A WRITTEN AGREEMENT

Generally, when a scientist is working for a large corporation, university, or nonprofit research center researching complex biotechnological issues, the scientist will have his or her own employment agreement. In some cases, there may not be an individual employment agreement. For example, some employers may have only issued to their employees an "intellectual property policy." Often, the employee will expressly agree to be bound by such policy. This policy may effectively function as an agreement; however, these policies are also discussed separately below.

Agreements can simplify the issues in a dispute between an employer and employee regarding ownership of discoveries and work product. From the employer's standpoint, such an agreement is obviously the preferred course as the common law rule generally places ownership in the hands of the employee. Consequently, it is almost always the employer who requests the execution of the agreement. Not surprisingly, such agreements generally are drafted to favor the employer. On the other hand, in new, high-risk companies or in nonprofit settings, it is not uncommon for employees to receive some form of royalty payment from their employer regarding laboratory discoveries and work product.

A. Common Terms and Conditions of Written Agreement

1. Assignment

The most important contractual term with regard to ownership of the discovery will be the assignment provision. The most effective assignment from the employer's point of view is one that assigns all inventions and discoveries, whether patentable or not, that are related to the employer's business or that the employee makes as a result of his employment. In addition, although the employer probably owns all tangible work product developed by the employee, the employer may include an express provision assigning title to such work product. An example of such a provision might read as follows:

> Employee hereby assigns all right, title and interest to any inventions, ideas, discoveries, concepts and improvements (''Inventions'') related to the Business, whether made during working hours or at home, and any Inventions made by Employee as a result of his employment. In addition, the Employee hereby assigns all right, title and interest to any writings, memoranda, research notebooks, drawings, sketches, computer programs, and manuals (''Work Product'') created in whole or part by Employee.

The first provision of this grant, assigning inventions related to the business, would encompass discoveries made by the employee related to the business, whether made during working hours or at home. Most likely, an employer would broadly define ''business'' to maximize the scope of this assignment. The second provision, assigning discoveries made as a result of the employment, would encompass those discoveries made through the employee's use of the employer's facilities, time, and materials to develop or conceive the discovery. Such provisions make it much more difficult for the employee to claim that the employee owns the discovery. Generally, an employee does not have his or her own business on the side, and such an agreement will not seem unreasonable.

An employer can go too far in drafting such a provision. Some states have enacted statutes that prohibit the employer from obtaining the rights to all of their employee's discoveries or inventions.[21] For example, Illinois and California have such laws.[22] The Illinois law declares as void against public policy any assignment of inventions that fall outside of the field of company business, that are developed by the employee outside the scope of work, and that are conceived of without use of company time and resources. In other jurisdictions, if the employer attempts to require assignment of all discoveries and inventions, a court might view the provision as unconscionable and, therefore, invalid or unenforceable. Arguably, such a clause is broader than is reasonably necessary to protect the legitimate business interests of the employer as it would encompass some inventions with no connection to the business of the employer. Some courts, recognizing that an employee has little bargaining power in negotiating some employment agreements, narrowly construe such assignment agreements against the employer.

One issue that often arises is the temporal scope of such an assignment. Clearly, the assignment should not cover discoveries or inventions made before the employment. One common technique used by employers to address this issue is to allow the employee to identify any inventions that have already been made by the employee at the time the employment agreement is executed. As a result, any invention on the list would be excluded from the assignment clause in the agreement. In the event of a dispute, however, the employer can use such a list as powerful evidence that a now valuable discovery was not invented by the employee before the date of employment.

With respect to inventions or discoveries allegedly made after the employment, similar issues may arise. An employer may argue that the invention or discovery was actually made during the recently terminated employment. Such a case will likely turn on the facts and circumstances of the case, particularly the similarity of the invention and the employee's work at the former employer. Some employers insert specific contractual provisions addressing inventions made after termination of employment. One approach employers may take is to include a term to the effect that the parties agree there is a "presumption" that any invention relating to the activities of the employee while working for the employer that are conceived or made within 1 year after termination of employment are to be assigned to the employer unless the employee can establish the contrary. The enforceability of such a presumption, i.e., the allocation of the burden of going forward with the evidence to rebut the presumption, might be seen in effect as an agreement about the rules of evidence to be applied by a court. Generally, parties cannot agree to alter the evidentiary rules of a court. Thus, the enforceability of such a presumption may be subject to challenge.

The final provision, assigning title to work product, may not be necessary in view of state and federal laws. Under state law, an employer providing paper and working materials to an employee should continue to own the tangible materials after the employee prepares the work product materials. As to any copyright in the work product, under the "work made for hire" doctrine, the employer owns the copyright in the work product, as a matter of law.

2. Disclosure of Discoveries and Confidentiality

Often, to facilitate the exploitation of a discovery or invention, the employer requires the employee to disclose promptly to the employer in writing all inventions and discoveries, whether patentable or not. Such a contractual requirement facilitates the avoidance of the absolute novelty bar to patentability in many foreign countries and the one-year statutory bar to patentability in the United States as a result of a public use, disclosure, or sale beyond one year.[23] It also serves as evidence that can be used in the event that there is an interference proceeding and the date of conception becomes an important issue. Disclosure allows management to evaluate the discovery and reduce the chance that a valuable discovery will go unnoticed and unexploited. To facilitate such disclosures, employers often have invention disclosure forms.

A typical disclosure clause might read as follows:

> Employee agrees to disclose fully and promptly to Employer all Inventions made or conceived by Employee, in whole or in part, during the term of this Agreement. Employee agrees to maintain in confidence the Invention. Except as specifically authorized in writing by the Employer, Employee shall not disclose, directly or indirectly, the Invention, or any information related thereto, to any third party.

This particular clause includes not only an obligation of disclosure to the employer, it also obligates the employee not to disclose the Invention to a third party.

In order to protect trade secrets and a patentable invention prior to the issuance of a patent, employers often expressly obligate the employee to honor confidentiality provisions. A typical confidentiality provision might read as follows:

> Employee recognizes that during the term of employment, Employee will receive, develop or otherwise learn various types of information that are secret and confidential. Except as authorized by Employer, Employee shall not disclose or use, directly or indirectly, either during or subsequent to employment with Employer, any confidential information of the Employer that Employee obtains during the course of the employment relating to the business of Employer of Inventions.

3. Cooperation Regarding Execution of Documents

Another provision commonly found in employment agreements requires the employee to cooperate in the execution all documents necessary for the employer to apply for and obtain patent protection in the United States and abroad and to protect ownership in the employer. Generally, this provision will also provide that the employer will bear all expenses related to the prosecution of the patent rights. Such a provision might provide as follows:

> At the request and expense of the Employer, and without additional compensation to the Employee, Employee agrees to make application in due form for United States and foreign letters patent on such Inventions, and will execute any and all instruments and documents and do any and all acts necessary or desirable in connection with such applications in order for the Employer to establish and perfect the entire right, title and interest in such Inventions and to also execute any and all instruments and documents necessary or desireable

> in connection with any continuations, renewals, reexaminations or reissues thereof or in the conduct of any related litigation or proceedings.

This provision would serve as the basis for the employer to seek specific performance in the event that a employee refused to cooperate in executing documents associated with obtaining a patent on the discovery. As for trade secrets, the employee cooperates in protecting the employer's right by agreeing to maintain the discovery in secrecy.

4. Royalty Payments

A less common provisions found in employment agreements pertaining to intellectual property rights is one that provides for payment of a royalty to the inventor. Although such a provision does not change the ownership of the intellectual property, it does provide meaningful rights to the inventor: a contractual right to a percentage or a share of the earnings or revenues resulting from the patented product or product made by a patented process. A common royalty provision might read as follows:

> Employer agrees to pay Employee a royalty of two percent (2%) on all Net Sales of Patented Products, whether manufactured by Employer, an assignee or a sub-licensee of Employer.

This provision also can be viewed as a form of consideration for the assignment by the employee of the inventions or discoveries made by the employee.

B. Intellectual Property Policies of the Employer

As discussed above, sometimes a employer and employee do not execute a formal two-party agreement with these particular provisions. Instead, the employer has a formal, written "intellectual property policy" or "research handbook" that includes similar provisions. Most major universities now have formal patent or intellectual property policies for use by professors and graduate students who conduct research.[24] In some cases, the employee executes an agreement binding the employee to the policies. In other cases, the employee never expressly agrees to be bound by the policies.

This absence of a signed agreement raises the question of the binding nature of these policies. Ironically, even in the absence of a written agreement incorporating the terms of a policy manual or handbook, employees have used these documents against employers in many employee-termination disputes. As a result, there is a considerable body of law regarding the enforceability of such policy manuals against an employer. Some courts consider the policies in determining if there was agreement between an employer and employee that the employer would not terminate the employee except for the reasons stated in the policy manual[25]; other courts have refused to find an agreement based on the policy manuals above.[26]

The effect of such policy manuals on intellectual property ownership has not been litigated to the same extent. However, generally speaking, under contract law, if the employee does not agree to be bound by these formal policies, i.e., does not sign an agreement to that effect, then a court might decide not to enforce the intellectual property policies against the employee. Of course, the employer can argue the research scientist was hired to invent. Interestingly, one court ruled that university researchers were "hired to invent."[27] On the other hand, if the employee considers the policies favorable, e.g., the provisions regarding payment of royalties, then the employee may want to enforce the policy against the employer. Certain facts improve the chances of enforcing such an agreement. For example, in some jurisdictions, if the employee can argue that he or she relied on the policy when accepting employment, then the court may be more likely to enforce the policies against the employer.[28]

V. THE EMPLOYEE OPERATING UNDER AN ORAL AGREEMENT

In some cases, an employee may not have a written agreement with his or her employer; however, the employee and the employer may have orally agreed that the inventions of the employee would be owned by the employer or agreed that the employee would own the inventions of the employee. As an initial matter, just because an agreement is oral does not mean it is automatically unenforceable. Some oral agreements are enforceable. Others are not because they fail to comply with what is known as the Statute of Frauds. The Statute of Frauds requires that certain agreements be in writing in order to be enforceable. For example, an agreement to convey an interest in land must be in writing to comply with the Statute of Frauds. Intellectual property, however, is treated as personal property, and there is no similar provision in the Statute of Frauds. regarding an agreement to convey personal property. However, other types of agreements are also subject to the Statute of Frauds. For example, an agreement that takes longer than one year to perform must be in writing to be enforceable under the Statute of Frauds. Consequently, because this type of agreement is subject to the Statute of Frauds, license agreements that last more than one year need to be in writing.[29]

Notwithstanding the possible enforceability of an oral agreement, it is another thing to prove that such an agreement was actually made and the specific terms of such an agreement. For example, juries have a natural suspicion when an agreement is not in writing.

Often, even though there is no written formal agreement, there may be correspondence that contains the material terms signed by the employer that the employee now wishes to enforce. Such an informal agreement may be enough to constitute a "writing" that is specific enough to remove the agreement from the Statute of Frauds and thereby make it enforceable. Another exception to the Statute of Frauds is where the party seeking to enforce the oral agreement can prove detrimental reliance on the oral agreement. For example, in some jurisdictions, if one party has materially changed its position in reliance on the promise, the Statute of Frauds will not prevent enforcement of an oral promise.

VI. OWNERSHIP RIGHTS OF INTELLECTUAL PROPERTY RESULTING FROM THE EFFORTS OF AN INDEPENDENT CONTRACTOR

Some research biotechnologists may not be employees but independent contractors. When one develops intellectual property as an independent contractor, it raises additional issues regarding ownership of intellectual property rights. For example, although copyright is not a form of intellectual property protection that protects biotechnology ideas, it is certainly important with regard to certain work product related to the ideas. Under federal copyright law, any rights of copyright will remain with the independent contractor absent an express assignment to the hiring party of such copyrights.[30] This may be important when the ownership of various research notes, drawings, and other writings are at issue. In addition, if a research scientist is an independent contractor, it provides an additional argument for the scientist that the hiring party should not obtain a shop right to practice the invention in the absence of an assignment of the invention. Some courts have limited the shop right doctrine strictly to employer-employee settings.[31] Other courts, however, have recognized that a shop right can arise in the context of an independent contractor relationship.[32]

A. Independent Contractor Defined

An independent contractor has been defined in various terms. One authority, the Restatement of Agency, which represents a "restatement" of the law by eminent judges and lawyers, defines an independent contractor as

> . . . a person who contracts with another to do something for him but who is not controlled by the other nor subject to the other's right to control with respect to his physical conduct in the performing of the undertaking. He may or may not be an agent.[33]

Another way to define an independent contractor is to say that the individual is not an employee. Generally, when the party for whom the services are to be performed has the right to control and direct the individual who performs the services not only as to the result to be accomplished by the work but also as to the details and means by which that result is accomplished. In short, the employer has the right to direct or control the manner in which the employee does his or her job. This does not mean that the employer actual does so; it is sufficient if the employer has the right to do so.[34] Often, the parties will agree that the individual is an "employee" or "independent contractor." However, the characterization or description of the relationship by the parties is not controlling. None of the factors controls the determination; rather all of the factors may be considered in making the determination.

The Internal Revenue Service has issued regulations regarding the definition of an "employee" that provides an idea of the types of factors common law courts consider when determining whether a biotechnologist qualifies as an independent contractor. The Internal Revenue Service has identified 20 factors that it uses in determining whether an individual is an employee under the common law rule.[35] None of the factors controls the determination; rather all of the factors may be considered in making the determination:

1. If the hiring party has the right to require compliance with instructions, then it is more probable that the individual is an employee.

2. If requirements are imposed that the individual work with an experienced employee, attend meetings, use particular methods, and correspond with the hiring party, thereby indicating that the services are to be performed in a particular manner or method, then, it is more probable that the individual is an employee.

3. If the services of the individual are integrated into the business operations of the hiring party, thereby suggesting that the success or continuation of the business depends to an appreciable degree on the performance of the services by the individual, then it is more probable that the individual is an employee.

4. If the services must be rendered personally by the individual, presumably the hiring party is interested in the methods used to accomplish the work as well as the result, then it is more probable that the individual is an employee.

5. If the individual hires, supervises and pays assistants pursuant to a contract whereby the individual is responsible only for the attainment of a result, then it is more probable that the individual is an individual contractor.

6. If the individual has a continuing relationship with the hiring party, that is, work is performed at frequently recurring although irregular intervals, then it is more probable that the individual is an employee.

7. If the hiring party establishes set hours of work for the individual, thereby indicating control, then it is more probable that the individual is an employee.

8. If the individual must devote substantially full time to the business of the hiring party this indicates control over the individual because the amount of time spent impliedly restricts the worker from doing other work. An independent contractor is free to work when and for whom he or she chooses.

9. If the work is performed on the employer's premises, thereby suggesting control, then it is more probable that the individual is an employee.

10. If the individual must peform the services in an order or sequence specified by the hiring party, thereby indicating control, then it is more probable that the individual is an employee.

11. If the individual is required to submit regular or written reports to the hiring party, thereby indicating a degree of control, then it is more probable that the individual is an employee.

12. If the individual is paid by the hour, weekly, or monthly, and not on by the job or on a straight commission, then it is more probable that the individual is an employee.

13. If the hiring party pays the individual's business and/or travel expenses, thereby indicating control over expenses, then it is more probable that the individual is an employee.

14. If the hiring party furnishes significant tools, materials, and other equipment, then it is more probable that the individual is an employee.

15. If the individual invests in facilities that are used to perform the services and are not typically maintained by an employee (i.e., an office), then it is less probable that the individual is an employee.

16. If the individual can realize a profit or suffer a loss as a result of the individual's services, that is, the individual is subject to real economic risk of economic loss, then it is less probable that the individual is an employee.

17. If the individual is working for more than one firm at the same time, it is less probable that the individual is an employee.

18. If the individual makes his or her services available to the general public on a regular and consistent basis, then it is less probable that the individual is an employee.

19. If the hiring party has the right to discharge or dismiss the individual because of a failure to obey instructions, even if the individual is producing a result that meets the contracts specifications, then it is more probable that the individual is an employee.

20. If the individual has the right to terminate the relationship with the hiring party at any time without incurring liability, then it is more probable that the individual is an employee.

As this long list indicates, depending on the facts and circumstances, the nature of the relationship between the hiring party and the individual biochemist may be an issue in an ownership dispute.

1. No Written Agreement

When there is no written agreement regarding the employment of an independent contractor, in much the same fashion as discussed above with the employee, the common law rules of contracts will control. Again, without an agreement to assign discoveries and inventions, the hiring party has to prove the independent contractor was hired to invent or that the hiring party has the shop right in the discovery or invention. As noted above, some courts have been willing to find a shop-right exists when the individual is an independent contractor; however, other courts have expressly limited the shop-right rule to the employer-employee relationship. Other legal theories may provide the hiring party with the basis to practice the invention notwithstanding an agreement. For example, the hiring party may allege an implied license exists to practice the invention or claim that the independent contractor is estopped to assert that the hiring party cannot practice the invention.[36] There is no doubt, however, that absent a writing, an independent contractor owns the copyright to all of his or her works of authorship produced during his or her independent contractor relationship. It may be that the hiring party is a joint author with joint ownership rights; but such joint rights will not deprive the independent contractor of his or her rights of ownership.

B. Common Terms and Conditions in Written Agreements

An agreement with an independent contractor will contain many of the same terms and conditions outlined above with regard to an employee. Perhaps the most common additional provision regarding the independent contractor will be a provision that expressly states the individual is an independent contractor. As noted above, such a provision is not always dispositive of the issue. If the independent contractor works for other entities doing similar research, then the hiring party may require specific restrictions to ensure the confidentiality of the research.

With regard to work product and writings covered by the federal copyright laws, the hiring party may insert a provision characterizing the work product generated by the independent contractor as a work made for hire:

> Contractor agrees that all forms of Work Product constituting copyrightable subject matter under the United States copyright laws shall constitute a "work made for hire," and that the Hiring Party shall be the owner of all rights, title and interest in and to such forms of Work Product, including all rights of copyright therein. To the extent any such Inventions and Work Product do not constitute a work made for hire, Consultant hereby assigns to the Hiring Party all of its right, title and interest to the Work Product under the United States copyright laws.

VII. JOINT OWNERSHIP

Often, several individuals working together discover a new biotechnology product or process. When more than one individual or corporation owns an invention or patentable discovery, with respect to patent rights, the co-owners are described by the law as tenants in common. Each tenant in common, or co-tenant, has an undivided interest in the property, including the right to the possession of the whole and the right to transfer his or her own interest. Another way tenancy in common arises is when the sole owner of the invention conveys to another an undivided interest in the invention.

Joint ownerships of patents have particular legal rules associated with them. Perhaps the most important one is that each owner becomes entitled to use the patented invention without requiring the consent of or accounting to the other.[37] In this regard, the U.S. Patent Code provides as follows:

> In the absence of any agreement to the contrary, each of the joint owners of a patent may make, use or sell the patent and invention without the consent of and without accounting to the other owners.[38]

Thus, no matter how small the interests of each joint owner, that particular joint owner can exercise the right to the patent without the consent of or the necessity of an accounting to any other joint owner.

Trade secrets are treated similarly to patents. For example, suppose two research scientists jointly discover a new biotechnology idea and decide to maintain it in confidence pending the issuance of a patent. When trade secrets are held by two co-discoverers, then each co-discoverer is a joint owner of the trade secret. Presumably, each research scientist is free to use the trade secret as that scientist desires unless the two scientists have agreed to maintain the discovery in confidence. However, so long as the two scientists agree to maintain the discovery in confidence, neither researcher may use or disclose the trade secret to the detriment of the other party or parties.[39] With respect to work product that qualifies for copyright protection, when a work is created by more than one author, the authors are co-owners in the joint work.[40] In the absence of an agreement to the contrary, all joint authors share equally in ownership of the work even though their respective contributions are not equal.[41] Like patents, the joint owners are characterized as tenants in common.[42]

Joint ownership obviously creates a number of problems for the exploitation of a patent or trade secret. For example, one joint owner alone cannot grant an exclusive license to practice a patent or trade secret; the joint owner can only grant a license as to the joint owner's own rights. As a result, when the parties enter into an agreement in advance, they usually place title to the discovery, including any patents or trade secrets with one entity. This allows the optimal exploitation of the discovery.

VIII. OWNERSHIP ISSUES IN COMMON BIOTECHNOLOGY RESEARCH AGREEMENTS

Because of the unique nature of the biotechnology industry and the location or employment of many important researchers at universities and nonprofit research centers, a number of unique agreements have been used to date in the biotechnology industry. For example, some corporations have entered into joint research agreements with universities, whereby the corporations fund research by professors at the universities. Other corporations enter into similar agreements with nonprofit research centers. This section will discuss some of these types of agreements and their effect on title.

A. Joint Research Agreements Involving Universities, Nonprofit Research Centers, and Corporations

Universities and other academic research centers are heavily involved in biotechnology research.[43] As a result, to use the talents of scientists associated with universities and research centers, corporations have entered into sponsored research agreements. A sponsored research agreement refers to an agreement whereby industry utilizes the research talent of a university, nonprofit corporation, or nonprofit research center to conduct research without having to hire the research scientists as employees.

One of the more famous such agreements is the $70 million agreement between Massachusetts General Hospital and Hoechst A.G. Although the hospital retains title to all inventions and there are no confidentiality obligations, Hoechst receives an exclusive license to market all potential inventions. In addition, to protect any potential patent rights, Hoechst has the right to review research manuscripts prior to submission for publication.[44] In similar sponsored research agreements, Monsanto and Harvard University and Monsanto and Washington University vest ownership in the universities and grant Monsanto an exclusive license to any patents.[45]

Sponsored research agreements between universities and industry are guided by the university's intellectual property policies, particularly patent policies. Because confidentiality is inconsistent with the values of academia,[46] universities do not consider trade secret protection as an alternative. This is because universities generally insist upon the publication of research results.[47]

Nonprofit research centers generally offer more flexibility than universities with regard to confidentiality and patent protection. This stems from the fact that the values of academia and free exchange of information may not necessarily be central to the nonprofit purposes of such centers. For example, a nonprofit center with an exempt purpose of facilitating the economically efficient exploitation and transfer of biotechnology might agree to treat a discovery as a trade secret in order to ensure a fair return on the sponsor's investment in research. In addition, as opposed to universities, a nonprofit research center is likely more willing to allow the sponsor to own the discovery or invention made using the funds of the agreement.

Ownership of discoveries and work product made in the research conducted at universities and nonprofit research centers can be complicated by another common source of funding, the federal government. To the extent government funding has been used to develop the invention, then federal law may cloud the title to the discovery. The extent of this cloud will be discussed below.

IX. OWNERSHIP OF INTELLECTUAL PROPERTY RIGHTS IN LABORATORY DISCOVERIES AND WORK PRODUCT FUNDED BY THE U.S. GOVERNMENT

A. Rights in Intellectual Property Made with U.S. Government Funds

Because of the magnitude of federally funded research in the area of biotechnology, the laws governing the rights in inventions and discoveries arising out of federally funded research has taken on an increasingly important role. Universities, nonprofit corporations, and research centers often enter into agreements with the federal government to obtain funding for biotechnology research. Ownership of discoveries made in research funded by the federal government is governed by statutes and regulations. One statute governs discoveries made with government funding in the context of patent rights. With respect to "data rights" and copyrights that might constitute work products, federal regulations govern ownership issues.

1. The Bayh-Dole Act: Patent Rights in Inventions Made with Federal Assistance

The Bayh-Dole Act, passed in late 1980, originally permitted small businesses and nonprofit organizations, such as universities, to elect to retain patent rights to technology developed under federally funded research.[48] Subsequently, the Act was amended to include all government contractors. The Act gives title to the contractor to inventions "which are conceived or first actually reduced to practice in the performance of work under a funding agreement" under the "funding agreement." With respect to any invention in which a contractor elects to retain title, however, the federal government shall retain a nonexclusive, worldwide, royalty-free license to the invention.[49]

One of the purposes of the Act was to establish a uniform policy in all agencies of the federal government regarding the disposition of patent rights in government-funded research. With few exceptions, the Act applies to all agencies of the federal government. According to the statute, the "policy and objective" of the Act was as follows:

> To use the patent system to promote the utilization of inventions arising from federally supported research or development; to encourage maximum participation of small business firms in federally supported research and development efforts; to promote collaboration between commercial concerns and nonprofit organizations, including universities; to ensure that inventions made by nonprofit organizations and small business firms are used in a manner to promote free competition and enterprise; to promote the commercialization and public availability of inventions made in the United States by United States industry and labor; to ensure that the Government obtains sufficient rights in federally supported inventions to meet the needs of the Government and protect the public against nonuse or unreasonable use of inventions; and to minimize the costs of administering policies in this area.[50]

a. Contractors covered by the Act Although the Act was originally limited to nonprofit organizations and small businesses, the amendments to the Act in 1986 changed that limitation:

> *Nothing in this chapter is intended to limit the authority of agencies to agree to the disposition of rights in inventions made in the performance of work under funding agreements with persons other than nonprofit organizations or small business firms in accordance with the Statement of Government Patent Policy issued on February 18, 1983*, agency regulations, or other applicable regulations or to otherwise limit the authority of agencies to allow such persons to retain ownership of inventions except that all funding agreements, including those with other than small business firms and nonprofit organizations, shall include the requirements established in paragraph

> 202(c)(4) and section 203 of this title. Any disposition of rights in invention made in accordance with the Statement or implementing regulations, including any disposition occurring before and after another section, are hereby authorized.[51]

The Statement of Government Patent Policy referred to in the Act reads as follows:

> *To the extent permitted by law, agency policy with respect to the disposition of any invention made in the performance of a federally-funded research and development contract, grant or cooperative agreement award shall be the same or substantially the same as applied to small business firms and non-profit organization* under Chapter 38 of Title 35 of the United States Code.
>
> In awards not subject to Chapter 38 of Title 35 Of the United States Code, any of the rights of the Government or obligations of the performer described in 35 U.S.C. 202-204 may be waived or omitted if the agency determines (1) that the interests of the United States and the general public will be better served thereby as, for example, where this is necessary to obtain a uniquely or highly qualified performer; or (2) that the award involves co-sponsored, cost sharing, or joint venture research and development, and the performer, co-sponsor or joint venturer is making substantial contribution of funds, facilities or equipment to the work performed under the award.
>
> In addition, agencies should protect the confidentiality of invention disclosure, patent applications and utilization reports required in performance or in consequence of awards to the extent permitted by 35 U.S.C. 205 or other applicable laws.[52]

Subsequently, in Executive Order No. 12591, President Ronald Reagan ordered that the head of each agency "promote the commercialization . . . of patentable result of federally funded research by granting to all contractors, regardless of size, the title to patents made in whole or in part with Federal funds in exchange for royalty-free use by our own behalf of the government." Thus, the benefits of the Act, including the ability to retain title to inventions, is now available to all entities conducting research using federal funds.

b. "Funding agreements" and their terms The Act defines the funding agreements that are subject to the Act as follows:

> The term "funding agreement" means any contract, grant, or cooperative agreement entered into between any Federal agency, other than the Tennessee Valley Authority, and any contractor for the performance of experimental, developmental, or research work funded in whole or in party by the Federal Government. Such term includes any assignment, substitution of parties, or subcontract of any type entered into for the performance of experimental development, or research work under a funding agreement as herein defined.[53]

This broad definition suggests that all federally funded "experimental, development or research work" will be covered by the Act. In the Federal Acquisition Regulations, known as the FARs, the regulations on patent rights under government contracts are based on the policies set forth in the Act as amended, and the subsequent Presidential Memorandum and Executive Order 12591.[54] Notwithstanding the breadth of federal contracts governed by the FARs, the patent rights provisions in the FARs apply to the same types of work contemplated by the Act. The patent rights in the FARs apply to

> . . . inventions made in the performance of work under a Government contract or subcontract thereunder *if a purpose of the contract or subcontract is the conduct of experimental, developmental or research work*, except to the extent statutory requirements necessitate different agency policies, procedures, and clauses as specified in agency supplemental regulations.[55]

Nonetheless, the patent rights in all federal government contracts should be handled as set forth in the Act. In dealing with the federal government, the research scientist should realize, however, that exceptions may exist to these patent policies for particular federal agencies. In this regard, each federal agency promulgates its own supplemental regulations to the FARs.[56]

Certain contractual terms and conditions of a funding agreement are specified in the Act. Among the provisions mandated by the Act are terms and conditions that require the following:

1. The contractor must disclose each invention to the federal agency within a reasonable time after it becomes known to the contractor personnel responsible for the administration of patent matters. In the event the invention it is not disclosed within such time, the government may receive title to the invention.

2. The contractor must make a written election to retain title to the inventions within 2 years after disclosure to the federal agency or within any additional time approved by the agency. When publication, on sale, or public use has initiated the 1-year statutory period in which patent protection can be obtained in the United States, the federal agency can shorten this period of time for making an election.

3. The contractor electing rights in an invention is required to agree to file a patent application prior to any statutory bar date that may occur under the U.S. patent code due to publication, on sale, or public use of the invention. The contractor so electing must also thereafter file a corresponding patent application in other countries in which it wishes to retain title within a reasonable time. To the extent that the contractor has not filed foreign patent applications within such reasonable time, the federal government may receive title to the invention in the United States or other countries as applicable.

4. With respect to an invention in which the contractor elects to retain ownership rights, the federal agency has a nonexclusive, nontransferable, irrevoca-

ble, paid-up license to practice or have practiced for or on behalf of the United States the invention throughout the world. The federal government may have additional rights, including the right to assign or have assigned patent rights as necessary for meeting the obligations of the United States under any treaty or other international agreement.

5. The federal agency will require periodic reporting on the utilization or efforts at obtaining utilization that are being made by the contractor or his licensees or assignees. This information, however, is to be treated by the federal agency as commercial and financial information obtained from a person and privileged and confidential and not subject to disclosure under 5 U.S.C. § 552.

6. Any U.S. patent application filed by the contractor must disclose that the invention was made using government funds and that the government has certain rights in the invention.

There are some contractual provisions that are required only for nonprofit organizations. First and most importantly, the nonprofit organization is required to share royalties with the inventor. This requirement ensures that even though ownership remains with the nonprofit corporation, the inventor will share in the monetary fruits of the invention. Second, the funding agreement must contain a prohibition of the assignment of rights to the invention in the United States without the approval of the pertinent federal agency. A requirement at the balance of any royalties or income earned by the contractor from the invention, after payment of expenses (including payments to the inventors) incidental to the administration of the invention, be utilized for the support of scientific research or education. Additionally, the nonprofit corporation, except where it proves infeasible after a reasonable inquiry, shall license the subject invention to small business firms.

c. Limitations on ownership rights *March-in Right.* Besides the U.S. government's right to a nonexclusive, nontransferable, irrevocable, paid-up license to the invention, the federal agency under whose funding agreement the invention was made also has "march-in rights." This is the right to require the contractor, an assignee or exclusive licensee of the subject invention to grant a license to a "responsible applicant" upon "terms that are reasonable under the circumstances."[57] The required license may be nonexclusive, partially exclusive, or exclusive in any field of use of the subject invention.

If the contractor, assignee, or exclusive licensee refused to grant such a license, then the federal agency can grant the license itself if the agency makes any of the following determinations:

> (1) the license is necessary because the contractor or assignee has not taken, or is not expected to take within a reasonable time, effective steps to achieve practical application of the subject invention in such field of use;

(2) such action is necessary to alleviate health or safety needs which are not reasonably satisfied by the contractor, assignee or licensees;

(3) such action is necessary to meet requirements for public use specified by Federal regulations and such requirement are not reasonably satisfied by the contractor, assignee or licensees; or

(4) action is not necessary because the agreement required by Section 204 has not been obtained or waived or because a exclusive licensee is in breach of its agreement obtained under Section 204.[58]

The contractor, assignee, or exclusive licensee can challenge this determination. Thus, even though the contractor has "title" to the invention, the contractor's ownership rights are highly qualified by the march-in rights of the government. It is difficult to estimate the risk of such action by an agency; however, to date, there are no reported cases involving such a determination, which arguably suggests that governmental agencies are not attempting to grant such licenses over the protests of the contractor or its assignees or licensees.

Various procedures govern the government's exercise of its march-in rights. For example, the agency must send the contractor written notice of the march-in. If the contractor opposes the march-in, the contractor must submit information or argument in opposition to the proposed march-in. If the information presented raises a genuine dispute over the material facts upon which the march-in is based, then the agency shall conduct a fact-finding hearing. At the hearing, the contractor may present evidence and may request that the hearing be closed to the public, including potential licensees.

Restrictions on transfer. There is also a restriction on any assignments and exclusive licenses of the invention. An assignment or exclusive license of the invention must include a provision that any products embodying the invention or produced using the invention will be manufactured "substantially" in the United States.[59] The federal agency may waive this requirement upon a showing that reasonable but unsuccessful efforts have been made to grant licenses on similar terms to potential licensees or that circumstances demonstrate that domestic manufacture is not commercially feasible.[60]

d. When an invention is made using government funds Sometimes there may be a dispute as to whether an invention was developed using government funding, thereby triggering the provisions regarding ownership in the Act. If the invention is conceived or reduced to practice in performance of work under a government contract, then the government obtains a license to the invention.[61] In one case, the contractor's failure to segregate the conception and reduction to practice of the invention from government funding led the court to conclude government funds were used to fund to develop the invention.[62]

B. Rights in Data and Copyrights in Federal Contracts

Another issue in utilizing government funds is the effect of the use of funding on the title to any work product associated with a discovery or invention. Generally speaking, the federal government statutes and regulations do not contemplate the creation and ownership of trade secrets through the use of government funds. Thus, this type of information is addressed under the subparts in the FARs entitled "Rights in Data and Copyright."[63]

The policies underlying the contractual provisions pertaining to data and copyrights developed using government funds are set forth in the regulations. In general, the government will not treat information developed using government funds as confidential. Rather the government seeks to disseminate such information so that it might be utilized. The policies of the federal government on data and copyrights provide as follows:

> It is necessary for the departments and agencies, in order to carry out their missions and programs, to acquire or obtain access to many kinds of data produced during or used in the performance of their contracts. Agencies require such data to: obtain competition among suppliers; fulfill certain responsibilities for disseminating and publishing the results of their activities; ensure appropriate utilization of the results of research, development, and demonstration activities including the dissemination of technical information to foster subsequent technological developments; and meet other programmatic and statutory requirements. Further, for defense purposes, such data are also required by agencies to meet specialized acquisition needs and ensure logistics support.
>
> At the same time, the Government recognizes that its contractors may have a legitimate proprietary interest (e.g., a property right or other valid economic interest) in data resulting from private investment. Protection of such data from unauthorized use and disclosure is necessary in order to prevent the compromise of such property right or economic interest, avoid jeopardizing the contractor's commercial position, and preclude impairment of the Government's ability to obtain access to or use of such data. The protection of such data by the Government is also necessary to encourage qualified contractors to participate in Government programs and apply innovative concepts to such programs. In light of the above considerations, in applying these policies, agencies shall strike a balance between the Government's need and the contractor's legitimate proprietary interest.[64]

To implement these policies, the FARs require that certain provisions be inserted into all government contracts that "require data to be produced, furnished, acquired or specifically used in meeting contract performance requirements."[65] The provisions are intended to specify the rights regarding the use, duplication, and disclosure of such data. These regulations contemplate different types of rights regarding data. For some data, the government has "unlimited rights." When the government has unlimited rights in data, it means that the government has the right to use or disclose the data in any manner and for any purpose.[66]

The FARs also contemplate restrictions on certain types of data. In this regard, the term *limited rights data* means data in which the government's rights in the data are restricted as set forth in the government contract. There are several alternative provisions addressing limited rights data. The clauses either allow the contractor to withhold the limited rights data or if the data are delivered to the government to generally restrict the government's right to disclose the information.[67] Computer software can be restricted like data. The term *restricted computer software* refers to computer software developed at private expense and that is a trade secret or is a published copyrighted computer software.[68] The contractor can protect limited rights data and restricted computer software by withholding such data from delivery to the federal government and delivering what is called "form, fit, and function data" in lieu thereof.

There are limits to data that can be classified as limited right data. The typical data rights clauses used for most government contracts vest the government with unlimited rights in certain data:

(1) Data first produced in the performance of a contract (except to the extent such data constitutes minor modifications to data that are "limited rights data" or "restricted computer software");

(2) Form, fit and function data delivered under the contract;

(3) Data (except as may be included with restricted computer software) that constitute manuals or instructional and training materials for installation, operation, or routine maintenance and repair of items, components or processes delivered or furnished for use under a contract; and

(4) All the data delivered under the contract other than limited rights data or the restricted computer software.[69]

In addition, to the extent that any of the data are published copyrighted data with a copyright notice, the federal government acquires them under a copyright license with an unlimited right to use.[70]

With regard to data used in providing services under government contracts, the federal government has recognized that a contractor utilizing data developed at private expense that embodies trade secrets or commercial or financial and confidential or privileged information needs to be protected. In the event that a contract involves co-sponsored research and development in which the contractor is required to make substantial contributions of funds or resources, and the contractor and government respective contributions are not readily segregable, then the contracting officer of the government may limit the acquisition of or acquire less than unlimited rights to any data developed and delivered under such contract.[71] However, the regulations require that, at a minimum, such government rights should assure use of the data for agreed-to governmental purposes.

C. Ownership Rights of Intellectual Property Developed by U.S. Government Employees

When a biotechnologist works for the federal government, the ownership issues are resolved under a different regulatory scheme. The ownership interests of inventions made by government employees is governed by Executive Order No. 10096. Pursuant to Regulations that have been issued under the authority of that Order,[72] federal agencies are empowered to make the final determination regarding the ownership of inventions made by government employees subject to the employee's right to appeal to the Under Secretary of Commerce for Economic Affairs.[73]

Executive Order No. 10096 applies to any inventions made by government employees after the date of the Order, January 23, 1950. It covers almost all executive branch departments, boards, and regulatory commissions.[74] It also applies to all civilian and military officers and employees, including any part-time consultants or employees.[75]

The inventions or discoveries covered by the Order are as follows:

> . . . any art, machine, manufacture, design, or composition of matter, or any new and useful improvement thereof, or any variety of plant, which is or may be patentable under the patent laws of the United States.[76]

Thus, the Order covers only those inventions within the United States Patent Code.[77]

One possible issue is the meaning of the word *made* with respect to "an invention made by a government employee." The Commissioner of Patents has held that *made* in fact means *conception* or *an actual reduction to practice* and is not limited to the earlier of these two events.[78] Thus, according to the Commissioner, if either event occurred with government funding, then the Order applies.

Under the current Order, each governmental agency initially reviews the case and determines the rights of the inventor and the government in the employee's invention or discovery.[79] As a result of this review, the agency will make one of the three following determinations:

(1) the government obtains the entire domestic right, title and interest in the invention;

(2) title left to the employee subject to the reservation of a non-exclusive, irrevocable royalty-free license to the government as the contribution of the government is insufficient to equitably justify complete assignment to the government, or the government has insufficient interest in invention; or

(3) the government should leave the entire right, title, and interest in the invention to the government employee.[80]

Once this determination is made, the employee can challenge an adverse result. If a government employee conducting biotechnological research wishes to retain title to his or her inventions, it will be necessary to negotiate through a maze of ''determinations'' and possibly appeals in order to secure title to the invention. Even then, however, the title will still be subject to a nonexclusive, irrevocable, royalty-free license in the invention to the government.

a. Government ownership of inventions The federal employee seeking to own his or her invention wants to avoid the situation in which the government ''shall'' obtain title. According to the Order, the federal government *shall* obtain all domestic right, title, and interest to an invention made by any government employee:

(i) during working hours; or

(ii) with a contribution by the Government of facilities, equipment, materials, funds or information, or of time of services of other government employees on official duty, or

(iii) which bears a direct relation to, or is made in consequence of the official duties of the inventor.[81]

The breadth of this provision is enlarged by the use of the disjunctive ''or'' in item (i) above, which means that an invention that satisfies any one of these three criteria shall be owned by the government. This is the first regulatory category of inventions made by government employees under the Order.

This broad category is further strengthened by a presumption that inventions made by certain federal employees fall within the category of inventions for which the federal government shall obtain the entire domestic right, title, and interest to the invention. This provision might be viewed as a broad regulatory equivalent of the ''hired to invent'' rule of the common law. Pursuant to the presumption, the government owns the invention when it is made by an employee who is employed or assigned the following tasks:

(i) To invent or improve or perfect any art, machine, design, manufacture, or composition of matter,

(ii) To conduct or perform research development work or both

(iii) To supervise, direct, coordinate, or review Government financed or conducted research, development work, or both, or

(iv) To act in a liaison capacity among governmental or non-governmental agencies or individuals engaged in such research or development work.[82]

This presumption may be rebutted by a showing of the facts and circumstances of the case. In addition, the Order provides that these presumptions "shall not preclude a determination that these facts and circumstances justify leaving the entire right, title and interest in and to the invention to the government employee." This suggests that the government can waive its rights even if the employee fell within the categories of the presumption.

b. Employee ownership of inventions After giving itself the benefit of this broadly worded presumption, the federal government allows those remaining government employees a similar rebuttable presumption: Any inventions made by the remaining government employees who are not employed or assigned the various tasks identified above are presumed to fall under a second regulatory category of inventions made by government employees. This second category is the category that encompasses inventions that the government employee will own. This regulatory category includes the following inventions made by government employees:

1. inventions in which "the contribution of the Government, as measured by the [three] criteria set forth above . . . , to the invention is "insufficient equitably to justify a requirement of assignment to the Government of the entire domestic right, title and interest in and to such invention;" or
2. inventions in which the Government has "insufficient interest" to obtain the entire right, title and interest therein (although the Government could obtain the invention because it falls under the first category of government employee inventions set forth above).

A victory for the employee is not a total victory. Even if the government employee's invention falls within this second category, the employee's title to such invention is "subject however, to the reservation to the government of a non-exclusive, irrevocable, royalty-free license in the invention with power to grant licenses for all government purposes."[83]

NOTES

1. Witte and Guttag, *Employee Inventions*, 71 J. Pat. Off. Soc'y. 467, 479 (1989).
2. U.S. Const. Art. 1, § 8 (italics added).
3. United States v. Dubilier Condensor Corp., 289 U.S. 178, 188 (1933).
4. See Chapter 5, which discusses inventorship.
5. Moore v. Regents of the University of California, 271 Cal. Rptr. 146, 793 P. 2d 479 (1990).

6. Moore v. Regents of the University of California, 202 Cal. App. 3d 1230, 249 Cal. Rptr. 494 (Cal. Ct. App. 1988), *rev'd*, 271 Cal. Rptr. 146, 793 P.2d 479 (Cal. 1990). Some commenators have supported the recognition of such a property right. Comment, *Spleen for Sale*: *Moore* v. *Regents of the University of California and the Right to Sell Parts of Your Body*, 51 Ohio St. L. J. 499 (1990); Danforth, *Cells, Sales and Royalties*: *The Patient's Right to a Portion of the Profits*, 6 Yale L. & P. R. 179, 193–95 (1988). Note, *Toward the Right of Commerciality*: *Recognizing Property Rights in The Commercial Value of Human Tissue*, 34 U.C.L.A. L. Rev. 207, 260–63 (1986).
7. Davis v. Alwac Int'l, Inc., 369 S.W.2d 797, 802 (Tex. Civ. App.—Dallas 1963; no writ); Dowse v. Federal Rubber Co., 254 F.2d 308, 309 (7th Cir. 1958).
8. United States v. Dublier Condensor Corp., 289 U.S. 178, 187 (1933); Solomons v. United States, 137 U.S. 342, 346 (1890).
9. Restatement of Agency, Second § 397 (1958).
10. See generally 35 U.S.C. §§ 101–118 (1982).
11. See generally 1 R. Milgrim, *Milgrim on Trade Secrets* § 5.02[4] (1990).
12. *Idem*, at § 5.02[4][b] at 5-62.
13. Solomons v. United States, 137 U.S. 342, 346 (1980). See also Standards Parts Co. v. Peck, 137 U.S. 342, 346 (1924); Heberd v. American Zinc, Lead & Smelting co., 161 F.2d 339, 342 (8th Cir. 1947).
14. 1 R. Milgrim, *Milgrim on Trade Secrets* § 5.02[4][a] n. 61 (1990).
15. Houghton v. United States, 23 F.2d 386, 390 (4th Cir. 1928).
16. D. Chisum, *Patents* § 22.03 [3] at 22-26-28 (1989).
17. United States v. Dublier Condenser Corp., 289 U.S. 178, 188–89 (1933).
18. 1 R. Milgrim, *Milgrim on Trade Secrets* § 5.02[4][c], at 5-65-66 (1990).
19. Francklyn v. Guilford Packing Co., 695 F.2d 1158, 1160–61 (9th Cir. 1983); Kurt H. Volk, Inc. v. Foundation for Christian Living, 534 F. Supp. 1059, 1083–84 (S.D.N.Y. 1982). Lukens Steel Co. v. American Locomotive Co., 197 F.2d 939, 941 (2d Cir. 1952) 237 F.Supp. 694, 703 (E.D. Wis. 1964), *aff'd*, 343 F.2d 534 (7th Cir. 1965).
20. 17 U.S.C. §§ 101 & 201(b) (1982).
21. See generally R. L. Gullette, State Legislation Governing Ownership Rights in Inventions Under Employee Invention Agreements, 62 J. Pat. Off. Soc'y 732 (1980).
22. Cal Lab. Code §§ 2870–72 (Deering 1990); Ill. Rev. Stat. ch. 140, ¶ 302 (1988).
23. 35 U.S.C. § 102(b) (1982).
24. Note, *Patent and Trade Secret Protection in University Industry Research Relationships in Biotechnology*, 24 Harv. J. Legis. 191, 223 (1987).
25. Duldulao v. St. Mary of Nazareth Hospital Center, 115 Ill.2d 482, 505 N.E.2d 314, 318 (1987); Toussaint v. Blue Cross & Blue Shield of Michigan, 408 Mich. 579, 613, 292 N.W.2d 880, 892 (1980); Woolley v. Hoffman-La Roche, Inc., 99 N.J. 284, 491 A.2d 1257, *modified*, 101 N.J. 10, 499 A.2d 515 (1985).
26. Vallone v. Agip Petroleum Co., 705 S.W.2d 757 (Tex. Civ. App.—Houston [1st Dist.] 1986, writ ref'd n.r.e.); Buffaloe v. United Carolina Bank, 89 N.C.App. 693, 366 S.E.2d 918 (1988).
27. Speck v. North Carolina Dairy Foundation, 311 N.C., 679, 319 S.E.2d 139 (1984).

28. Weiner v. McGraw-Hill, Inc., 57 N.Y.2d 458, 443 N.E.2d 441, 457 N.Y.S.2d 193 (1982).
29. Gate-Way, Inc. v. Hillgren, 82 F. Supp. 546, 554 (S.D. Cal. 1949), *aff'd per curiam*, 181 F.2d 1010 (9th Cir. 1959).
30. 17 U.S.C. § 201(b) (1982). See Community for Creative Non-Violence v. Reid, 109 S.Ct. 2166, 2179 (1989).
31. Mechmetals Corp. v. Telex Computer Prod., Inc., 709 F.2d 1287, 1291–92 (9th Cir. 1983); *Hobbs v. United States*, 376 F.2d 488, 495 (5th Cir. 1967); Crom v. Cement Gun Co., 46 F. Supp. 403, 404 (D. Del. 1942).
32. Francklyn v. Guilford Packing Co., 695 F.2d 1158, 1160–61 (9th Cir. 1983); Kurt H. Volk, Inc. v. Foundation for Christian Living, 534 F. Supp. 1059, 1083–84 (S.D.N.Y. 1982).
33. Restatement of Agency, Second § 2 (1958).
34. Often, the parties will agree that the individual is an "employee" or "independent contractor." However, the characterization or description of the relationship by the parties is not controlling.
35. These 20 factors were listed by the Internal Revenue Service in Revenue Ruling 87-41.
36. See Francklyn v. Guilford Packing Co., 695 F.2d 1158, 1160–61 (9th Cir. 1983); Kurt H. Volk, Inc. v. Foundation for Christian Living, 534 F. Supp. 1059, 1083–84 (S.D.N.Y. 1982).
37. 5 Walker on Patents, § 19:39, at 461 (3d. ed. 1986).
38. 35 U.S.C. § 262 (1982).
39. See 1 R. Milgrim, *Milgrim on Trade Secrets*, § 5.03(7).
40. 17 U.S.C. § 201(a) (1982).
41. Sweet Music, Inc. v. Melrose Music Corp., 189 F. Supp 655 (S.D. Cal. 1960).
42. 1 M. Nimmer, *Nimmer on Copyright*, § 6.09 (1990).
43. Note, *Patent and Trade Secret Protection in University-Industry Research Relationships in Biotechnology*, 24 Harv. J. Legis. 191, 191 (1987) (*citing* P. Daly, The Biotechnology Business 51 (1985).
44. *Idem*, at 287, n. 213.
45. *Idem*, at 228, n. 213.
46. Eisenberg, *Academic Freedom and Academic Values in Sponsored Research*, 67 Texas L. Rev. 1363, 1375 (1988).
47. Note, *Patent and Trade Secret Protection in University-Industry Research Relationships in Biotechnology*, 24 Harv. J. Legis. 191, 226 (1987) (*citing* P. Daly, the Biotechnology Business 51 (1985).
48. Act of December 12, 1980, Pub. L. No. 96-517, § 6, 94 Stat. 3015, 3019–3029 (1984), codified as amended at 35 U.S.C. §§ 200–211 (1989).
49. 35 U.S.C.A. § 202(c)(4) (Supp. 1990).
50. 35 U.S.C.A. § 201 (Supp. 1990).
51. 35 U.S.C.A. § 210(c) (Supp. 1990).
52. *Statement of Government Patent Policy*, February 18, 1983.
53. 35 U.S.C. § 201(b) (1982).
54. 48 C.F.R. § 27.302.

55. 48 C.F.R. § 27.300.
56. See 48 C.F.R. §§ 2.01 - 6302. (listing supplemental regulations for numerous federal agencies).
57. 35 U.S.C. § 203 (1982).
58. 35 U.S.C. § 203 (1982).
59. 35 U.S.C. § 204 (1982).
60. 35 U.S.C. § 204 (1982). 37 C.F.R. 401.14(i).
61. 48 C.F.R. § 27.300.
62. Pacific Technica Corp. v. United States, 3 U.S.P.Q. 2d 1168, 1190–94 (Cl. Ct. 1986), *aff'd in part*, 835 F.2d 871 (Fed. Cir. 1987).
63. 48 C.F.R. § 27.400 *et seq.*
64. 48 C.F.R. § 27.402.
65. 48 C.F.R. § 27.403.
66. 48 C.F.R. § 27.401.
67. 48 C.F.R. § 52.227-14.
68. 48 C.F.R. § 27.401.
69. 48 C.F.R. § 27.404.
70. *Idem.*
71. 48 C.F.R. 27.409.
72. These regulations have been promulgated at 37 C.F.R. §§ 501.1 - .10.
73. 37 C.F.R. § 501.8.
74. It covers all Executive Branch Departments, boards, regulatory commissions, and corporations owned by the United States in the Smithsonian Institution, but it does not include the Department of Energy for inventions under 42 U.S.C. § 2182.37 C.F.R. § 501.3(b).
75. 37 C.F.R. § 501.3(c).
76. 37 C.F.R. § 501.3(d).
77. For example, this means it would not cover sexually reproduced plant varieties under the Plant Variety Protection Act of 1970, 7 U.S.C. §§ 2321 *et seq.*
78. *In re* King, 3 U.S.P.Q. 2d 1747 (Comm'r pat. 1987).
79. 37 C.F.R. § 501.4.
80. Executive Order No. 10096, ¶ 37 C.F.R. § 501.6.
81. 37 C.F.R. § 501.6(a)(a).
82. 37 C.F.R. § 501.6 (a)(3).
83. 37 C.F.R. 501.6(a)(2).

7

Patent Infringement in the Field of Biotechnology

Charles E. Lipsey

Finnegan, Henderson, Farabow, Garrett & Dunner, Washington, D.C.

Amy L. Tsui Collins

Chiron Corporation, Emeryville, California

I. INTRODUCTION

The development of biotechnology patent law has advanced considerably since the U.S. Supreme Court decided in *Diamond* v. *Chakrabarty*[1] that a live, man-made microbe constituted patentable subject matter as a "manufacture" or "composition of matter" within the meaning of the U.S. patent law.[2] Since then, the U.S. Patent and Trademark Office (PTO) has ruled that claims drawn to modified plants, plant seeds, and plant tissue cultures are also patentable.[3] Further determined to be patentable are polyploid oysters produced by a certain claimed method[4] and the so-called transgenic mouse.[5]

The foregoing determinations spurred activity among companies engaged in biotechnological research and development. These companies are churning out patents covering such materials as deoxyribonucleic acid (DNA) fragments, recombinant vectors, transformed host cells, assay techniques, reagents, and products produced by recombinant DNA technology.

The remedies available to patent holders for infringement of their patents highlight the importance of patents as commercial assets. When a patent is found to have been infringed, the infringer is liable for damages, e.g., in an amount sufficient to compensate the patent holder for lost profits. Under certain circumstances to be discussed below, the patent holder may obtain up to three times the actual damages proved and possibly recover attorney fees (which are usually quite substantial). The ultimate remedy for patent infringement is an injunction prohibiting the infringer from practicing the patented invention until the patent expires.

The law of patent infringement, however, is generalized and applicable to all technologies. It was not designed specifically for biotechnology problems, and herein lies the source of many of the difficulties encountered in enforcing biotechnology patents. In the balance of this chapter, legal principles relating to patent infringement in general are discussed and, where possible, are illustrated with examples pertaining to patents in the biotechnology area.

II. PATENTS AS PROPERTY

The concept of ownership of a patent and the rights associated with such ownership can be explained and understood most readily by analogy to ownership of other property, such as a car or a piece of land. A car is considered personal property, as contrasted with a piece of land, which is considered real property.

Ownership of a car is evidenced by a certificate of title. This document sets forth the name of the owner and identifying characteristics of the car; e.g., the year, the model and make of the car, and a serial number. One can read the certificate of title and tell whether any particular car is owned by the person noted on the certificate. If the car in question meets the description in each portion of the certificate, i.e., is the same make, model, year, and serial number, then it is the car owned by the person named on the certificate.

Inherent in the ownership of a car is the right of the owner to exclude others from using the car without his or her authorization. In the event of unauthorized use, the owner can enlist the power of the state in which he resides, e.g., the police, to enforce his or her property right to eject others from his or her property, a right that is recognized and protected by the state.

Ownership of the car, however, does not automatically confer upon the owner the right to use the car himself or herself. The owner must still abide by the rules and regulations that are preconditions to the use of the car. These preconditions include, e.g., obtaining registration for the car and ensuring that it meets the state's safety standards.

Similarly, ownership of land can be established by a document, called a deed, that sets forth the name of the owner and an identifying description of the land. Traditionally, the land is described by "metes and bounds" defining the boundaries of the land. For example, a building lot may be legally described as follows:

> A building lot bounded on the north by K Street, on the east by 13th Street, on the west by 14th Street and on the south by I Street:

A person standing at position (1) in the illustration is indeed south of K Street, west of 13th Street, north of I Street and east of 14th Street. The person's

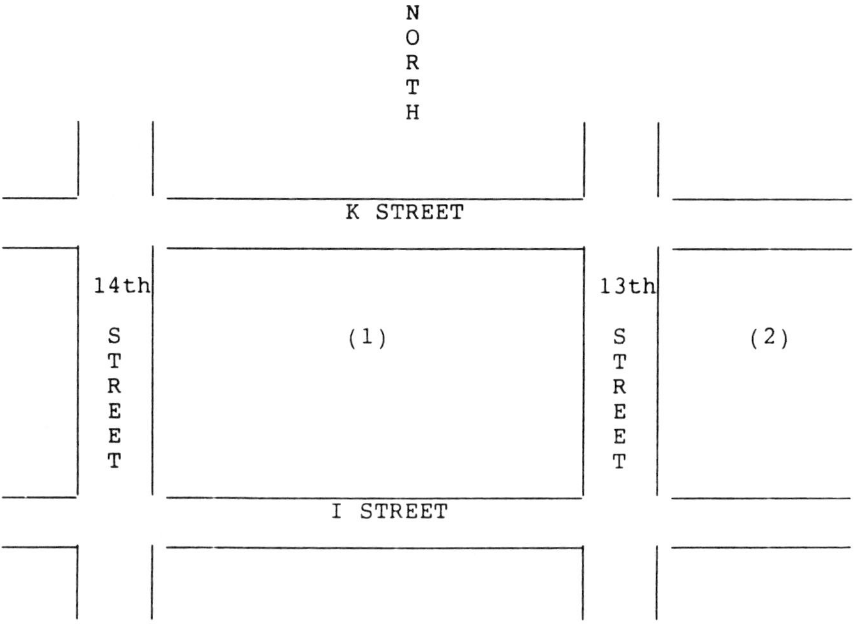

position is among those described in the deed and he or she is a trespasser. In contrast, the person standing at position (2), although south of K Street, north of I Street, and east of 14th Street, all as specified in the deed, does not meet the limitation in the deed that he or she be east of 13th street. Thus, this person is not a trespasser.

Just as ownership of a car entitles the owner to exclude others from using his or her car, ownership of land confers upon the land owner the right to exclude others from entry upon his or her land. Thus, the owner of the land can seek the assistance of the state to enforce his or her property right and eject those who enter upon the land without permission. Yet, ownership of the land does not entitle the owner to use the land in any way he or she wants. For example, the owner may still have to comply with applicable zoning laws.

The category of personal property may be further classified into tangible and intangible personal property. A car is tangible personal property. You can touch it, feel it, and drive it. In contrast, you cannot see, touch, smell, or feel intangible personal property, such as a corporation. Yet, as with other properties, real or personal, ownership of a corporation is evidenced by a document; i.e., a stock certificate, identifying the name of the corporation and the percentage ownership in terms of number of shares of the corporation. Ownership of a corporation carries with it the right to exclude others from taking control of the corporation and using it or, more appropriately, running it without the owners' permission. Again, the power of the state can be invoked to enforce the owners' property right.

With the above as background, a patent can be conceptualized as a deed to intangible personal property; i.e., an invention. Like the certificate of title to a car or the deed to a piece of land, a patent contains an identifying description of a patented invention describing the metes and bounds of the invention. These descriptions are called "claims," and are the focal point of any infringement analysis. In the simplest analysis, they work just like a deed to property—if the technology being used by someone else is described precisely by the words of the claim, then that person is an infringer.

By statute, ownership of a U.S. patent confers upon the owner a right to exclude others from making, using, and selling the patented invention in the United States. Any unauthorized making, using, or selling of a patented invention constitutes patent infringement and subjects the infringer to liability for damages.[6] Yet, as with other property, ownership of a patented invention does not confer upon the owner the unfettered right to use his or her invention. The owner still has to comply with governmental regulations, such as those imposed by the Food and Drug Administration (FDA) on the sale of drugs. Indeed, a patent owner may also be precluded from using his or her invention by a dominating patent owned by another.

Three types of patents are authorized by statute: utility patents,[7] plant patents,[8] and design patents.[9] Of the three types, only the utility patents and plant patents are of relevance in the biotechnology area; therefore, only those legal principles applicable to utility patents and plant patents will be discussed.

A utility patent may be granted for any "new and useful process, machine, manufacture, or composition of matter, or any new and useful improvement thereof," provided that the claimed invention meets all the statutory requirements and conditions for patentability.[10] A plant patent may be granted for any asexually reproduced distinct and new plant varieties "other than a tuber propagated plant or a plant found in an uncultivated state," provided that the novel plant meets all the conditions and requirements of patentability.[11]

Unlike ownership of a piece of land or ownership of a car, which remains with the owner for an indefinite period of time until the property is sold or otherwise disposed of, a patent for an invention is viable only for a fixed term. In the case of a utility patent or a plant patent, the term is 17 years from the date of issue. A further distinction between patent rights and other personal property rights resides in the fact that a patent right is enforceable only within the territorial confines of the country issuing the patent. A U.S. patent, therefore, is enforceable only in the United States, not in Canada or Europe.

In other respects, however, a patent has attributes of personal property. It can be assigned. It can be licensed exclusively or nonexclusively to cover the whole or a part of the United States.[12] Ownership rights in a patent can be enforced by invoking the power of the government to exclude others from making, using, and selling the patented invention; e.g., by bringing a patent infringement action against an alleged infringer for injunctive relief[13] or monetary damages in a U.S. District Court.[14]

III. STRUCTURE OF A PATENT

To understand patent infringement, one must understand the structure and function of the various parts of a patent as well as some of the substantive requirements for patentability. A patent contains two parts: a specification that constitutes a scientific description of the patented invention and a set of claims that constitute the legal description of the metes and bounds of the patented invention.

A. The Specification

The specification must contain a written description of the invention, the manner of making and using the invention, and the best mode of practicing the invention. These requirements are mandated by statute, 35 U.S.C. § 112, first paragraph, and are referred to, respectively, as the written description requirement, the enablement requirement, and the best mode requirement.[15] These requirements are absolute. Failure to comply with any one of these requirements renders the patent invalid. From the public policy point of view, the complete and enabling disclosure of an invention in the specification of a patent enriches the storehouse of technology available to the public and encourages improvements in the field. Such disclosure, therefore, is the quid pro quo that justifies the imposition on the public of an obligation not to use the technology without the patentee's permission for the limited time provided in the patent grant.

The validity of a patent may be subject to attack on several grounds other than those outlined above. Among other considerations, a patent must claim subject matter that is not identically disclosed or anticipated by prior art publications or patents.[16] Further, a patent claim must not describe an obvious variation of what was known in the art at the time the invention was made.[17]

B. The Patent Claims

A patent claim describes the metes and bounds of a patented invention without necessarily describing all of the details of the invention or how the invention works. The analogy to a real estate deed remains apt. In our previously offered example of a deed, the boundaries of the property are defined without stating whether the property is a church, an office building, or a park.

A claim generally begins with a preamble and ends with a substantive part, referred to as the body of the claim. Linking the preamble and the body of the claim is a transitional phrase that materially affects the scope of the claim.

Imagine, for example, a patent claim that reads:

> "An immunodiagnostic reagent comprising a monoclonal antibody specific for an antigenic determinant associated with an infecting organism."

The phrase "an immunodiagnostic reagent" is the claim preamble. It describes the environment in which the inventor contemplates that his or her invention will be used. The phrase "a monoclonal antibody . . . organism" is the body of the claim, describing the properties that any material must have to constitute an infringement of the patent. The word *comprising* is the transitional phrase that indicates whether the claimed invention includes just the subject matter that follows in the body of the claim or additional subject matter as well.

To pursue our analogy of patent claims to real estate deeds, two examples of possible infringement analyses are offered. If Company X sells in the United States a monoclonal antibody specific for the organism that causes strep throat attached to a radioactive label for use in detecting that organism, the thing being sold meets every limitation stated in the patent claim. Just as the person located at position (1) in our real estate example met all the limitations imposed by the deed and was, therefore, a trespasser, the product of Company X in this example meets all of the limitations imposed by the patent claim and is, therefore, trespassing upon, or "infringing," the patented property of another.

If, on the other hand, Company Y sells a product in the United States that is a labeled monoclonal antibody specific for an antigen found on the surface of human cancer cells used for the detection of cancer, it is not an infringer. Although their product meets most of the limitations in the patent claim, it does not fulfill the requirement that the antibody recognize an "infecting organism." Company Y is like the person at position (2) in our real estate example. Neither is trespassing upon, or infringing, the property of another.

Less obvious examples can be posited. What if the Company X product were sold for therapeutic rather than diagnostic use? Would it still infringe? What if the cancer detected by the product of Company Y was found to be caused by a virus? Does Company Y become an infringer?

Welcome to the world of patent litigation.

1. The Claim Preamble

The claim preamble points out the environment of the invention being claimed. A question that often arises in association with the preamble in an infringement action is whether it constitutes a positive limitation in a claim.

Let us consider the foregoing example where the antibody product of Company X is used as a *therapeutic* agent not as a diagnostic agent. The patent owner might contend that the preamble merely constituted a statement of intended use and did not impose any positive limitation on the scope of the claim. The infringer, in contrast, might contend that therapeutic antibodies were known in the prior art and that the preamble must limit the scope of the claim to immunodiagnostic reagents if the patent is to be valid.

An analogous issue arose in *Corning Glass Works* v. *Sumitomo Elec. U.S.A., Inc.*[18] In this case, the controversy centered around the interpretation and effect of the preamble of a claim asserted by Corning Glass Works (Corning) to be infringed.[19] The preamble read: "An optical waveguide. . . . " The accused infringer, Sumitomo Electric U.S.A., Inc. (Sumitomo), disregarded the preamble and contended that the asserted claim was anticipated by prior art that disclosed the structure described in the body of the claim.[20] Sumitomo contended that the subject preamble was merely a statement of intended use and not a limitation on the scope of the claim, and that the remainder of the claim completely defined the invention. Corning countered that the preamble was a positive limitation and must be interpreted in the manner defined in the specification.

The court of appeals for the Federal Circuit (Federal Circuit) was persuaded that not all optical fibers were suitable for use as optical waveguides and agreed with Corning. The Federal Circuit opined that the words of the preamble, which it found to be a positive limitation, gave "life and meaning" to the asserted claim because it required "the particular structural relationship defined in the specification for the core and cladding to function as an optical waveguide."[21]

The origin of the "life and meaning" test can be traced to *Kropa* v. *Robie*.[22] In *Kropa*, the Court reviewed 37 cases previously decided in which the significance of the claim preamble was disputed. Based upon these precedents, the court arrived at the conclusion that where the preamble was denied the effect of a claim limitation, the claims were drawn to structures, and the remaining portion of each of these claims following the preamble "was a self-contained description of the structure not depending for completeness upon the introductory clause."[23] The preamble was also denied the effect of a claim limitation in instances where the claims were drawn to products, and the introductory clause of each of these claims "mainly recited a property inherent in the old composition defined by the remaining part of the claim."[24] Under the above circumstances, the preamble "merely stated a purpose or intended use of that subject matter."[25]

In those cases where the preamble was given the effect of a positive claim limitation, the court in *Kropa* determined that the introductory phrase was essential to point out the invention defined by the claim. Under these circumstances, the preamble was considered "necessary to give life, meaning and vitality to the claims."[26]

In the *Corning* case, the Federal Circuit recognized that:

> . . . [n]o litmus test can be given with respect to when the introductory words of a claim, the preamble, constitute a statement of purpose for a device or are, in themselves, additional structural limitations of a claim.[27]

According to the Court, resolution of the matter can only be achieved by review of "the entirety of the patent to gain an understanding of what the inventors actually invented and intended to encompass by the claim."[28]

As a further illustration of a claim preamble, the phrase "A purified and isolated DNA sequence" would be considered the preamble in the following claim[29]:

> 1. A purified and isolated DNA sequence encoding erythropoietin, said DNA sequence selected from the group consisting of:
> (a) the DNA sequences set out in FIGS. 5 and 6 or their complementary stands;
> (b) DNA sequences which hybridize under stringent condition to the DNA sequences defined in (a).

The illustrated preamble is significant in terms of validity because it differentiates the balance of the claim from a naturally occurring DNA sequence encoding erythropoietin, which is not patentable subject matter under 35 U.S.C. § 101.

2. The Transitional Phrase

The transitional phrases that are most widely used in claims are the terms *comprises* or *comprising*, and the phrases *consists of* or *consisting of*, and *consists essentially of* or *consisting essentially of*. In the context of a claim, these terms and phrases constitute terms of art that have significant and definite meanings.

a. Comprises or comprising The term *comprises* or *comprising* as used in a claim includes the recited elements or steps but does not exclude other unspecified elements or steps. Either of these terms is considered an open-ended term that permits inclusion of unspecified ingredients even in major amounts.[30]

The use of one of these transitional phrases is illustrated in U.S. Patent No. 3,853,833, a patent relating to human growth hormone that was the subject matter of controversy in *Hormone Research Foundation, Inc.* v. *Genentech, Inc.*[31] Claim 3 of the patent asserted to be infringed recited:

> A method of producing a substance having growth-promoting activity which *comprises* [italics added]:
> (a) forming an unbridged polypeptide chain of amino acid residues in a sequence corresponding to FIG. 2 or 3 of the accompanying drawing;
> (b) generating sulfhydryl groups on the cysteine residues in said polypeptide chain;

(c) oxidizing said sulfhydryl groups under conditions effective to form disulfide bridges between said cysteine residues, thereby forming two intramolecular rings in the polypeptide chain.

The above-stated claim recites the essential steps of a claimed process. By use of the term *comprises,* the patented process does not exclude the possibility of the inclusion of additional steps. In other words, if a person practices the patented invention by performing all three recited steps of the above-recited process, without authorization from the patent holder, that person would infringe the subject patent, regardless of whether that person also performed additional steps in producing a substance "having growth-promoting activity."

The same principle applies when a claim recites a product. For example, in the following claim, the presence of other DNA segments is not excluded in a composition of matter (emphasis added)[32]:

As a composition of matter, a biologically functional recombinant plasmid capable of selection and replication in a prokaryotic cell *comprising* [italics added]: a first DNA segment . . . which segment is covalently joined . . . in vitro at its ends to the complementary ends of a second DNA segment. . . .

b. Consists of or Consisting of The phrase *consists of* or *consisting of* means including the recited features and no other unspecified features except for "impurities ordinarily associated therewith."[33] Both of these phrases are regarded as closed-ended terms and are construed to exclude from the scope of the claim a product or process that has elements or steps in addition to those expressly recited. In view of the limited scope of coverage provided by this transitional phrase, the phrase *consisting of* is not often used.

One cannot rely completely, however, on the phrase *consisting of* as absolutely closing the claim to additional elements. In *Special Metals Corp.* v. *Teledyne Industries, Inc.,*[34] e.g., where the claim to nickel-based alloys employed the transitional phrase *consisting of,* the court construed the patented invention to include two additional unspecified elements, carbon and boron.[35] Special Metals Corp. was able to persuade the court that the phrase "nickel base alloy" in the claim preamble would signify to a person skilled in the art that carbon and boron were necessary ingredients.[36]

The infringement analysis may be complicated by use of combinations of transitional phrases. Consider, for example, the following claim where the *consisting of* terminology appears in conjunction with *comprising* and the "Markush" language "selected from the group consisting of"[37]:

1. A recombinant DNA molecular *consisting of* [italics added] segments of DNA from different genomes which have been joined end-to-end outside of living cells and which have the capacity to infect some host and to be maintained therein, and the progeny thereof, *comprising* [italics added] a DNA sequence *selected from the group consisting of* [italics added]:

(a) the DNA inserts of Z-pBR322(Pst)/HcIF-2h (DSM 1700), Z-pBR322(Pst)/HcIF-SN35 (DS 1701), Z-pBR322(Pst)/HcIF-SN42 (DSM 1702) and Z-pKT287(Pst/HcIF-2h-AH6 (DSM 1703),

(b) DNA sequences which hybridize to any of the foregoing DNA inserts and which code on expression for a polypeptide of the IFN-α type, and

(c) DNA sequences which code on expression for a polypeptide of the IFN-α type coded for on expression by any of the foregoing DNA sequences and inserts,

said DNA sequences and inserts being operatively linked to an expression control sequence in said recombinant DNA molecule.

The Markush format will be discussed in greater detail below.

c. *Consists or Consisting Essentially of* The use of the phrase *consists essentially of* or *consisting essentially of* in a claim would include within the scope of the claim the recited features and other unspecified features that do not substantially affect the properties, functions, or characteristics of the invention.[38]

The term *consisting essentially of* was the subject of an appeal in *Ex parte Davis.*[39] The Examiner there had rejected certain claims to a sealing tape in an application by Davis on the basis of a single reference. A rejected claim, claim 13, read as follows[40]:

13. A sealing tape comprising a strip of backing material and a normally non-tacky coating thereon of a thermoplastic and solvent-applicable adhesive composition *consisting essentially of* [italics added]

Ethyl cellulose . one part
Staybelite . . . . . 1 to 3 parts
Hercolyn . . . . . . 0.5 to 1 part

According to the applicant, the claimed adhesive composition was normally nontacky and waterproof. This adhesive composition could be softened and rendered tacky "by the application of heat at temperatures from about 240°F to 280°F, or by the use of solvents such as ethyl alcohol, ethyl acetate, acetone, etc."[41] The composition of the cited reference, on the other hand, was pressure sensitive and tacky as the result of the presence of additional modifying agents not specified in Davis' claim.

The court determined that, in view of the meaning of *consisting essentially of* generally adopted by the PTO Examiners, the applicant's claim reciting three ingredients excluded the fourth ingredient of the cited reference. The court was persuaded that the tackifying "modifier" in the prior art materially changed the "fundamental character" of the three-ingredient composition claimed by the applicant.[42]

An example of a claim utilizing this transitional phrase can be found in U.S. Patent No. 4,703,008,[43] as follows:

> 7. A purified and isolated DNA sequence *consisting essentially of* [italics added] a DNA sequence encoding a polypeptide having an amino acid sequence sufficiently duplicative of that of erythropoietin to allow possession of the biological property of causing bone marrow cells to increase production of reticulocytes and red blood cells, and to increase hemoglobin synthesis or iron uptake.

Another example of the *consisting essentially of* terminology in a biotechnology patent can be found in claim 1 of U.S. Patent No. 4,766,075, the subject of dispute in *Genentech Inc.* v. *Wellcome Foundation, Ltd.*[44] This claim recited: "A DNA isolate *consisting essentially of* [italics added] a DNA sequence encoding human tissue plasminogen activator."

Although the foregoing examples were product claims, the phrase "consisting essentially of" can also be found in process claims. In *Ex parte Hoffman,*[45] the Board of Patent Appeals and Interferences (Board) held that there is nothing intrinsically wrong with the use of such language to modify method steps. The use of such a phrase renders the claim open only to the inclusion of steps that do not materially affect the basic and novel characteristics of the claimed method.

3. Body of the Claim

The body of the claim is usually what differentiates the patented invention from the prior art. By statutory mandate, the claim must particularly point out and distinctly recite the subject matter that the applicant regards as his or her invention.[46] This is basically a requirement for precision and definiteness in claim language so that the public will be on notice as to the types of actions that would constitute infringement of the patent. The bulk of infringement disputes revolve around interpretation of limitations in the body of the claim and will be the focus of the continuing discussion of infringement below.

IV. COMMON TYPES OF CLAIMS

Patents may be granted for processes, machines, articles of manufacture, or compositions of matter.[47] All patent claims must fall into one of these categories. Different types of claims may be treated somewhat differently in determining infringement. The basic types of patent claims are, therefore, briefly discussed here.

A. Product Claims

A product claim recites an invention that falls within the statutory classes of machines, articles of manufacture, compositions of matter, or an improvement thereof. A product may be claimed by characterization of its properties. For example, claim 1 of U.S. Patent No. 4,677,195 (the '195 patent), assigned to Genetics Institute, Inc. (GI), recites an erythropoietin characterized by its molecular weight and specific activity as follows[48]:

> 4. Homogeneous erythropoietin characterized by a molecular weight of about 34,000 daltons on SDS PAGE, movement as a single peak on reverse phase high performance liquid chromatography and a specific activity of at least about 160,000 IU per absorbance unit at 280 nanometers.

A product may also be claimed by characterizing the function the product is capable of performing. For example, a claim to a cell culture is drafted in this format in one of Genentech's patents[49]:

> 3. A cell culture capable of expressing human tissue plasminogen activator, obtained by transforming a mammalian cell line with a vector according to [another claim].

A product claim can be recited as *comprising* a component part or parts, as in the following composition claim of GI's '195 patent:

> 6. A pharmaceutical composition for the treatment of anemia *comprising* [italics added] a therapeutically effective amount of the homogeneous erythropoietin of claim 4 in a pharmaceutically acceptable vehicle.

A similar claim format was involved in *Intervet America, Inc.* v. *Kee-Vet Laboratories, Inc.*[50] The claim read[51]:

> A live vaccine effective against Infectious Bursal Disease in poultry upon a single administration to birds at the usual age of vaccination *comprising* [italics added] a live Infectious Bursal Disease virus belonging to the strain deposited at the ATCC under No. VR-2041.

When so written, product claims are usually interpreted as covering products containing the recited materials as well as other unspecified materials.

It is a cardinal principle of patent law that a product claim covers the product claimed regardless of the method by which it is made. In *Amgen, Inc.* v. *Chugai Pharmaceutical Co.*,[52] e.g., the District Court of Massachusetts found that the erythropoietin of Amgen, Inc. (Amgen), produced by recombinant techniques, infringed certain claims of U.S. Patent No. 4,677,195 assigned to Genetics Institute (GI). GI's claims covered a composition containing purified erythropoietin prepared by nonrecombinant techniques. In doing so, the court adhered to the principle that "[a] product patent gives the patentee the right to restrict the use and sale of the product regardless of how and by whom it was manufactured."[53]

Thus, Amgen's recombinant erythropoietin, even though not prepared by purification of the natural protein, as GI had done, was determined to have infringed GI's claims to "homogenous erythropoietin." On appeal, the Federal Circuit found GI's claims to be invalid for lack of enablement,[54] but the infringement analysis of the lower court serves to illustrate the point.

In the same vein, the lower court in *Scripps Clinic & Research Foundation* v. *Genentech, Inc.*[55] found, even without a trial, that Genentech's recombinant Factor VIII:C products literally infringed claims to Factor VIII:C in Scripps' reissue patent. Scripps' reissue patent disclosed the purification of natural human Factor VIII:C, in contrast to Genentech's product, which was produced by recombinant technology. Genentech had argued that their recombinant products should not be found to infringe Scripps' product claims by application of the so-called "reverse doctrine of equivalents," discussed later.

On appeal, the Federal Circuit ruled that consideration of additional evidence was required for determination of this issue and that it was necessary to have a trial in order to resolve it.[56] Although the district court's grant of summary judgment on the issue of infringement of Scripps' product claim was reversed, there appears to be a clear judicial inclination to find claims to purified natural products to be infringed by corresponding products made by recombinant DNA techniques.

B. Process Claims

Process claims are written in the form of the steps to be performed to effect the claimed process. For example, in U.S. Patent 4,677,195,[57] a process claim recites (with paragraphs inserted):

> 2. A method for purifying human erythropoietin comprising
>
> treating partially purified erythropoietin having a specific activity of at least about 80,000 IU per absorbance unit at 280 nanometers by reverse phase high performance liquid chromatography and
>
> eluting therefrom homogeneous human erythropoietin which moves as a single peak on reverse phase high performance liquid chromatography and is characterized by a specific activity of at least 160,000 IU per absorbance unit at 280 nanometers.

An example of an immunoassay process claim can be found in the patent involved in *In re Wands*,[58] in which the broadest claim on appeal read:

> 1. An immunoassay method utilizing an antibody to assay for a substance comprising hepatitis B-surface antigen (HBsAg) determinants which comprises the steps of:
>
> contacting a test sample containing said substance comprising HBsAg determinants with said antibody; and

determining the presence of said substance in said sample;

wherein said antibody is a monoclonal high affinity IgM antibody having a binding affinity constant for said HBsAg determinants of at least 10^9 M^{-1}.

A further example of a process claim is found in U.S. Patent No. 4,361,509, the subject matter of dispute in *Scripps Clinic and Research Found.* v. *Genentech, Inc.*[59] Claim 1 there recites:

1. An improved method of preparing Factor VIII procoagulant activity protein comprising the steps of
   (a) absorbing a VIII:C/VIII:RP complex from a plasma or commercial concentrate source onto particles bound to a monoclonal antibody specific to VIII:RP,
   (b) eluting the VIII:C,
   (c) absorbing the VIII:C obtained in step (b) in another adsorption to concentrate and further purify same,
   (d) eluting the absorbed VIII:C, and
   (e) recovering highly purified and concentrate VIII:C.

It is well established that a process patentee's power to exclude others extends only to the claimed process, in contrast to a product patentee's power, which extends to the product regardless of how it is made. As observed by the district court in *Amgen, Inc.* v. *Chugai Pharmaceutical Co.*[60]:

> A product patent gives the patentee the right to restrict the use and sale of the product regardless of how and by whom it was manufactured. A process patentee's power extends only to those products made by the patented process. A process patent thus "leaves the field open to ingenious men to invent and to employ other processes."
>
> A sale of a product made by a patented process does not itself infringe the patent; it is the unauthorized use of the process that infringes the patent.

C. Product-by-Process Claims

A product-by-process claim is a product claim in which the product is defined by the method of making it. In the past, a product-by-process claim had been interpreted to be infringed only by a product produced by following the same process described in the claim. In other words, an identical product made in a manner different from that specified in the claim was not found to infringe the patent claim.[61] In a single sentence of a recent case, however, the Federal Circuit appears to have swept away 100 years of well-settled law in this area. The court ruled that[62]:

> In determining patentability we construe the product as not limited by the process stated in the claims. Since claims must be construed the same way for

> validity and for infringement, the correct reading of product-by-process claims is that they are not limited to product prepared by the process set forth in the claims.

Under the circumstances, the Federal Circuit held that infringement of the Scripps *product-by-process* claims by Genentech's recombinant product may be considered at trial.[63] Whether the appellate ruling in this regard will withstand the test of time remains to be seen. Indeed, a different panel of the Federal Circuit has held otherwise in *Atlantic Thermoplastics Co.* v. *Faytex Corp.*[63b] Therein, the court held that:

> In light of Supreme Court caselaw and the history of product-by-process claims, this court acknowledges that infringement analysis proceeds with reference to the patent claims. Thus, process terms in product-by-process claims serve as limitations in determining infringement.

D. Jepson-Type Claims

Jepson-type claims are more popular in Europe than in the United States. These claims set forth in the preamble features that are old in the art and, in the body of the claim, the improvement sought to be patented. The following is an example of a Jepson-type claim (paragraphs added[64]):

> 19. In an immunometric assay to determine the presence or concentration of an antigenic substance in a sample of a fluid comprising forming a ternary complex of a first labelled antibody, said antigenic substance, and a second antibody said second antibody being bound to a solid carrier insoluble in said fluid wherein the presence of the antigenic substance in the samples is determined by measuring either the amount of labelled antibody bound to the solid carrier or the amount of unreacted labelled antibody,
>
> *the improvements comprising* [italics added] employing monoclonal antibodies having an affinity for the antigenic substance of at least about 10^8 liters/mole for each of said labelled antibody and said antibody bound to a solid carrier.

Another example of a Jepson-type claim is as follows[65]:

> 1. In an occult blood specimen test slide having a front panel, a rear panel, said front panel having one or more openings, sheet means carrying a test reagent between the front and rear panels underlying each of said openings, a hinged cover adapted to overlie a portion of the front panel and said openings and flap means in the rear panel opposite said openings and pivotable to expose the underside of the sheet, *the improvement comprising* [italics added]:

an area positioned on a portion of the sheet means facing the rear panel and isolated from the opening in the front panel, said area including a positive and negative monitor, said positive and negative monitors including the test reagent and said positive monitor additionally including a compound that reacts to environmental conditions in a manner similar to hemoglobin.

The elements stated in the preamble of a Jepson claim are considered positive limitations on the scope of the claim and must be present in an infringing product or process.

A cautionary note should be sounded in regard to Jepson-type claims. It has sometimes been held that the inventor's use of the Jepson claim format may constitute an admission that the subject matter described in the preamble of such a claim is in the prior art, apparently regardless of whether it is, in fact, prior art.[66] In drafting such claims, therefore, care should be taken that the inventor does not admit as prior art more than is absolutely necessary.

E. Markush Claims

The name *Markush claims* arose out of a chemical patent case, *Ex parte Markush*,[67] which sanctioned the claiming of a genus expressed as a group consisting of specified elements. The *Markush* decision authorized the claiming of such an artificial genus using the words "selected from the group consisting of A, B, and C." Often, this type of claim is resorted to when there is no generic term that will embrace these elements. To constitute a proper Markush group, the *Manual of Patent Examining Procedure* (MPEP) § 703.03(y) prescribes that the elements of the Markush group must ordinarily "belong to a recognized physical or chemical class or to an art-recognized class."

The courts have determined, however, that if all the elements of the Markush group belong to a subgenus that is not repugnant to scientific classification, then the claimed elements can be considered as part of a single invention, thereby meeting the requirement of unity of invention.[68] An example of a claim containing the Markush language, "selected from the group consisting of," is set forth below:

1. A purified and isolated DNA sequence encoding erythropoietin, said DNA sequence *selected from the group consisting of* [italics added]:
   (a) the DNA sequences set out in FIGS. 5 and 6 or their complementary strands; and
   (b) DNA sequences which hybridize under stringent conditions to the DNA sequences defined in (a).[69]

Another example of a Markush claim can be found in U.S. Patent No. 4,531,901 previously discussed, which reads:

4. The recombinant DNA molecule according to claim 1, wherein the expression control sequence is *selected from the group consisting of* [italics

added] a lac system, a trp system, major operator and promoter regions of phage λ, the control region of fd coat protein, and other sequences which control the expression of genes of prokaryotic or eukaryotic cells and their viruses.

A Markush claim is infringed by use of any one of the recited elements in the environment specified in the remainder of the claim.

F. Means Plus Function Claims

A *means plus function* claim refers to one that recites a structural element by reference to a means for performing a specified function. This type of claim is authorized by statute when a claimed invention is a combination made up of component parts.[70]

A *means plus function* claim was construed in *Palumbo* v. *Don-Joy Co.*,[71] an infringement action involving a patent[72] on a knee brace. Claim 1 asserted to be infringed recited:

> A dynamic patellar brace for preventing subluxation of a patella throughout the complete physiologic range of flexion and movement of the knee comprising:
> *means for* [italics added] bracing the patellar;
> *means for* [italics added] maintaining said patellar bracing means positioned laterally of the patella throughout the complete physiologic range of flexion and movement of the knee when the brace is in use; and
> *means for* [italics added] causing said patellar bracing means positioned laterally of the patella to apply a resultant force in the medial direction to the patella throughout the complete range of flexion and movement of the knee when the brace is in use.

Interpretation of *means plus function* language is made by reference to 35 U.S.C. § 112, sixth paragraph, which provides that

> . . . such claims shall be construed to cover the corresponding structure, material or acts described in the specification and equivalents thereof.

Drawing upon this language and upon precedents established by the court, the court expressly recognized in *Palumbo* that the patent claim must be construed "to cover *both* the disclosed structure *and* equivalents thereof."[73] To hold otherwise, "would be to nullify the [relevant] provision of § 112. . . . "

G. Independent and Dependent Claims

A claim may be written in independent, dependent, or multiple dependent form.[74] An independent claim recites all essential features of an invention. It stands or falls on its own. A dependent claim contains a reference "to a claim previously set forth [i.e., an independent or dependent claim] and specifies a further limitation of the subject matter claimed.''[75] A dependent claim is construed to incorporate all the limitations of the claim from which it depends.

A multiple dependent claim must contain a reference, in an alternative form only, to "more than one claim previously set forth and then specify a further limitation of the subject matter claimed."[76] A multiple dependent claim is also construed to incorporate all the limitations of the particular claim it referenced.

In the following examples, claim 1 is considered an independent claim, claims 2, 3, and 4 are considered dependent claims, and claim 5 is a multiple dependent claim.[77]

1. An apparatus for use in an immunoassay process for the detection of a target antigen in a liquid sample comprising:

   (a) a first member which is a porous membrane or filter and to which is bound an antibody against the target antigen, which member has upper and lower surface, the sample being applied to the upper surface, and wherein the antibody is bound within an area smaller than the area of the member to which the sample is applied; and

   (b) a second member which is a body of absorbent material having a surface over which the first member is placed and having capillaries therethrough in a direction generally transverse to the surface over which the first member is placed, which capillaries are in communication with the pores on the lower surface of the first member so as to draw liquid added to the upper surface which has permeated the first member into the capillaries of the member, the capillary communication between said first and second members having been established prior to, and maintained during, addition of liquids to the apparatus in the immunoassay process.

2. An apparatus according to claim 1 wherein the antibody is a monoclonal antibody.

3. An apparatus according to claim 1 wherein the membrane or filter is of glass or of nylon.

4. An apparatus according to claim 1 wherein the first member is separated from the second member by another porous member.

5. An apparatus according to claim 1, 2, 3, or 4 wherein the apparatus further comprises a container for the first and second members having an opening to permit addition of assay reagents to the upper surface of the first member.

V. INFRINGEMENT

A cause of action for patent infringement is created by the statute that states in part: "[e]xcept as otherwise provided in this title, whoever without authority makes, uses or sells any patented invention, within the United States during the term of the patent therefor, infringes the patent'' 35 U.S.C. § 271(a).

An action for infringement under 35 U.S.C. § 271(a) is limited to infringing activities ''within the United States.'' However, the patent has some extraterritorial effect when the patented invention is a process. Under 35 U.S.C. § 271(g), whoever ''without authority imports into the United States or sells or uses within the United States a product which is made by a process patented in the United States shall be liable as an infringer, if the importation, sale, or use of the product occurs during the term of such process patent.'' Some extraterritorial effect is also given U.S. patents by 35 U.S.C. § 271(f), which makes it an act of infringement to supply from within the United States all the material components of an invention patented in the U.S. for ultimate combination abroad.

As previously noted, patent infringement can be likened to trespass upon private property. Just as a determination of trespass is governed by reference to location of the trespasser in relation to the legal description of a piece of land, so too is infringement of a patent determined by reference to an allegedly infringing product or process in relation to the limitations in a patent claim. The initial phase of the analysis involves a comparison of the accused device with the literal language of the claim. This analysis leads to a conclusion on the issue of ''literal infringement.''

Historically, actions for patent infringement were brought in a court of equity seeking ''equitable'' relief, such as an injunction and an accounting for profits. Proceedings in a court of equity were intended to make the injured party whole and to do justice where monetary damages were insufficient to compensate the injured party. Over the years, there evolved an equitable ''doctrine of equivalents.'' The doctrine was applied in instances in which an allegedly infringing product or process did not clearly fall within the literal language of a claim but yet so closely resembled the patented invention that it ought, in fairness, to be found to have infringed the patent claim. In such instances, if an allegedly infringing device or process performed substantially the same function, in substantially the same way, to produce substantially the same result as the patented invention, the accused device or process was, for all intents and purposes, considered to be the same as the patented device and infringement was found.[78]

Coincidental with the evolution of the doctrine of equivalents was the evolution of the ''reverse doctrine of equivalents.'' Under the reverse doctrine of equivalents, a patent claim may literally describe an accused device or process, yet the accused device or process may be ''so far changed in principle from the patented article that it performs the same or similar function in a substantially different way.''[79] Under such circumstances, what appeared at first blush to be an infringement may ultimately be found not to constitute infringement. Needless to say, the doctrines of equivalents and reverse equivalents have injected a substantial amount of uncertainty into infringement analyses.

A. Literal Infringement

Literal infringement of a patented invention is established by a showing that every claimed limitation of the patented invention is present in the accused device or process. A literal infringement analysis is a two-step process. First, the claim that is said to be infringed, i.e., the asserted claim, must be construed without regard to the accused product to ascertain the meaning of the terms used. "A claim is construed in light of the claim language, the other claims, the prior art, the prosecution history, and the specification, *not* in light of the accused device."[80]

Once the asserted claim is properly construed, the device or process accused of infringement must be examined to determine whether each limitation of the properly construed claim is found in the accused product or process.[81]

1. Claim Interpretation

It can be appreciated that it is the *claims* and not the *specification* that define the invention. The threshold requirement in claim construction for infringement purposes, therefore, is examination of the claims at issue.[82] In defining an invention, however, the inventor may be his or her own lexicographer.[83] The inventor can exercise creativity in generating descriptive terms to explain the invention so long as the terms are defined, and so long as his or her definitions of the terms are not inconsistent with the meaning generally attributed to them by persons of ordinary skill in the art.

The inventors exercised this prerogative in creating the term *plasmid chimera* in United States Patent No. 4,237,224 to describe the concept of what is now called recombinant DNA.[84] In describing the invention, the patent states[85]:

> The process of this invention employs novel plasmids, which are formed by inserting DNA having one or more intact genes into a plasmid in such a location as to permit retention of an intact replicator locus and system (replicon) to provide a recombinant plasmid molecule. The recombinant plasmid molecule will be referred to as a "hybrid" plasmid or plasmid "*chimera*" [italics added].

To ascertain the true meaning of a disputed claim term, courts have resorted to the use of various types of evidence. "Claim interpretation involves a review of the specification, the prosecution history, the claims (including unasserted as well as asserted claims), and, if necessary, other extrinsic evidence, such as expert testimony."[86] Courts have also looked to the disclosures of prior art[87] and to the drawings.[88] In any event, claims "must be interpreted and given the same meaning for purposes of both validity and infringement analysis."[89]

In accordance with the current legal standard, where a claim is susceptible of two constructions, courts in an infringement setting will adopt the one interpretation that "will secure to the patentee his actual invention, rather than to adopt a construction fatal to the grant."[90] On the other hand, a patentee "cannot construe the claims narrowly before the Patent Office and later broadly before the Courts."[91]

a. Based upon literal or dictionary meaning When claims are construed, courts will generally give the words in a claim "their ordinary and accustomed meaning, unless it appears that the inventor used them differently."[92]

Yet, the ordinary meaning of disputed claim terms is not dispositive. "[R]esort must still be had to the specification and prosecution history to determine if the inventor used the disputed terms differently than their ordinary accustomed meaning."[93]

b. Based upon the specification The specification is generally viewed as a dictionary for the terms used in claims. It is, therefore, not surprising that the courts turn to the specification for interpretation of ambiguous claim terms. "[R]esort to certain extrinsic evidence (*i.e.*, the specification, prosecution history and other claims) is always necessary to interpret disputed claims."[94]

To say that a claim should be interpreted in light of the specification, however, is not to suggest that the meaning of the claim terms can be varied depending on the circumstances. This precept has been succinctly expressed by the Court in *White* v. *Dunbar*[95]:

> Some persons seem to suppose that a claim in a patent is like a nose of wax which may be turned and twisted in any direction, by merely referring to the specification, so as to make it include something more than, or something different from, what its words express. The context may undoubtedly be resorted to, and often is resorted to, for the purpose of better understanding the meaning of the claim; but not for the purpose of changing it and making it different from what it is.

There is often a confusion between looking to the specification to interpret the claims and reading limitations in the specification into the claims. The rule of law is that although the specification may be used for the purpose of interpreting claims, limitations described in the specification may not be read into the claims, either by the infringer for the purpose of avoiding infringement or by the patentee for the purpose of saving the claims from invalidation. For example, the court in *E.I. Du Pont De Nemours & Co.* v. *Phillips Petroleum Co.* stated[96]:

> It is entirely proper to use the specification to interpret what the patentee meant by a word or phrase in the claim [citations omitted]. But this is not to be confused with adding an extraneous limitation appearing in the specification, which is improper. By "extraneous," we mean a limitation read into a claim from the specification wholly apart from any need to interpret what the

patentee meant by particular words or phrases in the claim. "Where a specification does not *require* a limitation, that limitation should not be read from the specification into the claim."

It is in the nature of biological sciences that many of the terms used do not have the precise definitions to which we have become accustomed in other areas of science. This is not a fatal flaw in biotechnology patent claims as long as the claim language is as precise as the subject matter permits.

For example, in *Hybritech Incorporated* v. *Monoclonal Antibodies, Inc.*,[97] the district court held that claims specifying use of antibodies with an affinity of "at least about 10^8 liters/mole" were indefinite for failure to disclose "how infringement may be avoided because antibody affinity cannot be estimated with any consistency."[98]

The Federal Circuit disagreed, holding that a claim cannot be indefinite as long as the language used therein is as definite as it can be[99]:

> . . . if the claims, read in light of the specification, reasonably apprise those skilled in the art both of the utilization and scope of the invention, and if the language is as precise as the subject matter permits, the courts can demand no more.

The Federal Circuit was convinced that the record indisputably showed that calculating affinity was known in the art at the time of filing of the patent application, even though calculations were not precise or standard.

The permissible use of such claims in biotechnology patents necessarily results in some uncertainty as to exactly where the line lies between infringing and noninfringing products. With regard to the *Hybritech* patent, e.g., antibodies with measured affinities of 4.7×10^7 were found to satisfy the claim requirement for an affinity of "about 10^8". In so holding, the court relied in part on the art recognized error of a factor of 2 or 3 in affinity measurements.[100]

Interpreting patent claims in light of the specification has occasionally worked to the detriment of the patentee. In *Genentech Inc.* v. *Wellcome Foundation, Ltd.*,[101] e.g., the plaintiff, Genentech Inc. (Genentech), alleged that The Wellcome Foundation, Ltd. (Wellcome) infringed the claims of its patent, U.S. Patent No. 4,766,075 (the '075 patent), which recites, among other things, DNA sequences encoding tissue plasminogen activator (t-PA). The claims asserted to be infringed were claims 1, 3, and 8. Claims 1, 2, and 3 are reproduced below:

> 1. A DNA isolate consisting essentially of DNA sequence encoding human tissue plasminogen activator.

> 2. A recombinant expression vector containing a DNA sequence encoding human tissue plasminogen activator, wherein the vector is capable of expressing human tissue plasminogen activator in a transformed microorganism or cell culture.
> 3. A cell culture capable of expressing human tissue plasminogen activator, obtained by transforming a mammalian cell line with a vector according to claim 2.

Genentech asserted that the above claims encompassed variations in the t-PA molecule because of certain statements of such an intent made in the specification. The statements relied on were[102]:

> It will be understood that natural, allelic variations exist and occur from individual to individual. These variations may be demonstrated by (an) amino acid difference(s) in the overall sequence or by deletions, substitutions, insertions, inversions or additions of (an) amino acid(s) in said sequence.
>
> * * *
>
> The potential exists, in the use of recombinant DNA technology, for the preparation of various human tissue plasminogen activator derivatives, variously modified by resultant single or multiple amino acid substitutions, deletions, additions or replacements, for example by means of site directed, mutagenesis of the underlying DNA.

The court, however, interpreted these statements to mean that Genentech's asserted claims would be infringed only if the accused product was a DNA sequence that encoded *human* t-PA or a *naturally occurring allelic variant*. In construing Genentech's claims in this manner, the court determined that Wellcome's DNA sequences that were made by recombinant techniques and by biochemical modifications of the natural sequence did not infringe Genentech's claims.[103]

c. Based upon other claims The law is clear that when a patent claim does not contain a certain limitation and another claim does, that limitation should not be read into the former claim in determining infringement.

This concept is referred to as the doctrine of claim differentiation, and it stands for the proposition that "claims should be presumed to cover different inventions."[104] In the context of claim interpretation, therefore, "interpretation of a claim should be avoided if it would make the claim read like another one."[105] This concept is not applied rigidly, however. "If a claim will bear only one interpretation, similarity will have to be tolerated."[106]

This point of law is illustrated in *Scripps Clinic and Research Foundation* v. *Genentech, Inc.*,[107] in which Scripps brought an action for infringement against Genentech. The product claims asserted to be infringed were claims 24–29.[108]

Genentech contended that its factor VIII:C did not infringe Scripps' patent claims because it was produced by recombinant techniques, whereas Scripps' Factor VIII must be derived from human blood plasma. Scripps, on the other hand, argued that the term human VIII:C in the asserted claims meant that the claimed factor VIII:C only need to possess the same characteristics as that derived from humans.

The court examined Scripps' other product claims and found that none of them required Factor VIII:C to be derived directly from the blood stream. The process and product-by-process claims, on the other hand, all contained a limitation that Factor VIII:C be derived from "plasma" or commercial concentrate source."[109] Based upon the presence of such limitation in other claims, and upon the generally accepted principle of claim differentiation, the court concluded that Scripps' product claim to Factor VIII:C was not limited to a particular source and that Genentech's recombinant product infringed Scripps' asserted claims.

d. Based upon prosecution history The courts have also referred to the prosecution history of a patent as a guide to interpretation of ambiguous claims. "It is, of course, well settled that an invention is construed not only in the light of the claims, but also with reference to the file wrapper or prosecution history in the Patent Office."[110] The prosecution history of a patent includes, first of all, amendments to the specification and the claims made during the prosecution of the application that matured into the patent. Such amendments encompass both insertions of materials into and deletions of materials from the specification and the claims. The prosecution history further includes arguments made before the PTO in attempts to overcome objections or rejections asserted by the patent examiner. Particularly scrutinized are arguments presented to the PTO to overcome anticipation or obviousness rejections of the claimed invention on the basis of prior art disclosures.

The United States District Court for the District of Massachusetts in *Amgen, Inc.* v. *Chugai Pharmaceutical Co., Ltd.*[111] had occasion to review the prosecution history of the patent-in-suit.[112] The court was presented with the issue of whether the '008 patent embraced Chugai's process of manufacturing recombinant erythropoietin. The claims to be construed were claims 4 and 23, set out as follows:

> 4. A procaryotic or eucaryotic host cell transformed or transfected with a DNA sequence according to claim 1, 2 or 3 in a manner allowing the host cell to express erythropoietin.
>
> • • • •
>
> 23. A procaryotic or eucaryotic host cell transformed or transfected with a DNA sequence according to claim 7, 8, or 11 in a manner allowing the host cell to express [a polypeptide similar to erythropoietin].

If claims 4 and 23 were process claims, as Amgen urged, then Chugai's process of manufacturing recombinant erythropoietin outside the United States for sale in the United States might be found to infringe Amgen's claims (under 35 U.S.C. § 271[g]). If claims 4 and 23 were *product* claims, however, then Chugai's process would not infringe these claims unless Chugai was shown to have used Amgen's products in the United States.

In reviewing these claims, the court found the recitation of "[a] procaryotic or eucaryotic host cell transformed or transfected with a DNA sequence . . . in a manner allowing the host cell to express erythropoietin" to be ambiguous as to whether it was describing a product or a process.[113] In view of the ambiguity, the court decided that it must "look to the other factors of claim construction—in particular, the prosecution history—to construe these claims."[114]

Upon examination of the prosecution history, the court was convinced that Amgen's claims were product claims. The Court was persuaded that the interpretation of claims 4 and 23 urged upon the court by Amgen was similar to the process claims that were originally in the patent application[115] but that were canceled in response to the Examiner's rejection. Having canceled the process claims the court determined that Amgen could not construe the remaining product claims in a manner that would resurrect the subject matter surrendered.

Moreover, if during the prosecution of the patent application in the PTO, the applicant for patent argued that a claim term has a specific meaning, in order to obtain allowance of the claim, the term is usually given the same meaning when the patentee is asserting infringement of the claim.

e. Based upon expert testimony Since a claim phrase must be interpreted as it would be construed by a person of ordinary skill in the art, courts will also look to expert's testimony for interpretation of claim terms.[116]

In *Amgen, Inc.* v. *Chugai Pharmaceutical Co.*,[117] the Federal Circuit overturned the district court's finding of infringement by Amgen of claims 1 and 3 of U.S. Patent No. 4,677,195 (the '195 patent).[118] The Federal Circuit found that when properly construed, claims 1 and 3, each of which "contains the limitation that EPO have a specific activity of at least 160,000 IU/AU," were invalid for failure of the patent to provide an enabling disclosure.

Claim 1 recited:

> Homogeneous erythropoietin characterized by a molecular weight of about 34,000 daltons on SDS PAGE, movement as a single peak on reverse phase high performance liquid chromatography and a specific activity of at least 160,000 IU per absorbence unit at 280 nanometers.

The Federal Circuit concluded that the district court did not, however, err in deciding that the specific activity should be construed as defining in vivo activity when expert testimony from both sides showed that "to those skilled in the art, in the absence of an express statement in the patent, the claims would be construed to refer to *in vivo* rather than *in vitro* specific activity.''[119]

The point is also illustrated by the case of *Advanced Cardiovascular Systems, Inc.* v. *Scimed Life Systems Inc.*[120] There, the district court's grant of summary judgment was vacated on the basis that expert testimony would be particularly helpful in interpreting the meaning of a disputed term in the claims.

f. Other examples of claim interpretation based upon a combination of factors None of the above sources of reference by itself is dispositive for purposes of claim interpretation. Courts often look at the totality of the circumstances and make a determination based upon the facts of each case. The analyses in the following cases illustrate the interplay of the factors considered in interpreting claims.

SmithKline Diagnostics, Inc. *v.* Helena Laboratories Corp. In *SmithKline Diagnostics, Inc.* v. *Helena Laboratories Corp.*,[121] interpretation of the phrase ''similar to hemoglobin'' was at issue. The patented invention related to a specimen test slide and a method for detecting occult (hidden or invisible) blood in fecal matter. The test slide contained ''a piece of paper impregnated with a colorless compound, guaiac, which turns blue in the presence of a developing solution, such as hydrogen peroxide, and a catalyst, such as hemoglobin in the blood.'' A blue color, therefore, indicates that blood is present, thus, a ''positive'' result. Absence of blue, indicates the absence of blood, a ''negative'' result.[122]

The patent asserted to be infringed, U.S. Patent No. 4,365,970 (the '970 patent), provided for building positive and negative monitors separate from the test areas. The positive monitor contained a catalyst that ''must be a compound that reacts to environmental conditions in a manner similar to hemoglobin.'' The negative monitor, however, lacked the catalyst.[123] With use, a developing solution was added to the negative and positive monitors after it was applied to the fecal test areas. A blue color on the positive monitor indicated that the paper and solution were working. The absence of blue color on the negative monitors indicated that the slide ''has avoided contamination.''[124]

Claims 1 and 5 of the '970 patent both described the positive monitor as ''additionally including a compound that reacts to environmental conditions in a manner similar to hemoglobin.''[125]

The district court interpreted the claim term ''similar to hemoglobin'' as *excluding* hemoglobin itself. Based on this interpretation, the district court found the patented invention to be nonobvious and Helena's hemoglobin-containing slides to be noninfringing, either literally or under the doctrine of equivalents. The district court further stated that if the claims at issue were interpreted to cover hemoglobin, they would have been obvious and, therefore, invalid, in view of the prior art.

However, looking at the claim language, the specification, and the prosecution history of the '970 patent, the Federal Circuit disagreed with the district court and found the asserted claims to be valid and infringed. The Federal Circuit determined that the claim limitation at issue[126]:

> . . . does not identify specific catalysts to be included or excluded. Viewed in this manner, the limitation does not exclude hemoglobin; rather, it reflects the fact that a compound similar to hemoglobin may work better than hemoglobin itself.

Further, the Federal Circuit examined the specification of the '970 patent and concluded that the specification evidenced a clear intent by the inventors to include hemoglobin[127]:

> Since guaiac-based fecal occult blood tests are actually testing for the catalytic activity of hemoglobin in blood, the *positive monitor should employ either hemoglobin or a catalyst which would react to adverse environmental conditions in a manner similar to hemoglobin. Preferably*, the test slide of this invention employs *hemin*, a hemoglobin derived catalyst, as a catalyst in the positive monitor.

The Federal Circuit discounted Helena's argument that the patent "requires a control that deteriorates in the same way as blood deteriorates in the fecal sample," and that because hemoglobin does not deteriorate like blood, the patent claims cannot include hemoglobin.[128] The Federal Circuit reasoned that neither the claims nor the specification required the positive monitor catalyst to deteriorate like blood in a fecal sample.

The Federal Circuit then proceeded to look at the prosecution history of the '970 patent and used the teachings of the prior art to interpret the limitation of the claim at issue. The court noted that the limitation at issue was not in the original claim 1, and that claim 1 originally provided that the improvement comprised "a control area having a positive and a negative monitor that control area positioned on a portion of the sheet."[129]

The Examiner who examined the patent application had rejected claim 1 as obvious over U.S. Patents 3,996,006 to Pagano and U.S. Patent No. 4,175,923 to Friend. Friend disclosed a test product made of paper impregnated with guaiac, a section of which was impregnated with a blood component to form a built-in positive control. In use, the entire paper was sprayed with a developer. The paper was considered to be functioning properly if the portion impregnated with blood turned blue. The test paper was then dropped into a toilet bowl containing fecal matter. If the fecal matter contains blood, the remaining portion of the test paper turns blue.

The Pagano test slide did not contain built-in positive and negative monitors. During the prosecution the inventors argued that Friend did not teach the inclusion of a negative monitor. Subsequent to an interview with the Examiner, the Examiner entered into the record the following agreement: "Agreed to amendment of the claims as per Examiner's Amendment (Paper No. 9) to particularly recite the positive and negative monitors" (859 F.2d at 884). It was in Paper No. 9 that the inventors introduced the amendment, including the phrase "similar to hemoglobin."

The district court erroneously concluded that the inventors had amended claim 1 to avoid Friend as prior art, and that Friend disclosed the use of hemoglobin as the positive monitor catalyst. Therefore, the district court erroneously determined that claim 1 of the '970 patent must not include hemoglobin.

The Federal Circuit opined, however, that "Friend does not specifically disclose or claim a hemoglobin catalyst," but rather "Friend claims 'blood' as a substrate or composition for the positive monitor catalyst."[130] The Federal Circuit held that "Friend's teaching, although it includes hemoglobin as the catalyst, was not so restricted and an amendment excluding hemoglobin but including hemin (another blood component) would not have overcome Friend's broad disclosure of blood component catalysts." Hence, the district court's determination that the limitation at issue excludes hemoglobin was overturned by the Federal Circuit.

Hormone Research Foundation, Inc. v. *Genentech, Inc.* Use of the specification and prosecution history to interpret claims is further demonstrated in *Hormone Research Foundation, Inc.* v. *Genentech, Inc.*,[131] where the Federal Circuit affirmed a district court's interpretation of the phrase "corresponding to" in patentee's claim as meaning "identical to."

Hormone Research Foundations (HRF) was the owner of U.S. Patent No. 3,853,833 (the '833 patent), which related to human growth hormone. In a first patent application, the inventor described the solid-phase peptide synthesis of "a synthetic, 188 amino acid sequence exhibiting growth promoting properties," and indicated that he had identified the natural human growth hormone (HGH).[132] In a continuation-in-part application that matured into the '833 patent, the inventor conceded error in the original sequence filed in the parent case and proposed a different sequence, shown in a Fig. 2, containing 190 amino acids. Subsequently, however, HRF admitted that the sequence in Fig. 2 differed from natural HGH in several respects.[133]

Genentech produced Protropin and Protropin II by recombinant DNA techniques. Protropin differs from the structure of HRF's sequence depicted in Fig. 2 in several respects. First, Protropin has 192 amino acids instead of the 190 amino acids in Fig. 2 of HRF's '833 patent. Protropin has an additional methionine residue at the amino end of the sequence and an additional glutamine after position 68, like the natural HGH. Protropin also has glutamic acid instead of glutamine at position 73, aspartic acid instead of asparagine at position 106, and asparagine instead of aspartic acid at position 108.[134] The other Genentech product, Protropin II, has a structure that is identical to that of natural HGH.

HRF filed suit against Genentech for patent infringement. Several of HRF's patent claims, however, called for a protein with an amino acid sequence "corresponding to" that shown in Fig. 2 of its patent. Genentech defended on the ground that its Protropin products did not contain a product having the exact structure shown in Fig. 2 as required by the claims of the '833 patent.[135] The District Court agreed with Genentech on Genentech's interpretation of HRF's claims and found that Genentech's product did not meet all the limitations of HRF's claims. The district court interpreted the phrase "corresponding to" in the claims to mean "identical to," based upon "the most applicable dictionary definition.[136] Hence, the district court found no literal infringement of the patent claim.

The Federal Circuit affirmed the district court's result of a finding of no infringement by Genentech though through a different process, i.e., based upon the dictionary meaning of "corresponding" *in conjunction with the specification and prosecution history of the '833 patent.*[137] The Federal Circuit determined that the specification did not intend the phrase "corresponding to" to mean "similar to," as urged by HRF, by virtue of the following paragraphs in the specification[138]:

> The molecular structure illustrated in FIG. 1 was thought heretofore in the art *to correspond* to that of natural HGH, but as indicated earlier hereinabove, applicant has discovered that this is not correct. The correct structure of HGH has been determined by applicant and is illustrated in FIG. 2. It consists of a polypeptide chain of 190 amino acid residues, as compared to 188 amino acid residues in the polypeptide chain in FIG. 1.
>
> . . . .
>
> [W]hile there is *general similarity* between the overall conformations and structures illustrated in FIGS. 1 and 2, there are differences in the number and sequence of amino acid residues and in the size of the layer of the two intramolecular rings present in both structures.

The Federal Circuit opined that the above-cited passage from the specification "strongly suggests that the terms 'correspond' and 'similar [ ]' were not intended by the patentee to have the same meaning and that the former reflected true identity."[139]

The Federal Court further determined that statements made by HRF during prosecution of the application for the '833 patent evidenced the use of the phrase "corresponding to" to mean "identical to" rather than "similar to." Such statements include representation that the phrase "a structure corresponding to Fig. 2" in claim 12 was "directed solely to the compound of Fig. 2," was "specific to the structure of Fig. [ ] 2," was "limited to the structure [ ] shown in the drawing," and was "specific to the chemical formula of Fig. 2."[140]

Hence, in view of the specification and the prosecution history, the Federal Circuit concluded that the disputed term was used to connote "a material having the exact structure and conformation shown in Fig. 2 of the '833 patent"[141] and that Genentech's similar but not identical sequence did not literally infringe the claim.

2. Comparison of Accused Infringing Product or Process with Claims

There is usually little dispute about what the allegedly infringing product or process is. Accordingly, once the claim has been interpreted in the foregoing manner, the question of literal infringement or noninfringement usually follows quite readily. It is the patentee's burden, however, to introduce evidence that each of the limitations contained in the claim can be "read on"; i.e., describes the accused product or process. This may require the patentee to do tests or measurements on a specimen of the infringer's product in order to prove, e.g., what its specific activity or affinity is.

B. The Doctrine of Equivalents

1. Definition

Even where there is no literal infringement, infringement may still be found under the doctrine of equivalents if the limitation or limitations not literally present are there by equivalents.[142] The doctrine of equivalents was clearly set forth in *Graver Tank Mfg. Co., Inc.* v. *Linde Air Products Co.*[143] Under this doctrine, infringement is made out even in the absence of literal infringement when it is proved that the allegedly infringing product or process performs substantially the same function, in substantially the same way, to produce substantially the same result.

In *Graver Tank*, the U.S. Supreme Court articulated the public policy behind this doctrine; i.e., if others were permitted to imitate a patented invention by copying most but not all of the literal detail of the patented invention, it would "convert the protection of the patent grant to a hollow and useless thing."[144]

The doctrine of equivalents is an equitable doctrine judicially created to prevent pirating of a patentee's invention. This doctrine is said to create around a claim, "a penumbra which also must be avoided if there is to be no infringement."[145] In *Environmental Instruments Inc.* v. *Sutron Corp.*,[146] the Federal Circuit disavowed the district court's notion that "the doctrine of equivalents 'is not designed to permit a patent to exceed the scope of its claims,' " and ruled that "[t]he essence of the doctrine of equivalents is that it permits recovery for infringement where the accused device does *not* fall within the literal scope of the claims and is, therefore, outside their literal scope."[147]

Particularly enlightening in this regard is the court's decision in *Wilson Sporting Goods Co.* v. *David Geoffrey & Associates*,[148] where it states:

> To say that the doctrine of equivalents extends or enlarges *the claims* is a contradiction in terms. The claims—i.e., the scope of the patent protection *as defined by* the claims—remain the same and application of the doctrine *expands the right to exclude* to "equivalents" of what is claimed.

The doctrine of equivalents comes into play only when there is a determination of no literal infringement. In determining equivalency of a claimed element, the circumstances surrounding the purpose and function of the claimed element are examined[149]:

> What constitutes equivalency must be determined against the context of the patent, prior art, and the particular circumstances of the case. . . . Consideration must be given to the purpose for which an ingredient is used in the patent, the qualities it has when combined with other ingredients, and the function which it is intended to perform.

An important factor in the determination of equivalents is whether a person of ordinary skill in the relevant art would have known of the interchangeability of a feature of the patent claim that is absent from the allegedly infringing device with a feature that is present in the infringing device.[150] Understandably, when a person skilled in the art views an element in the patented device, e.g., a nail, to be interchangeable with another element in the allegedly infringing device, e.g., a screw, then a nail and a screw would likely be considered equivalent to each other.

Some courts have also viewed the status of a patent to be important in the determination of the range of equivalents to be accorded a patent claim. Such status pertains to whether the patent covers a pioneering invention or an improvement in a crowded art. The court in *Autogiro Co. of America* v. *United States*[151] provides the following example:

> A pioneer patent which occupies symbolically a six-inch circle will have three inches of equivalence if its range is fifty percent. An improvement patent occupying a two inch circle has only one inch of equivalence with the same range. Thus with relatively identical ranges, the scope of the patent provides the pioneer patent with absolutely a larger range of equivalence.

Of particular interest is that even though an equivalent must be found for every limitation of the patented claim,[152] the equivalent need only be somewhere in the accused device, not necessarily in a corresponding component.[153] As a corollary, a one-to-one correspondence of elements is not required for a finding of infringement. Claimed elements or steps may be combined in an allegedly infringing device without a loss of equivalence.[154]

To prove substantial identity between the claimed elements and the accused device under the doctrine of equivalents, such identity must be proven with respect to all three elements of the *Graver Tank* test; i.e., substantial identity of function, substantial identity in result, and substantial identity in the manner in which the result is achieved.[155] Such proofs cannot be subsumed in the proofs for literal infringement even though the proofs for literal infringement may bear on equivalence.[156]

The case of *Genentech Inc.* v. *Wellcome Foundation Ltd.*[157] serves to illustrate, albeit somewhat unsatisfactorily, the doctrine of equivalents analysis in biotechnology cases. The district court was there faced with a summary judgment motion brought by a patentee, Genentech Inc., against one of the alleged infringers, Genetics Institute, Inc. (GI), on the issue of infringement under the doctrine of equivalents.

Genentech's patents[158] covered (1) t-PA that was isolated and purified from melanoma and (2) the transformation of a mammalian cell line through genetic engineering techniques to produce a recombinant t-PA product. Human t-PA was known to be made up of 527 amino acids, and to be divided into five domains. First, an F domain that allows the t-PA to bind to fibrin; second, an E domain; third, a Kringle 1 (K1) domain; fourth, a Kringle 2 (K2) domain; and fifth, a serine protease domain. The K1, K2, and serine protease domains together constitute the enzymatically active portions of the t-PA molecule that cleave circulating plasminogen in the blood stream.

GI's allegedly infringing t-PA, FE1X, and met-t-PA, without question, served the same function as Genentech's t-PA; i.e., they stimulate the dissolution of fibrin clots by cleavage of plasminogen to plasmin. The met-t-PA was produced by substitution of a methionine residue for a valine at position 245 in the K1 region by biochemical techniques. GI's FE1X was produced from met-t-PA by deletion of the F and E domains of the protein.

GI argued that the FE1X cannot be equivalent to Genentech's t-PA because the FE1X does not contain the F domain for binding of the t-PA molecule to the fibrin clots. GI further contended that its FE1X exhibited an extended half-life in comparison with natural t-PA.

Genentech, on the other hand, argued that because FE1X required the presence of fibrin fragments for catalytic conversion of plasminogen to plasmin, the allegedly infringing product must accomplish the same function in the same way as the natural t-PA. Genentech further contended that GI's met-t-PA functioned enzymatically the same way as Genentech's t-PA despite the single amino acid substitution of valine by a methionine.

It appeared, therefore, that the patented product and the allegedly infringing products accomplished substantially the same function and obtained substantially the same result. The determination of infringement under the doctrine of equivalents, therefore, hinged on whether the patented product and the allegedly infringing product utilized the same means for cleavage of plasminogen to produce plasmin. The court determined that there were genuine issues of material fact that could not be determined without a trial. The issue, therefore, remains unresolved.[159]

2. Limitations on the Doctrine of Equivalents

a. Limitation by the prior art A claimed invention cannot be given a range of equivalents so broad that it would encompass a device or process that is in the prior art. The underlying rationale is that things that are in the prior art are already in the public domain and belong to the public. A patentee, therefore, has no right to remove such technology from the public domain.[160]

> The doctrine of equivalents exists to prevent a fraud on a patent, *Graver Tank Mfg. Co.* v. *Linde Air Prods. Co.*, 339 U.S. 605, 608, 85 U.S.P.Q. 328 (1950), *not* to give a patentee something which he could not lawfully have obtained from the PTO had he tried. Thus, since prior art always limits what an inventor could have claimed, it limits the range of permissible equivalents of a claim.

To aid in the determination of what range of equivalents to accord a claimed feature, a hypothetical claim can be drafted that covers the infringing device. The hypothetical claim is then examined to determine whether it is patentable over the prior art. If the hypothetical claim is not patentable over the prior art, then the patented claim cannot be given a range of equivalents that would cover the allegedly infringing device. If, however, the hypothetical claim is determined to be patentable over the prior art, then the patented claim could be given a range of equivalents that would cover the allegedly infringing device. In such an instance, the patented claim is infringed under the doctrine of equivalents.[161]

b. Prosecution history estoppel Prosecution history estoppel is a judicially recognized limitation on the doctrine of equivalents. This estoppel operates by limiting the range of equivalents that can be accorded a patented invention and is generally applied when amendments and arguments are made during prosecution of the patent application to overcome a prior art rejection asserted by the PTO.[162]

> The doctrine of equivalents is subservient to file wrapper estoppel. It may not include within its range anything that would vitiate limitations expressed before the Patent Office (citations omitted). Thus a patent that has been severely limited to avoid the prior art will only have a small range between it and the point beyond which it violates file wrapper estoppel. Similarly a patent which is a major departure from the prior art will have a larger range in which equivalence can function.

Thus, the doctrine of prosecution history estoppel precludes a patent owner from resurrecting a construction of a claim that was "surrendered during prosecution of his patent application."[163]

It is clear, however, that not all amendments and arguments introduced during prosecution result in application of prosecution history estoppel. If such were the rule, then the doctrine of equivalents would only be applied to those few claims that are not amended during prosecution. No rational basis can be advanced for such a rule. As articulated by the court in *Hughes Aircraft Co.* v. *United States*[164]:

> Amendments may be of different types and may serve different functions. Depending on the nature and purpose of an amendment, it may have a limiting effect within a spectrum ranging from great to small to zero.

The substance of the amendment, the reasons why certain subject matter was surrendered, and the circumstances surrounding the amendment are to be scrutinized. "[W]henever the doctrine is invoked, 'a close examination must be made as to, not only what was surrendered, but also the reason for such a surrender.' "[165] If amendments and arguments are made other than for the purpose of overcoming a prior art rejection, then prosecution history estoppel has not historically been applied. If, however, amendments are made to one element of the claimed invention to overcome prior art that does not affect a second element of the claimed invention, prosecution history estoppel will not normally be applied to prevent the patentee from asserting equivalents of the second element of the claimed invention.

Moreover, "[w]hen it comes to the question of which should control, an erroneous remark by an attorney in the course of prosecution of an application or the claims of the patent as finally worded and issued by the Patent and Trademark Office as an official grant, we think the law allows for no choice. The claims themselves control."[166]

C. The Reverse Doctrine of Equivalents

Application of the reverse doctrine of equivalents negates literal infringement. Under the reverse doctrine of equivalents, an allegedly infringing device can be so far changed in principle that it performs the same or similar function in a substantially different way, even though there may be literal infringement of the words of a claim.[167]

In *SmithKline Diagnostics* v. *Helena Laboratories Corp.*,[168] the accused infringer argued noninfringement of the '970 patent under the reverse doctrine of equivalents. Helena pointed out the alleged admission of one of the coinventors of the '970 patent that hemoglobin would not work and that hemoglobin operated in a substantially different way from the compounds SmithKline successfully used as positive monitor catalysts. The Federal Circuit, however, rejected Helena's argument and found that the inventor never stated that hemoglobin would not work and that the court had, in fact, interpreted claim 1 to include hemoglobin as one of the possible catalysts. Thus, hemoglobin "does not operate in a substantially different way from the compounds claimed—which include hemoglobin—and we reject Helena's argument based on the reverse doctrine of equivalents."[169]

In *Scripps Clinic & Research Foundation* v. *Genentech, Inc.*,[170] Genentech argued that its recombinant Factor VIII:C did not infringe Scripps' patent to Factor VIII:C by application of the reverse doctrine of equivalents. The Federal Circuit ruled that application of the doctrine of reverse equivalents "requires that facts specific to the accused device be determined and weighed against the equitable scope of the claims, which in turn is determined in light of the specification, the prosecution history, and the prior art."[171]

Genentech argued that its recombinant Factor VIII:C should be viewed as being changed in principle because the specific activities and purity that could be obtained by the recombinant technology exceeded those made available by Scripps' process. Scripps, however, disputed Genentech's assertions. In view of the questions of scientific and evidentiary fact that were in dispute and that were material to the issue of infringement, the Federal Circuit decided that a grant of summary judgment of infringement of Scripps' product claims was inappropriate and sent the case back for a trial on the issue.

IV. EXEMPTIONS FROM LIABILITY

A. The Common Law Experimentation Exemption

Although not written in the statute, there has been recognized historically a narrow exemption from liability for patent infringement for experimentation with a patented invention. In essence, experimentation with a patented invention in order to understand the principles upon which it works or for purely philosophical purposes or for amusement does not create liability for patent infringement.[172]

As the Federal Circuit observed in *Roche*, however, this is an exceedingly narrow exemption. It does *not*, as many academics and R & D Managers believe, exempt the experimentor from liability for all "research" uses of a patented invention. Specifically, once the experimentation turns to adapting the invention for use in the experimentor's business, the line has been crossed and liability for infringement may result.[173]

B. Statutory Exemption from Infringement

Under 35 U.S.C. § 271(e)(1), it is not an act of infringement "to make, use, or sell a patented invention . . . solely for uses reasonably related to the development and submission of information under a Federal law which regulates the manufacture, use or sale of drugs or veterinary biological products.''[174] This provision of the statute was enacted to overcome the Federal Circuit's decision in *Roche Products, Inc.* v. *Bolar Pharmaceutical Co.*[175] In *Roche*, the Federal Circuit determined that the manufacture, use, or sale of a patented invention during the term of a patent constituted an act of infringement even if it was for the sole purpose of conducting tests and developing information necessary to obtain regulatory approval. Congress enacted § 271(e)(1) under the Drug Price Competition and Patent Term Restoration Act of 1984, Pub. L. 98–417, 98 stat. 1585, to allow manufacturers of generic drugs to be positioned to enter the market immediately upon the expiration of the patent covering the drug.

The U.S. Supreme Court, in *Eli Lilly & Co.* v. *Medtronic, Inc.*,[176] affirmed the decision of the Federal Circuit holding that the protection accorded by this section of the statute extends to certain medical devices that are regulated by the same federal law.

Courts have varied in their interpretation of § 271(e)(1) in regard to the phrase ''solely for uses reasonably related to the development and submission of information . . . .'' In *Scripps Clinic & Research Foundation* v. *Genentech, Inc.*,[177] Genentech argued that its use of Factor VIII:C was exempted from infringement under 35 U.S.C. § 271(e)(1) because, ''though not solely for purposes related to FDA testing, [they] bear some reasonable relationship to such purposes'' (666 F. Supp. at 1396). The district court, however, disagreed on the ground that this interpretation of the statute, eliminating the express statutory limitation ''solely for'' would vitiate § 271(e)(1). The court found that because Genentech's use of Factor VIII:C was not solely related to the purpose of obtaining FDA approval, its sales and uses of Factor VIII:C ''serving multiple purposes unrelated to meeting FDA requirements, clearly lie beyond the protection of § 271(e)(1).''[178]

In this regard, Genentech's uses of Factor VIII:C included uses related to the preparation of Genentech's application for a European patent and performance of the Genentech-Cutter agreement to develop a process for manufacturing Factor VIII:C on a commercial scale. Moreover, Genentech received substantial sums from its business partner, Cutter, in exchange for supplying Cutter with Factor VIII:C and for an agreement to share proprietary rights with Cutter. Furthermore, the agreement contemplated the marketing of recombinant Factor VIII:C outside the United States before the expiration of the Scripps patent (666 F. Supp. at 1396).

In *Amgen, Inc.* v. *Chugai Pharmaceutical Co.*,[179] the district court found no infringement by the defendant, Chugai, because Chugai had demonstrated that its shipment of two DNA samples to its business partner, GI, was for the purpose of conducting stability testing that was ''solely for uses reasonably related to the development and submission of information,'' under reporting requirements of the federal drug laws, and that GI's transfer of a small quantity of host cells transfected with a cDNA sequence encoding EPO to Chugai was de minimus and did not constitute infringement.

The evidence showed that the FDA requires DNA stability data in connection with the regulatory approval process. Chugai sent two samples of DNA to GI, requesting GI to clone the EPO gene and perform DNA sequence analysis. Evidence further showed that the two DNA samples sent were incapable of producing EPO because the cloning vectors used would not permit expression of EPO. Evidence also showed GI made no use of the information generated as a result of the stability testing other than to send it to Chugai. Evidence also showed that the stability data were submitted to the FDA. Based on Chugai's evidence that the sole purpose of stability testing was to obtain FDA approval, and based on Amgen's lack of evidence to the contrary, the district court found no infringement under the circumstances.

In a recent decision from the Northern District of California, *Intermedics, Inc.* v. *Ventritex, Inc.*,[180] the court granted protection of the exemption under § 271(e)(1) to the allegedly infringing manufacturer even though the manufacturer "intends to commercialize the device "before" the expiration of the allegedly infringed patent. The court reasoned that in enacting Section 271(e)(1), Congress wanted "to create a legal environment that would enable new, medically beneficial, cost-competitive products to reach the general marketplace in meaningful volume just as soon as the undistorted operation of the patent laws would permit (i.e., as soon as the 17-year life of relevant patents expired)."[181]

The court ruled that the availability of the exemption turns on actual uses, not on the "purposes," of the party doing the using.[182] Thus, the court would not rely on defendant's intent as a basis for denying defendant entitlement to § 271(e)(1) defense.

The patentee further argued that the infringer had engaged in a variety of activities unrelated to obtaining FDA approval, which activities allegedly demonstrated that the use of the accused product was not solely for uses reasonably related to FDA approval. The court, however, interpreted § 271(e)(1) as being applicable to only those acts that would constitute infringement but for the availability of this exemption. Because only acts of manufacture, use, or sale of a patented invention constitute infringement but for this exemption, "[i]t is these kinds of acts, only, that must be 'solely for uses reasonably related' to generating data for submission to the FDA."[183] The court characterized this section of the statute as Congress telling the public[184]:

> You may commit acts of *infringement* only so long as *those* acts are solely for uses reasonably related to gaining FDA approval to market your product. If you engaged in infringing activities for other uses, the exemption will not protect you. But if you are engaged in *non-infringing* acts for other uses, you do not lose the benefits of this statutory amendment.

Thus, the exemption is not lost when an infringer has engaged in noninfringing acts whose uses fall outside the protection of § 271(e)(1).[185]

VII. REMEDIES FOR PATENT INFRINGEMENT

The remedies for infringement are provided by statute in 35 U.S.C. § 281, which provides that "[a] patentee shall have remedy by civil action for infringement of his patent." The remedies available include an injunction, as provided by 35 U.S.C. § 283,[186] damages, as provided by 35 U.S.C. § 284,[187] and attorney's fees, as provided by 35 U.S.C. § 285.[188]

A. Preliminary Injunction

A court is authorized by statute to grant injunctions to prevent an alleged infringer from making, selling, and using the claimed invention. The grant or denial of an injunction resides within the discretion of the court, based upon a balancing of equities.[189]

In order to obtain a *preliminary* injunction pursuant to 35 U.S.C. § 283, i.e., an injunction to prevent further infringement during the pendency of the lawsuit, a party must establish a right thereto in view of four factors: (1) a reasonable likelihood of success on the merits; (2) irreparable harm if the injunction is not granted; (3) the balance of hardships not tipping in the infringer's favor; and (4) the impact of the injunction on the public interests.[190]

Traditionally, if a patent has been previously adjudicated as valid, the prior adjudication would be considered as evidence demonstrating a likelihood of success on the merits. Irreparable harm is often presumed where there is a strong showing of infringement and the validity of the patent.

B. Damages

Damages for patent infringement are provided by 35 U.S.C. § 284.[191] The purpose of section 284 is to ensure that "the patent owner would in fact receive full compensation for 'any damages' he suffered as a result of the infringement."[192]

If a patentee is able to prove the amount of profits lost as a result of the infringement, the entire amount of lost profits may be recoverable. In the event the amount of lost profits cannot be proved, then the patentee is entitled to at least a reasonable royalty for the use made of the invention by the infringer, together with interest and costs as fixed by the court.[193]

1. Lost Profits

In order to obtain lost profits, a patent owner must prove by a preponderance of the evidence that, but for the infringing activity, he or she would have made the sales made by the infringer or charged higher prices for its patented products, or both.[194] It is not necessary for the patent owner to prove causation as a certainty or "negative every possibility that the purchaser might not have bought another product other than his absent infringement."[195] A reasonable probability that the patent owner would have made the sales the infringer made suffices.[196] In determining profits, fixed costs are excluded.[197]

Moreover, a patentee is not required to prove the exact amount of damages and a trial court is permitted to approximate, if necessary, the amount to which the patent owner is entitled.[198] In particular, when damage in fact is clear, any doubts regarding the amount must be resolved against the infringer.[199] This is particularly so when the accused infringer destroyed evidence of the extent of its wrongdoing.[200]

Thus, the question of whether a patentee is entitled to lost profits is not based upon a subjective case-by-case analysis. Rather, an objective standard of "reasonable probability" serves the purpose.[201] A reasonable probability that the patentee would have made the sale but for the infringing sale can be inferred in two instances. First, when the patent owner has proved: (1) a demand for the patented product during the period of infringing sales; (2) an absence of acceptable noninfringing substitutes; (3) the patent owner had the marketing and manufacturing capability to meet the demand for the products covered by the patent; and (4) the amount of profit the patent owner would have made but for the infringing sales.[202]

Second, in a two-supplier market where the patentee and the infringer are the only suppliers in the market, one may infer that the patentee would have made the infringer's sales or charged higher prices but for the infringing competition.[203] Where there are more than two suppliers, proof that the other competitors, other than the accused infringer, did not sell the patented invention entitles the patent owner to lost profits. Moreover, the patent owner may introduce proof of its market share and base its lost profits on that market share, leaving the balance of the infringer's sales to a reasonable royalty determination.[204]

When the inference of "reasonable probability" is made out under either of these tests, then "the inference approaches conclusiveness."[205] The burden then shifts to the infringer to show that it is unreasonable to infer that the patentee would have made all of the infringing sales but for the infringement.

Definition of "Acceptable Noninfringing Substitutes" In considering what is deemed an "acceptable noninfringing substitute," the Federal Circuit has held that[206]:

> The alleged acceptable noninfringing substitute must not have a disparately higher price than or possess characteristics significantly different from the patented product.

Moreover, ''[i]f purchasers are motivated to purchase because of particular features of a product available only from the patent owner and infringers, products without such features would obviously not be *acceptable* noninfringing substitutes.''[207]

In this regard also, ''[a] technology that was immature and more expensive than the patented technology during the time of infringement cannot have been an acceptable noninfringing substitute.''[208] Evidence that a patented product filled a long-felt need of its users and enjoyed commercial success also indicates a lack of an acceptable noninfringing substitute.[209]

b. Lost Profits May Include Those Resulting from Sales of Unpatented Products Under the ''entire market value rule,'' a patent owner may recover the lost profits resulting from sales of unpatented components, if he or she can show that he or she could reasonably anticipate the sale of the unpatented components together with sale of the patented components.[210]

Under *Lam, Inc.* v. *Johns-Manville Corp.*,[211] lost profits as a result of price erosion is compensable. In such instances, a patent owner is forced to reduce the price of its patented products in response to the infringer's competition.[212] At the same time, a patent holder who proves that he or she would have increased prices had the infringer not been in competition with the patentee, can also collect damages under a price erosion claim.[213]

2. Reasonable Royalty

A reasonable royalty is an amount ''which a person, desiring to manufacture and sell a patented article, as a business proposition, would be willing to pay as a royalty and yet be able to make and sell the patented article, in the market, at a reasonable profit.''[214] The courts have determined that the amount of reasonable royalty after infringement cannot in all fairness be the same as a reasonable royalty before infringement. Otherwise, the infringer would have created a ''heads-I-win, tails-you-lose'' situation.

In *Georgia-Pacific Corp.* v. *U.S. Plywood Corp.*,[215] the court set forth certain factors a willing buyer and a willing seller would likely consider in determining what a reasonable royalty rate should be. These include[216]:

> [the parties'] relative bargaining strengths; the anticipated amount of profits that the prospective licensor reasonably thinks he would lose as a result of licensing the patent as compared to the anticipated royalty income; the anticipated amount of net profits that the prospective licensee reasonably thinks he will make; the commercial past performance of the invention in terms of public acceptance and profits; the market to be tapped; and any other economic factor that normally prudent businessmen would, under similar circumstances, take into consideration in negotiating the hypothetical license.

Hence, the reasonable royalty rate does not reflect the infringer's actual profits but rather the parties' expectations and bargaining positions at the time of the first infringement.[217]

In *SmithKline Diagnostics, Inc.* v. *Helena Laboratories Corp.*,[218] the Federal Circuit affirmed the district court's determination of a reasonable royalty rate of 25% of Helena's total sales of infringing goods. SmithKline had urged a royalty rate of 48% and Helena had urged a rate of 3%. The Federal Circuit held that "[a] court is not restricted in finding a reasonable royalty to a specific figure put forth by one of the parties."[219]

C. Personal Liability of Corporate Officers

Corporate officers and directors can be held personally liable for patent infringement if they participate in, induce, and approve acts of patent infringement by a corporation.[220]

D. Prejudgment Interest

The purpose of an award of prejudgment interest is to ensure that the patent owner is placed in as good a position as he or she would have been in had the infringer not infringed.[221] Thus, the patent owner is not only entitled to damages but also to "foregone use of the money between the time of infringement and the date of judgment."[222] In determining prejudgment interest, the court has discretion in deciding the rate of interest and whether the interest should be compounded.[223]

E. Attorney's Fees

The recovery of attorney's fees by the prevailing party in exceptional cases is authorized by statute.[224] Whether a case can be considered exceptional must, in the first instance, be determined by the district court, bearing in mind that the purpose of section 285 is to prevent gross injustice. Examples of exceptional cases are instances in which there is willful infringement, misconduct during litigation, vexatious or unjustified litigation, or a frivolous suit. In other words, there must be some finding of unfairness, bad faith, or inequitable conduct on the part of the unsuccessful party.[225]

F. Increased Damages

Increased damages of up to three times the amount found or assessed is authorized by statute. 35 U.S.C. § 284 provides in part:

> Upon finding for the claimant the court shall award the claimant damages adequate to compensate for the infringement . . .

> When the damages are not found by a jury, the court shall assess them. In either event the court may increase the damages up to three times the amount found or assessed.

If infringement is willful, increased damages may be awarded at the discretion of the trial judge.[226] One of the purposes of allowing increased damages is to deter willful patent infringement by punishing the willful infringer.

It is well-settled law that where a potential infringer has actual notice of another's patent rights, he or she has an affirmative duty of due care. That duty of due care may require that the potential infringer, who has knowledge of another's patent covering the infringing product, obtain competent legal advice of counsel before infringing or continuing to infringe.[227] Where the infringer does not have prior knowledge of another's patent covering the infringing device, the infringer cannot be held liable for willfulness for not seeking legal advice before commencement of infringement.[228]

Failure to obtain legal advice is not, however, necessarily determinative of willful infringement. However, such failure constitutes one of the factors supporting a finding of willfulness. The test is whether, under all circumstances, a reasonable person would prudently conduct himself or herself with any confidence that a court might find the patent invalid or not infringed.[229]

VIII. CLOSING WORDS OF ADVICE

From the foregoing, it is hoped that the reader has perceived that patents can be extremely valuable assets. The reader will, undoubtedly, also have perceived that the line between infringing and noninfringing conduct can be exceedingly murky. A determination of infringement or noninfringement often cannot be made simply by reading the patent claims. In many cases, the prosecution history of the patent must be reviewed before firm judgments can be made about what the patent claims mean and what is likely to constitute an infringement under the doctrine of equivalents.

There are potentially dramatic penalties in the event you are perceived to be a willful infringer. Damage awards of tens of millions of dollars are not uncommon in patent infringement litigation. Attorney fees in excess of a million dollars are not uncommon. An assessment of attorney fees or increased damages for willful infringement can constitute a huge economic penalty.

While reliance on advice of counsel is a legitimate defense to a charge of willfulness, such advice is often not obtained in a timely manner. In most cases, the question of willfulness will be determined as of the time infringement began. In view of the very narrow exemption for experimental uses of patented inventions, infringement may be found to have occurred at a much earlier stage of the development process than most scientists and managers might think. Failure to obtain infringement clearance at an early stage of the development process may, therefore, have dramatic adverse economic consequences.

What can you do to minimize your exposure? Make an effort to know the patent literature relevant to your field of endeavor, get legal advice regarding potentially infringed patents, and get that advice early.

NOTES

1. 447 U.S. 303, 65 L. Ed. 2d. 144, 100 S. Ct. 2204, 206 U.S.P.Q. 193 (1980). Note, however, that at n.9, the Supreme Court commented in passing that as early as 1873, a patent had been granted to Louis Pasteur on a yeast that was "free from organic germs of disease."

   In fact, the Court Appeals for the Federal Circuit had ruled, in 1977, that a claim to a biologically pure culture of a microorganism was not unpatentable and was not excludable from 35 U.S.C. § 101 simply on the ground that the microorganism was a living thing. *In re* Bergy, 563 F.2d 1031, 195 U.S.P.Q. 344 (C.C.P.A. 1977).
2. 35 U.S.C. § 101 provides:

   > Inventions patentable
   >
   > Whoever invents or discovers any new and useful process, machine, manufacture, or composition of matter, or any new and useful improvement thereof, may obtain a patent therefor, subject to the conditions and requirements of this title.

3. See *Ex parte* Hibberd, 227 U.S.P.Q. 443 (Bd. Pat. App. Int. 1985).
4. *In re* Allen, 2 U.S.P.Q.2d 1425 (Bd. Pat. App. Int. 1987).
5. U.S. Patent No. 4,736,866, issued April 12, 1988, to Harvard University as assignee. In actuality, this patent claims a transgenic nonhuman mammal, not limited to a mouse.
6. 35 U.S.C. § 271 provides in part:

   > (a) Except as otherwise provided in this title, whoever without authority makes, uses or sells any patented invention within the United States during the term of the patent therefor, infringes the patent.

7. 35 U.S.C. § 101.
8. 35 U.S.C. § 161.
9. 35 U.S.C. § 171.
10. 35 U.S.C. § 101.
11. 35 U.S.C. § 163. Note that the Plant Variety Protection Act, 7 U.S.C. § 2321 *et seq.*, provides some protection, albeit not a patent right, for sexually reproduced plant varieties.
12. 35 U.S.C. § 261.
13. 35 U.S.C. § 283.

14. 35 U.S.C. § 284.
15. 35 U.S.C. § 112, first paragraph provides:

> The specification shall contain a written description of the invention, and of the manner and process of making and using it, in such full, clear, concise, and exact terms as to enable any person skilled in the art to which it pertains, or with which it is most nearly connected, to make and use the same, and shall set forth the best mode contemplated by the inventor of carrying out his invention.

16. See 35 U.S.C. § 102.
17. See 35 U.S.C. § 103.
18. 868 F.2d 1251, 1256 (Fed. Cir. 1989).
19. Claim 1 of one of the patents-in-suit, U.S. Patent No. 3,659,915 recites:

> An optical waveguide comprising
> (a) A cladding layer formed of a material selected from the group consisting of pure fused silica and fused silica to which a dopant material on at least an elemental basis has been added, and
> (b) a core formed of used silica to which a dopant material on at least an elemental basis has been added to a degree in excess of that of the cladding layer so that the index of refraction thereof is of a value greater than the index of refraction of said cladding layer, said core being formed of at least 85 percent by weight of fused silica and an effective amount up to 15 percent by weight of said dopant material.

20. Prior art discloses "a substantially transparent luminescent glass in the form of a fiber comprised of a doped silica core . . . having a sheath of silica." Corning Glass Works v. Sumitomo Elec. U.S.A., Inc., 868 F.2d 1251, 1256, 9 U.S.P.Q.2d 1962 (Fed. Cir. 1989).
21. *Idem*, 868 F.2d at 1257.
22. 187 F.2d 150, 88 U.S.P.Q. 478 (C.C.P.A. 1951).
23. *Idem*, 187 F.2d at 152, 88 U.S.P.Q. at 480.
24. *Idem*, 187 F.2d at 152, 88 U.S.P.Q. at 480–81.
25. *Idem*, 187 F.2d at 152, 88 U.S.P.Q. at 481.
26. *Idem*.
27. Corning Glass Works v. Sumitomo Elec. U.S.A., Inc., 868 F.2d 1251, 1257 (Fed. Cir. 1989).
28. *Idem*.
29. U.S. Patent No. 4,703,008, issued on October 27, 1987, to inventor Fu-Kuen Lin, and assigned to Kiren-Amgen, Inc.
30. See, *e.g.*, *In re* Davis, 80 U.S.P.Q. 448, 450 (Bd. App. 1948); Water Technologies Corp. v. Calco Ltd., 850 F.2d 660, 7 U.S.P.Q.2d 1097 (Fed. Cir.) *cert. denied*, 488 U.S. 968, 109 S. Ct. 498 (1988).
31. 904 F.2d 1558, 15 U.S.P.Q.2d 1039 (Fed. Cir. 1990), *cert. dismissed*, 111 S. Ct. 1434, 113 L. Ed.2d 485 (1991).
32. U.S. Patent No. 4,468,464, issued on August 28, 1984, to inventors Cohen & Boyer, and assigned to Stanford University.

33. *Ex parte* Davis, 80 U.S.P.Q. 448 (Bd. App. 1948).
34. 717 F.2d 128, 219 U.S.P.Q. 953 (4th Cir. 1983).
35. The disputed claim reads as follows (emphasis added):

    2. A highly workable nickel base alloy *consisting of* from about 18 to about 20 percent of chromium, from about 11 to about 13 percent of cobalt, from about 5.5 to about 7.0 percent molybdenum, from about 1.9 to about 2.2 percent of aluminum, from about 2.9 to about 3.2 percent titanium, from about 0.9 to about 1.2 percent of tungsten, and the balance of nickel except for inconsequential amounts of impurities.

36. Special Metals Corp. v. Teledyne Industries, Inc., 219 U.S.P.Q. at 955.
37. U.S. Patent No. 4,531,901, issued on July 23, 1985, to inventor Weissmann and assigned to Biogen N.V.
38. See, e.g., Water Technologies Corp. v. Calco Ltd., 850 F.2d 660, 7 U.S.P.Q.2d 1097, 1102, (Fed. Cir.), *cert. denied*, 488 U.S. 968, 109 S.Ct. 498 (1988).
39. 80 U.S.P.Q. 448 (Bd. App. 1948).
40. *Idem*, at 448.
41. *Idem*, at 449.
42. *Idem*, See also *In re* Garnero, 412 F.2d 276, 162 U.S.P.Q. 221 (C.C.P.A. 1969).
43. Issued on October 27, 1987, to inventor Fu-Kuen Lin and assignee Kiren-Amgen, Inc.
44. 14 U.S.P.Q.2d 1363, 1367 (D. Del. 1990).
45. 12 U.S.P.Q.2d 1061 (Bd. Pat. App. Int. 1989).
46. 35 U.S.C. § 112, second paragraph.
47. 35 U.S.C. § 101.
48. In Amgen, Inc. v. Chugai Pharmaceutical Co., 927 F.2d 1200, 18 U.S.P.Q.2d 1026 (Fed. Cir. 1991), however, the Federal Circuit determined that the limitation "at least about 160,000 IU per absorbance unit" was not enabled.
49. U.S. Patent No. 4,752,603, one of the patents-in-suit in Genentech Inc. v. Wellcome Foundation Ltd., 14 U.S.P.Q.2d 1363 (D. Del. 1990).
50. 887 F.2d 1050, 1052, 12 U.S.P.Q.2d 1474 (Fed. Cir. 1989).
51. U.S. Patent No. 4,530,831 owned by Intervet America, Inc.
52. 706 F. Supp. 94, 103 (D. Ma. 1989).
53. *Idem*, at 104.
54. Amgen, Inc. v. Chugai Pharmaceutical Co., 927 F.2d 1200, 18 U.S.P.Q.2d 1015 (Fed. Cir. 1991).
55. 666 F. Supp. 1379, 3 U.S.P.Q.2d 1481 (N.D. Cal. 1987), *modified in part* on other grounds, 6 U.S.P.Q.2d 1018 (N.D. Cal. 1988).
56. 927 F.2d 1565, 18 U.S.P.Q.2d 1001 (Fed. Cir. 1991).
57. Issued on June 30, 1987, to inventors Hewick and Sechra, and assigned to Genetics Institute, Inc.
58. 858 F.2d. 731, 8 U.S.P.Q.2d 1400 (Fed. Cir. 1988).
59. 666 F. Supp. 1379, 1385, 3 U.S.P.Q.2d 1481 (N.D. Cal. 1987), *modified on reconsideration*, 678 F. Supp. 1429, 6 U.S.P.Q.2d 1018 (N.D. Cal. 1988).
60. 706 F. Supp. 94, 107, 9 U.S.P.Q.2d 1833 (D. Ma. 1989).

61. See, e.g., Scripps Clinic and Research Foundation v. Genentech, Inc., 666 F. Supp. 1379, 1386–87 (N.D. Cal. 1987) *aff'vd in part*, *rev'd in part* and *vacated in part* on other grounds, 927 F.2d 1565, 18 U.S.P.Q.2d 1001 (Fed. Cir. 1991).
62. 927 F.2d at 1583.
63. One of the product-by-process claims in issue in Scripps Clinic and Research Foundation v. Genentech, Inc., 927 F.2d 1565, 18 U.S.P.Q.2d 1001 (Fed. Cir. 1991), read as follows:

> 13. Highly purified and concentrated human or porcine VIII:C prepared in accordance with the method of claim 1.

63b. 23 U.S.P.Q.2d 1481, 1491 (Fed. Cir.), *reh'g-denied*, 23 U.S.P.Q.2d 1481 (Fed. Cir. 1992).
64. U.S. Patent No. 4,376,110 entitled "Immunometric Assays Using Monoclonal Antibodies," issued to Gary S. David and Howard E. Greene and assigned to Hybritech Incorporated, as discussed in Hybritech Inc. v. Monoclonal Antibodies, Inc., 802 F.2d 1367, 231 U.S.P.Q. 81, 83 (Fed. Cir. 1986), *cert. denied*, 480 U.S. 947, 107 S. Ct. 1606 (1987).
65. U.S. Patent No. 4,365,970 assigned to SmithKline Diagnostics, Inc., as discussed in SmithKline Diagnostics, Inc. v. Helena Laboratories Corp., 859 F.2d 878, 881, 8 U.S.P.Q.2d 1468, 1470 (Fed. Cir. 1988).
66. See, e.g., *In re* Fout, 675 F.2d 297, 213 U.S.P.Q. 532 (C.C.P.A. 1982).
67. 1925 C.D. 126, 340 O.G. 839.
68. *In re* Harnisch, 631 F.2d 716, 206 U.S.P.Q. 300, 305 (C.C.P.A. 1980).
69. U.S. Patent No. 4,703,008, issued on October 27, 1987, to inventor Fu-Kuen Lin and assignee Kiren-Amgen, Inc.
70. 35 U.S.C. § 112, 6th paragraph provides:

> An element in a claim for a combination may be expressed as a means or step for performing a specified function without the recital of structure, material, or acts in support thereof, and such claim shall be construed to cover the corresponding structure, material, or acts described in the specification or equivalents thereto.

71. 762 F.2d 969, 226 U.S.P.Q. 5 (Fed. Cir. 1985).
72. U.S. Patent No. 4,296,744.
73. Palumbo v. Don-Joy Co., 762 F.2d at 974.
74. 35 U.S.C. § 112, 3d paragraph.
75. *Idem.*
76. *Idem.*
77. U.S. Patent No. 4,632,901 to Valkirs et al., issued December 30, 1986, and assigned to Hybritech Incorporated.
78. See Graver Tank Mfg. Co. v. Linde Air Products Co., 339 U.S. 605, 608, 70 S. Ct. 854, 85 U.S.P.Q. 328, *reh'g denied*, 340 U.S. 845 (1950).
79. *Idem*, 85 U.S.P.Q. at 330.
80. SRI Int'l v. Matsushita Elec. Corp. of America, 775 F.2d 1107, 1120, 227 U.S.P.Q. 577, 584 (Fed. Cir. 1985) (emphasis in original).

81. Hormone Research Foundation, Inc. v. Genentech, Inc., 904 F.2d 1558, 15 U.S.P.Q.2d 1039 (Fed. Cir. 1990).
82. McGill Inc. v. John Zink Co., 736 F.2d 666, 672, 221 U.S.P.Q. 944 (Fed. Cir.) *cert. denied*, 469 U.S. 1037, 105 S. Ct. 514 (1984).
83. ZMI Corp. v. Cardiac Resuscitator Corp., 844 F.2d 1576, 1580, 6 U.S.P.Q.2d 1557 (Fed. Cir. 1988), *remanded to*, 11 U.S.P.Q. 1634 (D. Or. 1989), *aff'd*, 15 U.S.P.Q.2d 1398, 1990 W.L. 28066 (Fed. Cir. 1990).
84. Issued to inventors Stanley N. Cohen and Herbert W. Boyer on December 2, 1980, and assigned to Stanford University.
85. U.S. Patent No. 4,237,224, Col. 2, lines 4–10.
86. Hormone Research Foundation Inc. v. Genentech Inc., 904 F.2d 1558 at 1562, 15 U.S.P.Q.2d 1039 at 1043 (Fed. Cir. 1990), *cert. dismissed*, 111 S. Ct. 1434, 113 L. Ed.2d 485 (1991).
87. See, e.g., Control Components, Inc. v. Valtech, Inc., 609 F.2d 763, 770, 204 U.S.P.Q. 785, *reh'g denied*, 616 F.2d 892 (5th Cir. 1980), *cert. denied*, 449 U.S. 1022 (1980); to a subsequently filed patent (see, e.g., General Electric Co. v. United States, 572 F.2d 745, 760, 198 U.S.P.Q. 65 (Ct. Cl. 1978).
88. See, e.g., Autogiro Co. of America v. United States, 384 F.2d 391, 155 U.S.P.Q. 697 (Ct. Cl. 1967).
89. SmithKline Diagnostics, Inc. v. Helena Laboratories Corp., 859 F.2d 878, 882 (Fed. Cir. 1988).
90. Smith v. Snow, 294 U.S. 1, 14, 79 L. Ed. 721, 729 (1934).
91. Autogiro Co. of America v. United States, 384 F.2d 391, 155 U.S.P.Q. 697, 704 (Ct. Cl. 1967).
92. ZMI Corp. v. Cardiac Resuscitator Corp., 844 F.2d 1576, 6 U.S.P.Q.2d 1557 (Fed. Cir. 1988), *remanded*, 11 U.S.P.Q.2d 1634 (D. Or. 1989), *aff'd*, 1990 W.L. 28066 (Fed. Cir. 1990); Envirotech Corp. v. Al George, Inc., 730 F.2d 753, 759, 221 U.S.P.Q. 473, 477 (Fed. Cir. 1984); General Electric Co. v. United States, 572 F.2d 745, 753, 198 U.S.P.Q. 65, 71 (Ct. Cl. 1978).
93. ZMI Corp. v. Cardiac Resuscitator Corp., 844 F.2d at 1579.
94. Moeller v. Ionetics, Inc., 794 F.2d 653, 656, 229 U.S.P.Q. 992, 994 (Fed. Cir. 1986).
95. 119 U.S. 47 (1886).
96. 849 F.2d 1430, 1433, 7 U.S.P.Q.2d 1129 (Fed. Cir.) *cert. denied*, 488 U.S. 986, 109 S. Ct. 542 (1988).
97. 802 F.2d 1367, 231 U.S.P.Q. 81 (Fed. Cir. 1986), *cert. denied*, 480 U.S. 947, 107 S. Ct. 1606, 94 L. Ed.2d 792 (1987).
98. *Idem*, 231 U.S.P.Q. at 94.
99. *Idem*.
100. Hybritech Incorporated v. Abbott Laboratories, 4 U.S.P.Q.2d 1001 (C.D. Cal. 1987), *aff'd*, 849 F.2d 1446 (Fed. Cir. 1988).
101. 14 U.S.P.Q.2d 1363, 1367–68 (D. Del. 1990).
102. Genentech Inc. v. Wellcome Foundation, Inc., 14 U.S.P.Q.2d 1363, 1367–68 (D. Del. 1990), referring to the '075 patent at column 5, lines 13–18 and 21–27.

103. *Idem*, at 1368–1370.
104. Autogiro Co. of America v. United States, 384 F.2d 391, 155 U.S.P.Q. 697, 708 (Ct. Cl. 1967).
105. *Idem*.
106. *Idem*.
107. 666 F. Supp. 1379 (N.D. Cal. 1987).
108. Claims 24–29 of the U.S. Reissue Patent No. RE 32011, issued on October 22, 1985, recite as follows:

> 24. A human VIII:C preparation having a potency in the range of 184 to 1172 units per ml. and being substantially free of VII:RP.
>
> 25. A human VIII:C preparation of claim 24, wherein the VIII:C concentration is at least 160,000 fold purified relative to VIII:C in plasma.
>
> 26. A human VIII:C preparation of claim 24, wherein the ratio of VIII:C to VIII:RP is greater than 100,000 times the ratio of plasma.
>
> 27. A human VIII:C preparation of claim 24, wherein said VIII:C is isolated from VIII:C/VIII:RP and 90–100 percent of the VIII:RP has been removed.
>
> 28. A human VIII:C preparation having a specific activity greater than 2240 units/mg.
>
> 29. A human VIII:C preparation of claim 28 wherein the potency is in the range of 184 to 1172 units/ml.

109. Scripps Clinic and Research Foundation v. Genentech Inc., 666 F. Supp. 1379 at 1389.
110. Graham v. John Deere Co., 383 U.S. 1, 33, 86 S. Ct. 684, 148 U.S.P.Q. 459 (1966).
111. 706 F. Supp. 94, 9 U.S.P.Q.2d 1833 (D. Ma. 1989).
112. United States Patent No. 4,703,008, entitled "DNA Sequences Encoding Erythropoietin," issued on October 27, 1987, and assigned to Kiren-Amgen, Inc. ("the '008 patent").
113. *Idem*, at 108.
114. *Idem*.
115. A representative process claim surrendered is as follows:

> 69. A process for the production of a polypeptide having part or all of the primary structural conformation and one or more of the biological activities of naturally-occurring erythropoietin, said process comprising:
>
> growing, under suitable nutrient conditions, procaryotic or eucaryotic host cells transformed or transfected with a DNA vector according to claim 62, and isolating desired polypeptide products of the expression of DNA sequences in said vector.

116. Autogiro Co. v. United States, 384 F.2d 391, 397, 155 U.S.P.Q. 697, 702 (Ct. Cl. 1967); Fromson v. Advance Offset Plate, Inc., 219 U.S.P.Q. 1137, 1142 (Fed. Cir. 1983).

117. 927 F.2d 1200, 18 U.S.P.Q.2d 1016 (Fed. Cir.) *cert. denied*, 1991 U.S. Lexis 4341 (1991).
118. Entitled "Method for the Purification of Erythropoietin and Erythropoietin Compositions," issued to Dr. Rodney Hewick and assigned to Genetics Institute, Inc. ("GI").
119. Amgen, 927 F.2d at 1216, 18 U.S.P.Q.2d at 1029.
120. 887 F.2d 1070, 12 U.S.P.Q.2d 1539 (Fed. Cir. 1989).
121. 859 F.2d 878 (Fed. Cir. 1988).
122. *Idem.* at 880.
123. *Idem.*
124. *Idem.*
125. Claim 1 of the '970 patent, asserted to be infringed, reads as follows:

    In an occult blood specimen test slide having a front panel, a rear panel, said front panel having one or more openings, sheet means carrying a test reagent between the front and rear panels underlying each of said openings, a hinged cover adapted to overlie a portion of the front panel and said openings and flap means in the rear panel opposite said openings and pivotable to expose the underside of the sheet, the improvement comprising: an area positioned on a portion of the sheet means facing the rear panel and isolated from the openings in the front panel, said area including a positive and negative monitor, said positive and negative monitors including the test reagent and said positive monitor additionally including a compound that reacts to environmental conditions in a manner similar to hemoglobin.

126. 859 F.2d at 883.
127. *Idem.*
128. *Idem.*
129. *Idem*, at 884.
130. *Idem*, at 885.
131. 904 F.2d 1558, 15 U.S.P.Q.2d 1039 (Fed. Cir. 1990).
132. *Idem*, 904 F.2d at 1560, 15 U.S.P.Q.2d at 1040.
133. These differences were:

    (1) HGH has one additional amino acid in the chain. (Specifically, glutamine is included at a position after the 68th amino acid.); and
    (2) HGH has slightly different amino acids in positions 73, 106, and 108. (Specifically, there is glutamic acid instead of glutamine at position 73, aspartic acid instead of asparagine at 106, and asparagine instead of aspartic acid at 108.)

134. *Idem*, 15 U.S.P.Q.2d at 1041.
135. Examples of claims asserted to be infringed and containing the "corresponding to" language are (with emphasis added):

    A method of producing a substance having growth-promoting activity that comprises:

a. forming an unbridged polypeptide chain or amino acid residues *in a sequence corresponding to Fig. 2 or 3* of the accompanying drawing;
b. generating sulfhydryl groups on the cysteine residues in said polypeptide chain;
c. oxidizing said sulfhydryl groups under conditions effective to form disulfide bridges between said systeine residues, thereby forming two intramolecular rings in the polypeptide chain.

3. A method of producing a substance having growth-promoting activity which comprises:
a. Forming an unbridged polypeptide chain of amino acid residues *in a sequence corresponding to (i) the portion of the sequence of Fig. 2* of the drawing from positions 86 to 190 or (ii) the said portion (i) combined with any fraction of the remaining portion from position 85 to position 1;
b. generating sulfhydryl groups on the cysteine residues in said polypeptide chain;
c. oxidizing said sulfhydryl groups under conditions effective to form a disulfide bridge between said cysteine residues across positions 181 and 188; thereby forming an intra-molecular ring in the polypeptide chain.

12. A composition of matter consisting essentially of a synthetic, biologically active substance which has a structure corresponding to Fig. 2 of the accompanying drawing.

136. Hormone Research, 904 F.2d at 1563, 15 U.S.P.Q.2d at 1043.
137. *Idem.*
138. *Idem.*
139. *Idem.*
140. *Idem.*
141. *Idem*, 904 F.2d at 1563–64, 15 U.S.P.Q.2d at 1044.
142. Becton Dickinson & Co. v. C.R. Bard, Inc., 922 F.2d 792, 17 U.S.P.Q.2d 1097 (Fed. Cir. 1990).
143. 339 U.S. 605, 70 S. Ct. 854, 85 U.S.P.Q. 328, *reh'g denied*, 340 U.S. 845 (1950).
144. *Idem*, 85 U.S.P.Q. at 330.
145. Autogiro Co. of America v. United States, 384 F.2d 391, 155 U.S.P.Q. 697 (Ct. Cl. 1967).
146. 877 F.2d 1561, 11 U.S.P.Q.2d 1132 (Fed. Cir. 1989).
147. *Idem*, 11 U.S.P.Q.2d at 1136.
148. 904 F.2d 677, 684, 14 U.S.P.Q.2d 1942, 1948 (Fed. Cir.), *cert. denied*, 111 S. Ct. 537, 112 L. Ed.2d 547 (1990).
149. Genentech Inc. v. Wellcome Foundation Ltd., 14 U.S.P.Q.2d 1363, 1371 (D. Del. 1990).
150. See, e.g., *Graver Tank.*
151. 384 F.2d 391, 155 U.S.P.Q. 697, 705 (Ct. Cl. 1967).
152. See, e.g., Pennwalt Corp. v. Durand - Wayland Inc., 833 F.2d 931, 4 U.S.P.Q.2d 1737 (Fed. Cir. 1987).

153. Corning Glass Works v. Sumitomo Elec. U.S.A., Inc., 868 F.2d 1251, 1259, 9 U.S.P.Q.2d 1962 (Fed. Cir. 1989).
154. See, e.g., Sun Studs, Inc. v. ATA Equipment Leasing, Inc., 872 F.2d 978, 989, 10 U.S.P.Q.2d 1338, *vacated in part*, 11 U.S.P.Q.2d 1479, *reh'g granted*, 882 F.2d 1583, *reinstated*, *reh'g denied*, 892 F.2d 73 (Fed. Cir. 1989).
155. Nestier Corp. v. Menasha Corp.-Lewisystems Division, 739 F.2d 1576, 222 U.S.P.Q. 747 (Fed. Cir. 1984), *cert. denied*, 470 U.S. 1053, 105 S. Ct. 1756 (1985).
156. *Idem*, 222 U.S.P.Q. at 749.
157. 14 U.S.P.Q.2d 1363, 1370–71 (D. Del. 1990).
158. United States Patent No. 4,766,075 (the '075 Patent) and United States Patent No. 4,752,603 (the '603 Patent).
159. Genentech Inc. v. Wellcome Foundation Ltd., 14 U.S.P.Q.2d 1363, 1370–71 (D. Del. 1990).
160. See, e.g., Wilson Sporting Goods v. David Geoffrey & Associates, 904 F.2d 672, 14 U.S.P.Q.2d 1942, 1948 (Fed. Cir. 1990).
161. *Idem*, 14 U.S.P.Q.2d at 1949.
162. Autogiro Co. of America v. United States, 384 F.2d 391, 155 U.S.P.Q. 697, 705 (Ct. Cl. 1967).
163. Hughes Aircraft Co. v. United States, 717 F.2d 1351, 219 U.S.P.Q. 473, 481 (Fed. Cir. 1983).
164. *Idem*, at 481.
165. Hormone Research, 904 F.2d at 1564, 15 U.S.P.Q. at 1044, citing Loctite Corp. v. Ultraseal Ltd., 781 F.2d 861, 871, 228 U.S.P.Q. 90, 96 (Fed. Cir. 1985).
166. Intervet America Inc. v. Kee-Vet Laboratories, Inc., 887 F.2d 1050, 1054, 12 U.S.P.Q.2d 1474, 1477 (Fed. Cir. 1989).
167. See, e.g., SRI Int'l v. Matsushita Electric Corp. of America, 775 F.2d 1107, 1123–24 (Fed. Cir. 1985); Texas Instruments, Inc. v. United States Int'l Trade Commission, 846 F.2d 1369, 1371, 1372, 6 U.S.P.Q.2d 1886 (Fed. Cir. 1988), *reh'g denied*, 7 U.S.P.Q.2d 1414 (1988).
168. 859 F.2d 878, 889 (Fed. Cir. 1988).
169. *Idem*, at 890.
170. 927 F.2d 1565, 18 U.S.P.Q.2d 1001 (Fed. Cir. 1991).
171. 927 F.2d at 1581, 18 U.S.P.Q.2d 1014.
172. See, e.g., Roche Products, Inc. v. Bolar Pharmaceutical Co., 733 F.2d 858, 221 U.S.P.Q. 937 (Fed. Cir. 1984), *cert. denied*, 469 U.S. 856, 105 S. Ct. 183, 83 L. Ed.2d 117 (1984).
173. *Idem*.
174. 35 U.S.C. § 271(e)(1) provides:

> It shall not be an act of infringement to make, use, or sell a patented invention (other than new animal drug or veterinary biological product) (as those terms are used in the Federal Food, Drug, and Cosmetic Act and the Act of March 4, 1913) which is primarily manufactured using recombinant DNA, recombinant RNA, hybridoma technology or other processes involving site specific genetic manipulation techniques) solely for uses reasonably related to the development

and submission of information under a Federal law which regulates the manufacture, use, or sale of drugs or veterinary biological products.

175. 733 F.2d 858, 221 U.S.P.Q. 937 (Fed. Cir. 1984), *cert. denied*, 469 U.S. 856, 105 S. Ct. 183, 83 L. Ed.2d 117 (1984).
176. 879 F.2d 849, 11 U.S.P.Q.2d 1649 (Fed. Cir. 1989), *aff'd* 110 S. Ct. 2683 (1990).
177. 666 F. Supp. 1379 (N.D. Cal. 1987).
178. *Idem*, at 1396.
179. 13 U.S.P.Q.2d 1737 (D. Mass. 1989), *aff'd in part, vacated in part*, 97 F.2d 1200 (Fed. Cir. 1991).
180. 20 U.S.P.Q.2d 1422 (N.D. Cal. 1991).
181. *Idem*, at 1424.
182. *Idem*, at 1429.
183. *Idem*, at 1427.
184. *Idem*, at 1427–28 (emphasis in original).
185. *Idem*, at 1428.
186. 35 U.S.C. § 283 provides:

> The several courts having jurisdiction of cases; under this title may grant injunctions in accordance with the principles of equity to prevent the violation of any right secured by patent, on such terms as the court deems reasonable.

187. 35 U.S.C. § 284 provides:

> Upon finding for the claimant the court shall award the claimant damages adequate to compensate for the infringement but in no event less than a reasonable royalty for the use made of the invention by the infringer, together with interest and costs as fixed by the court.
>
> When the damages are not found by a jury, the court shall assess them. In either event the court may increase the damages up to three times the amount found or assessed.
>
> The court may receive expert testimony as an aid to the determination of damages or of what royalty would be reasonable under the circumstances.

188. 35 U.S.C. § 285 provides: "The court in exceptional cases may award reasonable attorney's fees to the prevailing party."
189. Smith Int'l Inc. v. Hughes Tool Co., 718 F.2d 1573, 1578, 219 U.S.P.Q. 686, 690 (Fed. Cir.), *cert. denied*, 464 U.S. 996, 104 S. Ct. 493 (1983).
190. Hybritech Incorporated v. Abbott Laboratories, 849 F.2d 1446, 1451, 7 U.S.P.Q.2d 1191, 1195 (Fed. Cir. 1988).
191. 35 U.S.C. § 284 provides:

> Upon finding for the claimant the court shall award the claimant damages adequate to compensate for the infringement but in no event less than a reasonable royalty for the use made of the invention by the infringer, together with interest and costs as fixed by the court.
>
> When damages are not found by a jury, the court shall assess them. In either event the court may increase the damages up to three times the amount found or assessed.

The court may receive expert testimony as an aid to the determination of damages or of what royalty would be reasonable under the circumstances.

192. SmithKline Diagnostics, Inc. v. Helena Laboratories Corp., 926 F.2d 1161, 1164, 17 U.S.P.Q.2d 1922, 1924 (Fed. Cir. 1991), citing General Motors Corp. v. Devex Corp., 461 U.S. 648, 103 S. Ct. 2058, 654–55, 217 U.S.P.Q. 1185, 1188 (1983).
193. 35 U.S.C. § 284. See also Fromson v. Western Litho Plate & Supply Co., 853 F.2d 1568, 1574, 7 U.S.P.Q.2d 1606, 1612 (Fed. Cir. 1988).
194. Marsh-McBirney, Inc. v. Montedoro-Whitney Corp., 882 F.2d 498, 505, 11 U.S.P.Q.2d 1794 (Fed. Cir. 1989) *vacated*, 112 L. Ed.2d, 111 S. Ct. 775, *reinstated in part*, 939 F.2d 969 (Fed. Cir. 1991); see also Gyromat Corp. v. Champion Spark Plug Co., 735 F.2d 549, 554, 222 U.S.P.Q. 4, 8 (Fed. Cir. 1984); and Panduit Corp. v. Stahlin Bros. Fibre Works, Inc., 575 F.2d 1152, 1156, 197 U.S.P.Q. 726, 730 (6th Cir.), *cert. denied*, 439 U.S. 856 (1978).
195. Kaufman Co. v. Lantech, Inc., 926 F.2d 1136, 17 U.S.P.Q.2d 1828 (Fed. Cir. 1991).
196. See, e.g., Bio-Rad Laboratories, Inc. v. Nicolet Instrument Corp., 739 F.2d 604, 616 222 U.S.P.Q. 654, 663 (Fed. Cir.), *cert. denied*, 469 U.S. 1038 (1984).
197. Paper Converting Machine Co. v. Magna-Graphics Corp., 745 F.2d 11, 22, 223 U.S.P.Q. 591 (Fed. Cir. 1984).
198. Del Mar Avionics, Inc. v. Quinton Instruments Co., 836 F.2d 1320, 1327, 5 U.S.P.Q.2d 1255 (Fed. Cir. 1987) (citing Bio-Rad Laboratories, Inc. v. Nicolet Instrument Corp., 739 F.2d 604, 616, 222 U.S.P.Q. 654 (Fed. Cir. 1984); see also Yarway Corp. v. Eur-Control U.S.A., Inc., 775 F.2d 268, 275, 227 U.S.P.Q. 352 (Fed. Cir. 1985).
199. Ryco, Inc. v. Ag-Bag Corp., 857 F.2d 1418, 1427, 8 U.S.P.Q.2d 1323 (Fed. Cir. 1988).
200. Beatrice Foods Co. v. New England Printing and Lithographing Co., 899 F.2d 1171, 14 U.S.P.Q.2d 1020 (Fed. Cir. 1990).
201. Kaufman Co. v. Lantech, Inc., 926 F.2d 1136 at 1141, 17 U.S.P.Q.2d at 1828 at 1831.
202. Panduit Corp. v. Stahlin Bros. Fibre Works, Inc., 575 F.2d 1152, 1156, 197 U.S.P.Q. 726, 729–30 (6th Cir. 1978).
203. Lam, Inc. v. Johns-Manville Corp., 718 F.2d 1056, 1065, 219 U.S.P.Q. 670, 675 (Fed. Cir. 1983).
204. State Indus., Inc. v. Mor-Flo, 883 F.2d 1573, 12 U.S.P.Q.2d 1026 (Fed. Cir. 1989), *cert. denied*, 110 S. Ct. 725 (1990).
205. *Idem.*
206. Kaufman Co. v. Lantech, Inc., 926 F.2d at 1142, 17 U.S.P.Q. at 1832.
207. SmithKline Diagnostics, Inc. v. Helena Laboratories Corp., 926 F.2d at 1166, 17 U.S.P.Q.2d at 1926 (Fed. Cir. 1991). See also Kaufman, 926 F.2d at 1142–43, 17 U.S.P.Q.2d at 1833 ("[a] product that is lacking the advantages of that patented can hardly be termed a substitute acceptable to the customer who wants those advantages.")
208. *Kaufman*, 926 F.2d at 1142, 17 U.S.P.Q.2d at 1832.

209. Panduit Corp. v. Stahlin Bros. Fibre Works, Inc., 575 F.2d 1152, 1162, 197 U.S.P.Q. 726, 734 (6th Cir. 1978).
210. See Paper Converting Machine Co. v. Magna-Graphics Corp., 745 F.2d 11, 23, 223 U.S.P.Q. 591, 599 (Fed. Cir. 1984); Cori Corp. v. Wilco Marsh Buggies & Draglines, Inc., 761 F.2d 649, 656, 225 U.S.P.Q. 985, 989 (Fed. Cir.), *cert. denied*, 474 U.S. 902, 106 S. Ct. 230, 88 L. Ed. 2d 229 (1985); Tektronix, Inc. v. United States, 552 F.2d 343, 351, 213 Ct. Cl. 257, 193 U.S.P.Q. 385, 393 (1977), *modified*, 213 Ct. Cl. 307, 557 F.2d 265 (1977), *cert. denied*, 439 U.S. 1048, 1990 S. Ct. 724, 98 L. Ed.2d 707 (1978).
211. 718 F.2d 1056 at 1065, 219 U.S.P.Q. 670 at 675 (Fed. Cir. 1983).
212. Yale Lock Mfg. Co. v. Sargent, 117 U.S. 536, 553 (1886).
213. Kalman v. Berlyn Corp., 914 F.2d 1473, 16 U.S.P.Q.2d 1093, 1102 (Fed. Cir. 1990); TWM Mfg. Co. v. Dura Corp., 789 F.2d 895, 902; 229 U.S.P.Q. 585 (Fed. Cir.), *cert. denied*, 479 U.S. 852, 107 S. Ct. 183 (1986).
214. Panduit Corp. v. The Stahlin Bros. Fibre Works, Inc., 575 F.2d 1152, 1157, 1158, 197 U.S.P.Q. 726 (6th Cir. 1978); see also Trans-World Mfg. Corp. v. Al Nyman & Sons, Inc., 750 F.2d 1552, 1568; 224 U.S.P.Q. 259 (Fed. Cir. 1984).
215. 318 F. Supp. 1116, 1120; 166 U.S.P.Q. 235 (S.D.N.Y. 1970), *modified*, 446 F.2d 295, 170 U.S.P.Q. 369 (2d Cir.) *cert. denied*, 404 U.S. 870 (1971).
216. 318 F. Supp. at 1121. See also H. M. Stickle v. Heublein, Inc., 716 F.2d 1550, 1562, 219 U.S.P.Q. 377, 385 (Fed. Cir. 1983).
217. State Indus. Inc. v. Mor-Flo Indus, Inc., 883 F.2d 1573, 1580; 12 U.S.P.Q.2d 1026 (Fed. Cir. 1989), *cert. denied*, 493 U.S. 1022, 110 S. Ct. 725 (1990).
218. 926 F.2d 1161, 17 U.S.P.Q.2d 1922 (Fed. Cir. 1991).
219. 926 F.2d at 1168, 17 U.S.P.Q.2d at 1928.
220. See, e.g., Orthokinetics, Inc. v. Safety Travel Chairs, Inc., 806 F.2d 1565, 1579, 1 U.S.P.Q.2d 1081, 1091 (Fed. Cir. 1986).
221. General Motors Corp. v. Devex Corp., 461 U.S. 648, 655; 217 U.S.P.Q. 1185 (1983).
222. *Idem*, 461 U.S. at 656, 217 U.S.P.Q. at 1188.
223. *Idem*, at 461 U.S. 655–57.
224. 35 U.S.C. § 285 provides: "The court in exceptional cases may award reasonable attorney fees to the prevailing party."
225. See, e.g., Bayer Atktiengesellschaft v. Duphar International Research B.V., 738 F.2d 1237, 1242, 222 U.S.P.Q. 649, 652 (Fed. Cir. 1984); Hughes v. Novi American, Inc., 724 F.2d 122, 125, 220 U.S.P.Q. 707, 710 (Fed. Cir. 1984); Stickle Heublein, Inc., 716 F.2d 1550, 1564, 219 U.S.P.Q. 377, 387 (Fed. Cir. 1983); Orthopedic Equipment Co. v. All Orthopedic Appliances, Inc., 707 F.2d 1376, 1384, 217 U.S.P.Q. 1281, 1287 (Fed. Cir. 1983); Stevenson v. Sears, Roebuck and Co., 713 F.2d 705, 713, 218 U.S.P.Q. 969, 976 (Fed. Cir. 1983); Advanced Transformer Company v. Levinson, 837 F.2d 1081, 1085, 5 U.S.P.Q.2d 1600, 1603 (Fed. Cir. 1988).
226. Ryco, Inc. v. Ag-Bag Corp., 857 F.2d 1418, 1429, 8 U.S.P.Q.2d 1323, 1332 (Fed. Cir. 1988).

227. Rolls-Royce, Limited v. GTE Valeron Corp., 800 F.2d 1101, 1109, 231 U.S.P.Q. 185, 191 (Fed. Cir. 1986).
228. Jurgens v. McKasy, 927 F.2d 1552, 18 U.S.P.Q.2d 1031 (Fed. Cir.), *cert. denied*, 1991 U.S. Lexis 5205 (1991).
229. See Central Soya Co. v. George A. Hormel & Co., 723 F.2d 1573, 1577, 220 U.S.P.Q. 490, 492 (Fed. Cir. 1983); Ryco, Inc. v. Ag-Bag Corp., 857 F.2d 1418, 1428 (Fed. Cir. 1988).

8

Intellectual Property Policies, Research Agreements, Consulting Agreements, and Conflicts of Interest

Katherine L. Chapman

University of Texas Southwestern Medical Center at Dallas, Dallas, Texas

I. OVERVIEW OF INTELLECTUAL PROPERTY POLICIES—WHAT THEY HAVE IN COMMON

A. Introduction

In the absence of any discernible history of the matter, it seems safe to surmise that university intellectual property policies arose from two sources: One was the faculty member concerned, or even insistent, that something be done with his or her scientific discovery; the other was the university's fear of losing its ownership rights in the intellectual property through either inadvertence or ignorance. Only later did the prospect of producing income from intellectual property become a reality on university campuses. As a consequence, relatively few universities have had policies for decades[1]; the majority began writing policies in the 1960s with a rush developing in the 1970s.[2] Most policies have been revised one or more times and have much in common with one another, although there remain a few interesting differences that are discussed in this chapter.

B. Common Features of University Intellectual Property Policies

1. Preamble

Most universities introduce their policy on intellectual property (or inventions, patents, and technology transfer or inventions and licensing as the policy may be called) with a statement of philosophy regarding intellectual property, emphasizing the institution's missions, as well as its objectives in identifying, developing, and commercially exploiting scientific inventions, discoveries, and creations of faculty members. Most of the preambles also express the universities' concerns for and obligations to the public. A few express a lesser but realistic concern for receiving a fair share of any monetary return on commercialization of its faculty's intellectual property. Many preambles are elegant:

> [The] University, with primary concern for the education and training of students and the advancement, preservation and dissemination of knowledge, recognizes that the various activities of its faculty, students, and employees may produce patents and inventions which further support these goals. The multiplicity and diversity of fields of endeavor, the complexities of technologies and the speed with which change presently occurs mandates the establishment of a coherent system which manages both the process of discovery and the fruits of those discoveries.[2a]

Or "[I]deas or creative works produced at the University should be used for the greatest possible public benefit . . . [and the] . . . public benefit should take precedence over profit making, either by the University or the individual scholar."[3] Others are more practical:

> [T]he University does not pursue research in anticipation of profits. The principal goals of research remain the discovery of new knowledge and its dissemination. Since some of the by-products of research may be of commercial value . . . , it needs an appropriate mechanism to recognize developments, protect them and bring them to the marketplace.[4]

In restating their academic missions, many universities make it clear that their faculty do not carry out research for the purpose of inventing, but if inventions arise, the universities believe it is their responsibility to develop them for the benefit of the public, the university, the inventor, and the research sponsor, if any.

Statements explaining the source of the university's legal ownership of the intellectual property (*if* the university will own it) may also appear in the preamble of the policy, such as

> [W]hen University support makes the enterprise possible or when it provides extra or special support, either with money, facilities, or equipment, for the development of ideas or the production of works, it is reasonable for the University to participate in the fruits of the enterprise. . . .[5]

Or "The University recognizes that inventions . . . may result from the activities of University employees in the course of their duties or through the use of University facilities, funds, or resources."[6] There are variations within each university regarding what gives them a legal interest in an individual's intellectual property. The most common indications are employment, use of the university's facilities, equipment, supplies and an invention falling within the scope of the individual's university-based research regardless of the time or place of invention.

Some of the preambles also briefly state the policy's applicability to certain individuals (i.e., individuals who are generally obligated to disclose inventions to the university):

1. Faculty, students, and employees;[7]
2. Faculty, staff, fellows, students, and other individuals in the university community;[8]
3. Faculty, staff, and other persons receiving compensation from the university for services rendered, as well as students, graduate students, and others, whether compensated or not, who work on any project under university control.[9]

2. Categories of Intellectual or Tangible Property

Many university intellectual property policies list in varying ways the categories or types of intellectual policy included in or excluded from the coverage of the policy. Brown University uses the most common term, *invention*, in their policy, defining it to be all inclusive: "[I]nvention is intended not only to include patentable inventions, but also proprietary information relating to useful articles of commerce, including but not limited to computer programs, secret processes, etc."[10] Device, product, or method is the listing Duke University uses for the inventions it requires to be reported.[11] Auburn University, on the other hand, states specifically that their

> [P]olicy covers patentable subject matter and patents but does not extend to copyrightable or copyrighted material. Patentable subject matter involving the act of invention covers: 1. Processes, machines, compositions of matter and other articles of manufacture. . . . 2. Design. 3. Plants.[12]

An example of exclusions from the terms of an intellectual property policy is:

> Articles submitted to or published by scholarly and professional journals. . . . [C]lass notes produced in connection with regularly scheduled courses of instruction. . . . [S]cholarly and professional books, texts, works of art, musical compositions and the like. . . . [S]ponsored grants or contracts may contain specific provisions regarding ownership, copyright, or royalty income privileges related to materials generated under the contract. These conditions are binding on the University and the author or producer. . . .[13]

Although many universities recognize their ownership of copyrightable works, trademarks, know-how, and tangible research products or materials,[14] these categories are not generally included in intellectual property policies. The "tangible" properties are most often handled in policies similar to, but separate from, the intangible or intellectual property policies. This distinction illustrates the growing sophistication of university intellectual property management. In earlier years, many universities recognized and handled only inventions. If any of the tangible property issues arose, they were "forced" into the provisions of the patent policy.[15] However, today many universities have separate policies for each kind of tangible and intangible property because the handling of each is complex in its own way.

Inventions and patents are the categories of intellectual property terms that are present in virtually all policies; many policies do not define the terms but impliedly adopt the definitions used by the U.S. Patent and Trademark Office. Other institutions have authored their own definitions. For instance, the University of Illinois defines the term *invention* as "any device, process, material, plant product, or other innovation."[16] Stanford defines it as "a novel and useful idea relating to processes, machines, manufactures, and compositions of matter. It may cover such things as new or improved devices, systems, circuits, chemical compounds, mixtures, etc."[17] The term *patent* is defined by some but not all of the policies. Those who do define it use similar definitions, such as

> A patent is a grant issued by the U.S. Patent and Trademark Office giving an inventor the right to exclude all others from making, using or selling the invention within the United States, its territories and possessions, for a period of 17 years from the patent grant.[18]

Copyrightable works or "materials subject to copyright" are clearly defined by the University of Pennsylvania, in its copyright policy as

> 1. Books, journals, articles, texts, glossaries, bibliographies, study guides, laboratory manuals, syllabi, tests, and proposals. 2. Lectures, musical or dramatic compositions, and unpublished scripts. 3. Films, film strips, charts, transparencies, and other visual aids. 4. Video and audio tapes and cassettes. 5. Live video or audio broadcast. 6. Programmed instruction material. 7. Computer programs. 8. Other copyrightable material.[19]

The majority of the university policies at least mention works subject to copyright. Some set up a specific, detailed policy and procedure on how they are handled; others are more simplistic, stating only that such works should be disclosed to the appropriate university office.

Computer software is another category of intellectual property (or tangible property as some universities call it[20]) that appears in most university policies.

Often it is on the list of works subject to copyright—in the copyright policy—as well as being the subject of its own policy.[21] It is clearly defined by Northwestern University:

> ''Computer software'' is used to designate computer programs, in the broadest sense, and is meant to include user's manuals and other explanatory material that accompany computer programs, and computerized databases. Also included are microcodes, subroutines, operating systems, high level languages application programs in whatever form expressed (machine or assembly language, source or object code) or embodied (chip architecture, ROM, disk or tape storage, program listings).[22]

Tangible research property appears to be a generic term that includes biological materials and a number of the other similar terms for tangible properties. The common feature of all of the applicable terms is that they are innovations that cannot be readily protected by patenting or copyrighting. (Some universities include computer software in their definition of tangible research property,[23] but most do not.) One university defines this category of intellectual property as

> tangible (or corporeal) items produced in the course of research projects supported by [the university] or by external sponsors. TRP includes such items as: Biological materials[,] computer software[,] computer databases[,] circuit diagrams[,] engineering drawings[,] integrated circuit chips[,] prototype devices[, and] equipment.[24]

Another institution, focusing more on biotechnological research states: '' 'biological materials' [include] clones, plasmids, gene fragments, DNA probes, hybridomas or monoclonal antibody cell lines[.] . . . [25]''

Know-how is actually defined in very few policies, though it may be very valuable to the university if licensed by itself or if licensed simultaneously with patents. The few that do define know-how can be illustrated by the following: ''Most biomedical/biotechnological research produces innovations that . . . include . . . laboratory techniques (know-how).''[26] ''[R]oyalty payments [may be made] to the University in consideration of the 'know how' or accumulated unpublished research data furnished to the company.''[27] ''[L]aboratory techniques, or 'know-how' . . . [are] almost never worth attempting to patent[,]''[28] as stated above as one of the characteristics of know-how. Harvard Medical School makes the excellent point that know-how can be valuable in a scientific sense, in benefiting society by advancing medicine, and in helping create links with industry to bring innovation into use and support research.[29]

3. Coverage of Intellectual Property Policies

The individuals who are covered by intellectual property policies are introduced in the section on the Preamble (Sect. I.P.1.). Although it is commonplace for the policies to apply to scientists and physician faculty members and staff, the policies may differ in applicability to students (undergraduate or graduate), postdoctoral fellows, trainees, visiting faculty, and, in a few policies, parttime faculty. Every faculty member should review the institutional policy to determine if it is applicable to him or her.

4. Ownership

Ownership of the intellectual property is dealt with in every policy reviewed by the writer, either explicitly or in some indirect manner.

Most universities provide that the institution is the owner of the intellectual property developed by the individuals to whom the policy is applicable. For instance, Duke University provides: "Inventions resulting from research or other work conducted by University employees in whole or in part on University time or with significant use of University funds or facilities *shall be considered the property of the University*."[30] Likewise, it is the policy of the University of Pennsylvania that "[A]ny invention or discovery which may result from work carried out on University time . . . is the property of the University."[31] When, as is evident in these policies, the university will own certain inventions and discoveries, the policy also places on the scientist-faculty member certain duties: (1) the duty to assign. For instance, one policy states: "The University may require inventors to execute an assignment and related documents necessary to further patenting and development of an invention. . . . "[32]; and another says, "University personnel shall execute all documents necessary to implement the University's Patent Policy . . . including but not limited to . . . assignments of inventions."[33] (2) the duty to cooperate. The University of Nebraska's formal document accepting an offer of a patentable discovery from an inventor states that

> the University . . . will . . . pursue protection *with the help of the originators*. . . . "[34] Another policy even more explicitly provides in its section on "Seeking a Patent" that "the inventor shall cooperate fully in such effort, including execution of all necessary or desirable agreements, applications, assignments and other forms and instruments.[35]

There continue to be a few universities who allow for ownership of certain intellectual property by the faculty member rather than the university. Stanford University and Washington University are well-known examples. Stanford's policy on inventions, patents, and licensing provides: "Unlike industry and many other universities, Stanford's invention rights policy allows all rights to remain with the inventor if possible."[36] The exception to this policy illustrates another type of ownership of intellectual property within the university community—the sponsor of the research from which the invention arose. "Some [sponsored research] agreements permit the University to retain title . . . some provide for the reverse or defer allocation of rights.''[37] Other universities, such as the University of Illinois at Urbana-Champaign, provide that its staff will negotiate with a research sponsor who ''desire[s] special terms such as those related to ownership of . . . patents. . . . ''[38]

The majority of universities insist upon ownership of its scientists' inventions by way of an assignment, but to make sure that one is knowledgeable on this point, the scientist should carefully read the policy or inquire of the appropriate university administrator.

5. Publication Rights of Scientist or Academic

One of the prime misconceptions about patenting of university inventions is that a scientist cannot publish and also file for patent protection. Since misconception can discourage disclosure of inventions by the faculty member, many universities make some statement in their policy regarding the compatibility of publication and patenting. Harvard Medical School says that it is dedicated to encouraging ''the faculty's customary manner of disseminating research results, while providing the means to preserve rights to such property.''[39] This policy also explains why the publication and patenting can be accomplished without forfeiting either. The University of Pennsylvania does much the same thing by restating its major function of advancement and dissemination of knowledge, and explaining the way publication and patenting can coincide. At this point in the policy, there also is a discussion of appropriate delay in publication under sponsored research agreements,[40] which is discussed in the Sponsored Research Agreement section (Sect. II) of this chapter.

With or without explicit statements in the intellectual property policy, universities see the right to publish as an integral part of academic freedom and make every effort to allow publication as usual while protecting the discovery through patenting.

6. Administrative Arrangements for Management of Intellectual Property

Most policies contain explanations of how the specific university administratively manages the intellectual property. Some institutions have a single office that handles intellectual property from the time of first disclosure to final licensing.[41] Other universities have more than one office involved in the process. ''Patent administration at Stanford is handled by two groups: the Intellectual Property Administration (IPA) staff as part of the Sponsored Projects Office, and the Office of Technology Licensing.''[42] Each office has separate functions in a university with this administrative structure, and it is incumbent upon the faculty member to become aware of the functions of each by referring to the university's written policy or visiting the offices.

Auburn University is an example of a university who uses a nonprofit organization to manage its intellectual property. Inventions are initially disclosed to the Secretary-Director of the Auburn Research Foundation[43]; and the University of Nebraska and University of Washington are examples of universities that "may develop its inventions through the university's foundation or Research Corporation or other licensing agents."[44] The scientist must be aware of the options the university has since certain rights of the inventor, including the amount of royalties, can be changed by the handling of the invention by a non-profit organization, especially if the non-profit organization is not connected in any way with the university.

7. Disclosure of Intellectual Property to Administration

It is the rule in intellectual property policies that the faculty member–inventor is obligated to disclose an invention or discovery to the appropriate university administrative office or committee, regardless of who may have rights to the invention. Even Stanford University, which generally allows invention rights to remain with the inventing faculty member, requires that "[A]n invention disclosure must be prepared and submitted for each invention conceived or first actually reduced to practice under a University contract or grant with patent terms."[45] This obligation of the inventor is significant because it is one of the only ways that universities have of finding out about scientific discoveries and there may be an imminent publication, which could forfeit rights to the invention if no patent application is filed. If the faculty member's laboratory notebooks are not available, the written disclosure may serve as evidence of inventorship if a controversy develops over who was the first inventor.

Timing of disclosure to universities of inventions and discoveries developed by faculty members are dealt with in intellectual property policies in one or two ways. The policy is either silent[46] or states no more than the admonition to disclose *promptly* after the scientist knows he or she has made a discovery or invention.[47] Some policies do not mention a time frame for making the disclosure to the administration but state the U.S. law regarding the time for filing a patent application—within 1 year after publication, offer for sale, or public use.[48]

Virtually all policies provide for the manner of reporting of inventions to the university. Either the university has its own invention disclosure form that the faculty member must complete[49] or he or she needs to provide a written description of the invention or discovery in order to fulfill the policy's disclosure requirement.[50] The key is for the disclosure to be clear and complete enough for it to serve as the focus of a patentability search, technical evaluation, assessment of commercial feasibility, and perhaps legal evidence of conception and reduction to practice of the invention.

8. Institution's Internal Evaluation Process of Intellectual Property Disclosure

Most policies clearly provide for a university committee to provide some evaluation of the invention disclosed by the faculty member. Some of the committees start with "initial screening, followed when indicated by a detailed evaluation of the invention. This may be done through an internal review process, or by referral to an external organization [either related to the university or not,] that manages the evaluation."[51] Other policies seem to favor review by staff alone[52] or with committee responsibilities being broader and more policy based, including "review . . . of decisions relating to, or policies affecting, [an] invention or discovery.[53]

Many of the policies do not list the factors that will be considered in the evaluation of the invention disclosed to the university. Some spell them out in detail, such as the University of Illinois at Urbana-Champaign. That policy states: After submission, "an evaluation is made to answer the following questions: Does the University have rights in the invention? Is the invention patentable? Can the invention be developed commercially?"[54] Factors evaluated by the administration of the University of Virginia are whether the invention is the product of university research as defined in the institution's patent policy; whether the university will refer the invention to an outside organization to determine the patentability and marketability of the invention, and, if so, to which outside organization the invention will be referred.[55] Disclosures at the University of Pennsylvania are examined for technical merit, patentability, and commerciability.[56] Although the third or fourth factor evaluated may vary with the institution, virtually every institutional policy that states factors by which inventions are evaluated, lists legal protection available and commercial potential as significant factors. Whatever factors the university selects to evaluate, it will need the cooperation (i.e., time and thought) of the faculty member, and if there is a sponsor, the sponsor may well want to be involved (under confidentiality agreements).

9. Availability of Legal Protection for Intellectual Property

Different types of intellectual property that may arise at the university are subject to different types of legal protection.

a. Inventions Inventions have been previously defined in this chapter. However, in brief, an invention is a novel and useful discovery relating to processes, devices, manufactures, and compositions of matter. Inventions are generally subject to legal protection by patent—a grant to the inventor or assignee by the U.S. government of the right to exclude other people from making, using, or selling the invention in the United States for 17 years. Patenting is a complex procedure that most often is undertaken by a patent agent attorney representing the university and/or inventor. Some inventions do not meet the complex criteria for patentability. If this occurs, the invention may switch to another category of intellectual property, "know-how," which is valuable, proprietary information of a technical nature that may be licensed to industry despite its lack of patent protection.

b. Know-How "Know-how is factual knowledge not capable of precise, separate description, but that when used in an accumulated form, after being acquired as the result of trial and error, gives to the one acquiring it an ability to produce something that he or she otherwise would not have known how to produce with the same accuracy or precision found necessary for commercial success."[57] Often know-how can be valuable to the university, and as pointed out in the Harvard Medical School's *Guide to Protecting and Managing Intellectual Property*, it can be "valuable in a scientific sense, in benefitting society by advancing medicine and in helping create links with industry to bring innovation into use and support research."[58] The key to legally protecting and licensing know-how is in keeping it confidential. Often this can be done without violating the institution's no secrecy in research policy and while protecting the scientist's right to publish, since there are times that it is a scientist's preference not to publish certain information, such as laboratory techniques or that knowledge that one generally thinks is hidden away in one's mind. A faculty member who has valuable scientific information, which he or she thinks may be know-how, should consult with the appropriate administrator on campus to determine how best to protect it.

c. Copyrightable Works The term *copyrightable works* has been discussed and illustrated in Section I.B.2. on Categories of Intellectual or Tangible Property. The term itself discloses the type of legal protection that is available—copyright protection. Copyright law protects original works of authorship from copying, while at the same time denoting ownership and control of the work. A long list of works (as listed above in Categories of Intellectual or Tangible Property) are subject to copyright protection with minimal effort, so the scientist should know his or her institution's policy and be quick to seek advice.

Many universities claim ownership to copyrightable works created for institutional purposes. They also must retain the right to perform the obligations to research sponsors or other third parties.

It should be noted that many university policies exclude from university ownership and from university-initiated legal protection certain works that are traditionally faculty owned. For instance, rights to textbooks, other books, and journal articles and similar works are retained by the faculty member (and not claimed by the university), so that any royalties derived from these are retained by the faculty member.

d. Computer Software Computer Software is unusual in that it is subject to several types of legal protection depending on its nature and specific characteristics. Copyright protection is the most common form of legal protection used for computer software. However, the possibility for patenting now exists. The determination of how to protect will depend on the concerns, commercial aspirations of, and desire to control future use and distribution by the faculty member and university. As with all university intellectual property, the institution's policy must be followed and any obligations to research sponsors must be met. Some universities have computer software distribution mechanisms on their campuses. If the faculty member desires to have the software distributed for research purposes only, this may be an avenue to investigate if the service is offered at the university.

e. Tangible Research Materials Tangible research materials are by-products of research activities that are in a tangible, as opposed to intangible, intellectual property form. They are also discussed in Section I.B.2. on Categories of Intellectual or Tangible Property. Some tangible research materials or products, such as certain biological materials, may be of commercial value even if patent or copyright protection is unavailable. The commercial value may be enhanced by controlling its distribution and having recipients contractually agree not to disclose the tangible research materials further and not to use for other than research purposes. This can successfully be done without damaging the traditional scientific exchanges of tangible materials or products. The materials may be transferred for commercial uses pursuant to a license-like agreement in exchange for royalties.

Few intellectual property policies explicitly state how the university goes about obtaining legal protection for inventions within the various categories of intellectual property. All forms of protection can be perfected by the nonlawyer university administrator except for the patenting of inventions that (although done by a few rare scientist-inventors themselves) generally requires patent agents or patent attorneys. Accordingly, the protection of copyrightable works, computer software, know-how, and tangible research materials can usually be done internally. Preparation and prosecution of patent applications on inventions can be done by in-house patent attorneys or agents or by legal counsel outside the institution retained by the university. In many areas of biotechnology, it is necessary to use outside legal counsel because of the scientific complexity. Often these attorneys have doctorates in various fields that allow them to converse readily with the faculty members and the patent examiners in the U.S. Patent and Trademark Office. If the institution has significant numbers of invention disclosures in that area they may have such a person in-house).

10. Sponsor's Right to Intellectual Property

As illustrated by the lengthy discussion of Sponsored Research Agreements later in this chapter (Sect. II), there are many, many issues regarding the rights of sponsors of research at the university. Many universities are developing policies on sponsored research issues that are separate and apart from their intellectual property policies.[59] Some allow their departments and units to develop their own sponsored project policies so long as they do not conflict with general university policy.[60] Even those universities with separate policies on sponsored research often briefly state in their intellectual property policies the basic obligations to sponsors and the effect on the intellectual property rights of the faculty member and university.

a. Commercial Sponsor The relationship between the university and the sponsor is controlled by the written agreement between the two that is virtually always made subject to the university's policies. The intellectual property policies generally state the specific or unique rules that apply to commercial sponsors of university research, such as, ownership of inventions arising from the commercially sponsored research remains with the university.[61] The policies usually go a step further to explain that the commercial sponsor may have an option to license or a right of first refusal to the inventions that arise under the agreement. Whether the license can be exclusive or nonexclusive is also frequently stated.[62]

b. U.S. Government as Sponsor The other type of sponsor mentioned in virtually every intellectual property policy is the U.S. government.[63] The government funds many university research grants, so it has been very important for universities to provide in their policies for handling of inventions that arise under that research grant. Since the passage in 1980 of the Patent and Trademark law Amendment Act,[64] commonly called P.L. 96-517, universities have been allowed to elect to retain ownership over most inventions made in the course of government-sponsored research. In order to take advantage of the law, universities must, among other things, report the invention to the granting agency, file and prosecute a patent application, actively seek licensees, and grant the government a royalty-free license to use the invention for governmental purposes. Many policies go this far in explaining the obligations to a federal government sponsor in order to correct what is still a misapprehension by many scientists—that federal agencies retain rights to inventions.

c. Other Sponsors Only occasionally do the intellectual property policies mention sponsors other than industry or government. However, occasionally other sponsors are mentioned if only to say that research sponsored by nonprofit foundations or private individuals (as contracts, not gifts) are controlled by the terms of the specific contract or agreement.[65] Traditionally, reporting the disclosure of the invention to the administration and the progress in licensing is all that is required by these other sponsors because they are not usually in a position to commercialize an innovative discovery. It is interesting that in recent times even some of these private individuals and foundations have begun to ask for options to license inventions arising from the research they fund. These sections of the policies serve as a vivid reminder to the faculty member that he or she must be aware of the obligations to the sponsor of his or her research regardless of whether the sponsor is an individual, corporation, foundation, or other entity.

11. Licensing of Intellectual Property

Most intellectual property policies include brief statements about licensing of inventions owned by the university. Most begin with a statement of goals, stressing the need to benefit the public and further the missions of the university.[66] Some state goals in terms of exclusive or nonexclusive licensing, such as, Yale University.

> The objective of the university is to achieve broad access to the results of its research. Therefore, as a general policy the university wishes to grant nonexclusive licenses. . . . Exclusive licenses, however, may be granted when there is clear evidence that this is the most effective route for arranging public access to the benefits of the invention.[67]

Many universities have their own in-house licensing offices, often dedicated solely to that purpose. These universities' intellectual property policies state that these in-house offices will handle licensing of inventions, often seeking assistance and cooperation of the inventor and, perhaps, of the inventor's department.[68]

A number of universities state in their intellectual property policies that the university may elect (on university-owned inventions) to pursue the development of inventions through a related foundation or through a patent management company.[69] This type of provision merely notifies the academic scientist of the options for licensing the university may, in its discretion, choose between.

12. Distribution of Compensation Received for Intellectual Property

One of the issues of prime concern to the scientist is how the rewards of marketing are distributed. The bulk of the intellectual property policies state explicitly in the policy how the income will be divided. Most use net income or net patent revenue as their base for distribution, deducting certain costs borne by the university in getting the invention to the point of actual sales.[70] The costs deducted from the income received from licensees vary in their terminology but are similar in their intent. The goal of most universities is to recover their not insubstantial costs in evaluating, patenting (which entails searching, preparing, filing, and prosecuting) and licensing the invention. Virtually all universities seek to recover out-of-pocket costs, such as outside patent attorney and U.S. Patent and Trademark Office fees directly attributable to the specific invention. Many also include in the determination of costs to be recovered: a generic statement of administrative expenses,[71] or a specific percentage, such as 15% for administrative costs.[72] Many policies state that licensing costs will be deducted[73]; others use terms such as "patent promotion costs."[74] One policy sets a ceiling (to be adjusted annually based on constant 1984 dollars) for the "direct costs associated with patent prosecution" that may be deducted from royalty income.[75]

Table 1 Examples of Division Between the Inventor and University or Related Foundation from Net Income

Example	Inventor	University	Related foundation
1	50% of the first $200,000 of cumulative NI; 25% of cumulative NI thereafter	Remaining net income for research[76]	
2	40% of the first $20,000 revenue, 35% of the second $20,000 revenue, 30% of any additional revenue		Any surplus derived from the patent program is used to support "appropriate university activities" with consideration given to departments originating the successful patents[77]
3	50% of net royalties	50% of net royalties with first consideration given to support of research[78]	

The actual percentages of the net income that go to the inventing scientist as his or her personal income and to the university vary greatly by institution (and may have been altered within the institution more than once). It can generally be said that most universities are generous to the inventor, hoping to both reward and motivate him or her; and that most universities use their share of royalty income to further research within their institution.

Some examples of the division between the inventor and university from net income (NI) (i.e., after deductions for costs) for each invention are given in Table 1.

Some universities provide additional information in their policy regarding how the university's share of royalty will be distributed. For example, Harvard provides that

> The university's share of such net income will ordinarily be distributed as follows: The first $32,500 to the Faculty and academic department from which the . . . invention originated for support of the research activities of the inventor. The next $67,500 to be divided 1/2 as above and 1/2 for the general use of the academic department from which the . . . invention

> originated. The additional income shall be allocated between the Central University and the Faculty involved. . . . The needs of the academic department and the research areas within which the . . . invention originated shall always be considered in formulating the details of such allocation.[79]

One university even gives its inventor-faculty members an incentive to donate a portion of his or her personal 25% of the net proceeds from commercializing his or her invention to the university for use in research. If the donation takes place, the university will add an equal amount into that university research account.[80]

Few university intellectual property policies explicitly state how equity or shares in a company (received in lieu of cash royalty payments for licensing technology) will be distributed. Of those few that do, the allocation is generally in the same manner as that university's policy provides for cash royalties and fees.[81]

In order to determine which distribution rules apply to which types of intellectual property, the scientist must look to the specific university policy. If copyrightable works are included in the policy's definition of intellectual property, the distribution of royalties will be the same as it is for inventions. If the university separates copyrightable works from the intellectual property policy, then one should look for the specific royalty distribution policy for copyrightable works. The same is true for computer software, tangible research materials or products, or know-how.

13. Dispute Resolution

A number of university intellectual property policies provide for a method of dispute resolution. Most have procedures that involve internal resolution by committees and senior administrators.[82] A few have final determinations made by the board of regents.[83] Others provide for final resolution of a dispute by arbitration, either outside the university according to a set of specific rules such as the Rules of the American Arbitration Association,[84] or inside the university by an ad hoc arbitration panel named by the inventor and university.[85]

14. Recommendation to Scientists

The one recommendation that can be made to the scientist without hesitation is to read and understand the institution's intellectual property policy. Although there are common clauses among the policies, each policy has unique features that the scientist will be deemed to know and that it will be necessary to know in order to fully and appropriately exploit one's intellectual property for the good of the public, the university, the inventor, and the sponsor of research, if any.

II. RESEARCH AGREEMENTS

A. Introduction

Everyone has seen the statistics showing the peak and decline of federal funding for university research. Few have realized that as the federal funding decreased, industry's contribution has increased, albeit slowly and small. Even optimists do not expect federal funding to grow in the 1990s; virtually all university scientists and research administrators anticipate an increase in funding by commercial firms even though the ultimate potential is limited (i.e., "[I]ndustrial funding does not have the potential to replace federal funding as the major source of financing university research.")[86] Simultaneously, industry has been searching for new sources of innovative ideas in order to improve its competitiveness in the world.

1. The Need for Research Agreements

University and industry, for these reasons, are seeking creative methods of cooperating. If universities can continue to discover the new knowledge and commercial firms to apply it to the real world, at the same time that they communicate with each other and share information, the new creative relationships formed will be beneficial to both entities. Many universities recognize the potential for these types of relationships with industry. Some state outright in their policies that they have

> a long tradition of welcoming support for research projects from commercial sponsors . . . [believing] [s]upport aids our pursuit of new knowledge and, through the cooperative interactions it spawns, often facilitates the practical applications of such knowledge to the benefit of society.[87]

One institution was *mandated* by its Board of Regents to develop

> closer cooperation between the university and the industrial sector . . . [for] more effective application of new knowledge and new technology for the benefit of society.[88] The same university states in its policy that research activities with industry increase the likelihood that educational and research activities at the university will reflect current societal needs.[89] Most policies are realistic, stating that "industrially supported research agreements . . . provide valuable support.[90]

Universities are hopeful, yet realistic, perhaps surprisingly so, knowing that to industry a competitive position is critical. Wary of "too good to be true" offers from industry, they are still optimistic about the authentic and sincere offers to join in cooperative research with commercial firms. Of all the types of research relationships between universities and commercial firms or industry that have evolved through the years, sponsored research agreements are the most frequently used (though called other names). "These relationships are based on a *quid pro quo*: the [industrial] sponsor provides financial support of specific research in exchange for certain rights to use the results. . . ."[91]

2. Mutual Benefit as Goal

The emphatic need is for mutual benefit, a balance reflecting both parties' interest. If the parties can negotiate, there is no reason, even with the differing values, goals, and cultural realities of academic institutions and commercial firms, that the relationship cannot be a "win-win" situation for both parties. In fact, one of the authorities in this field, Donald R. Fowler, General Counsel of the California Institute of Technology, concluded from a 1983 survey that "no overwhelmingly important impediments seem to control or dominate the university–industry research relationship."[92]

3. Types of Research Agreements

With these ideas and concerns in mind, universities have developed many forms of research agreements, most of which have the same aims and, often, common provisions or clauses. Two representative agreements are (a) sponsored research agreements and (b) collaborative research agreements. These names are descriptive of the function of the agreements to be discussed, but are not necessarily the names most often used by universities. Industrial research agreements and sponsored projects are examples of synonymous terms for sponsored research agreements. Collaborative agreements can also be called cooperative research agreements or collaborative research agreements. Specifically excluded from this discussion are the wide variety of other agreements universities can enter into with industry: biological testing agreements, materials (or tangible research products) transfer agreements, confidential and limited use agreements, patent or know-how license agreements, consortium agreements, or consulting agreements.[93] Also excluded are traditional grants and gifts generally thought to be funds for research that, although there are stipulations attached, are not negotiated in the sense that a contract is.

The sponsored research agreement involves two independent parties, both bringing to the negotiation table desire for what they want to give to the relationship and what they want to receive from the relationship. Fundamentally, the commercial firm wants to give money to the university in exchange for the results of a particular researcher's defined research project; the university and its researcher wants to receive funds to carry out research of academic interest, the results of which the commercial sponsor will receive some nature of legal right to. In a sponsored research agreement, the research is done by one party, the university. In a true *collaborative* research agreement, both parties carry out research either separately or together at the industry or university laboratory and share methods and results.

B. Clauses or Provisions in Sponsored Research Agreements

1. Nature of Contractual Agreement

It is very important for the scientist to understand the nature of a research agreement. First of all, it *is* a binding legal agreement. Many scientists think the agreement is merely an informal "gentlemen's agreement" that does not in any way bind the individuals or institutions. This naive thinking can be extremely dangerous, leading the scientist to unintentionally and inadvertently breach the agreement, putting himself or herself and the institution that employs the scientist at risk contractually, financially, and legally. It is imperative that each scientist take the obligations of the sponsored or collaborative research agreement very seriously. The agreement must be read, studied, understood, and followed closely during the period of performance. Taking the agreement seriously is both intelligent and necessary; its essential quality is commitment—an agreement binding under penalty of law.

2. Scope of Research to be Performed

After a scientist understands the gravity of the university's entering into a research agreement with a commercial firm on the scientist's behalf, it becomes imperative for the scientist to describe accurately the scope of the proposed research program or research project.

As a matter of course in the life of university researchers, funds to finance research projects are too limited to properly finance the quantity and quality of work that the researchers believe to be necessary. Consequently, the scientist may exaggerate the scope of research that he or she can provide to the commercial firm. Not intending to misrepresent the desired scope of the research project, the scientist may overstate the potential results of the research project and embroider on the details of how the research will be carried out. This is a realistic part of the competition for very limited funds. However, if and when a commercial firm decides to accept the university's offer to fund the research, the researcher (and university research administrator) must assure that the description of the research program or project is as specific, precise, and accurate as is humanly possible. Donald R. Fowler stated this point very realistically: "Failure to have a well-defined scope can lead to embarrassing and costly disputes over who owns the patent rights to inventions made by the researcher . . . during the term of the research agreement but not necessarily as part of it."[94] To protect the researcher from his or her own enthusiasm, the university research or contract administrator must remind researchers that after a commercial firm indicates interest in the researcher's proposal for research, the proposal must be carefully analyzed and the boundaries of the research must be described specifically. Speculation on research that may come to fruition *after* termination of the agreement must be removed; language pulling in research of colleagues, explicitly or implicitly, must be deleted; only accurate and precise descriptions of the actual scientific research and work to be done should be included, since there will ultimately be many repercussions from this, the subject matter of the agreement. Rights to patents or technology will often turn on the statement of the scope of the research program. Many sponsored research agreements contain statements such as: Sponsor has an option to an exclusive license on any inventions or discoveries *that arise under the terms of the defined Research Program* during the Term of the Agreement. If the "Research Program" is not clearly and factually defined, the parties—sponsor and university—will certainly disagree on what can arise from it. This is only one example. There are others, all of which emphasize that there is no substitute for precision and clarity in the description of the research program.

3. Performance Obligations

a. Sponsor Obligations (Monetary) University researchers tend to think of the sponsor's performance obligations under the sponsored research agreement as solely monetary, not only because the researcher needs the funds but because the budget is one part of the agreement that may be studied and analyzed over and over before being presented to the sponsor and is often an issue for hard bargaining with the sponsor. Although not the only obligation of the sponsor, it is an important part of the agreement. If the researcher's estimate is incorrect, the research can suffer and the mutual benefit of the relationship may dissolve. Important aspects of the funding provision of which the university research administrator and the researcher must be aware are the following.

Start Date for Funding. When the draft or drafts of the sponsored research agreement are circulated, a starting date for funding is identified. It may be the same or different from the effective date of the agreement. It may be a "date certain," as attorneys say when they mean a specified date (e.g., September 1, 1993), a particular period of time after the happening of an event (e.g., 30 days after execution of the agreement[95]), or at the time of an occurrence itself (e.g., at the time of execution of the agreement). It is vitally important to identify this date properly because that is when the first check is due to the university. In many institutions, receipt of that first check is extremely significant because that is the first date on which many necessary tasks under the research agreement can be done (or started), such as advertising for a new position that may be necessary to carry out the research program or ordering special equipment or supplies, all of which may take a considerable time to complete before the actual work can begin.

Term of the Agreement. Another difficulty in dealing with research sponsors regarding funding is setting the "term" of the agreement. *Term* means the agreed upon length of time the research program will be performed, or the time in between the effective date (i.e., starting date) of the agreement and the termination date (i.e., ending date). There are advantages and disadvantages of setting the term for a standard 1 year, although the same can be said about the longer terms for agreements. This writer's experience is that the researcher, despite the downside to be discussed, almost without exception wants as long a term for the research project as can be negotiated with the sponsor. The researcher is more attuned to the needs of (a) the university laboratory for stability in and foreseeability of funding and (b) the scientific research to progress, reveal new knowledge, and meet the faculty member's goals.

Two authorities in sponsored research agreements, Donald R. Fowler[96] and Bernard D. Reams, Jr.,[97] advise that some commercial sponsors of research may prefer longer-term funding if the research is advancing in accordance with the plans and desires of the sponsor.

One method assured to protect the commercial sponsor who makes a long-term agreement to fund research is suggested by Fowler[98] and Reams.[99] Reams advises to "provide in the contract that the university must meet . . . performance standards prior to [continuation of] financial support."[100] Fowler advises negotiating a provision whereby long-term funding of a whole research project by a commercial sponsor could be enhanced (for both parties) by setting up checkpoints (agreed upon by the parties) for research projects. "Gates" or "targets," as he calls them, can be used to assure that the researchers are continuing to meet objectives at the appropriate intervals in the research project, after the company has obligated itself for long-term consistent funding.[101]

Payments and Indirect Costs or Overhead During the Term (i.e., Duration) of the Agreement. It is vitally important for both the university and commercial sponsor to have the agreement spell out clearly the number and schedule of payments to be made and the exact amount of each payment. It should also be stated explicitly whether the university's indirect costs (or overhead) and fringe benefits has been added into the payment amounts scheduled in the agreement, or if indirect costs are addressed elsewhere and must be added in by the sponsor to the precise payment amounts listed in the agreement. It is very important for the sponsor to be told about indirect costs and fringe benefits at the earliest point in discussions whether it be in the cover letter to the commercial firm accompanying the sponsored research agreement, the budget originally drawn up by the scientist and sent to the sponsor, or set forth in the payment or funding clause in the first draft of the sponsored research agreement. Some universities have rigid policies on the indirect cost rate. Others may negotiate the indirect cost rate, taking into consideration the value of the research project and of graduate training, the precedent-setting nature of such action, and the financial implication. The recommendation made to the University of Minnesota Board of Regents by an administrative committee advised that

> [T]he university should operate in a manner of reasonable fairness, which would balance the value of training graduate students, transferring technology, and promoting research, against the financial consequence of not fully recovering indirect costs.[102]

Regardless of when it is brought up or where it is placed, it is imperative that both parties know precisely the amount of total funding and of each periodic payment (including the indirect costs) and the exact days when payments are due in the university's offices.[103] If the university is to bill or invoice the sponsor for the periodic payments, the parties must spell out in the agreement precisely how the invoicing will work (e.g., based on estimates of expenses or some other basis) in order to avoid confusion later or inadvertent breach of the agreement.

Cost Index for Inflation. For long-term sponsored research agreements, such as 5 years or more, or short-term agreements in times of varying high inflation, the university will want to include in the sponsored research agreement a cost index for inflation. This will allow an amount arrived at by the use of a particular defined formula to be added to the periodic payment, if appropriate. Reams (1986) gives examples of some of these cost index terms. The Massachusetts General Hospital–Hoechst A. G. Contract of 1980 specifically provided for an adjustment of payments "to reflect changes in the value of the dollar between 1981 and the time[s] of payment . . . "[104] (i.e., 1983–1990). That agreement used the CPI–Medical Care Professional Services Component Index published by the United States Bureau of Labor Statistics.[105] Other indexes are available, such as the GNP (Gross National Product) Deflator Index, which was defined in the Washington University–Monsanto Biomedical Research Agreement of 1982, as "the quarterly revised Implicit Price Deflator for the Gross National Product as reported by The United States Department of Commerce, Bureau of Economic Analysis."[106] In all cases, the index must be applied in accordance with the specific procedures set forth in the agreement. The most important point to remember is to make certain that both parties understand the index and how it is to be applied.

b. Sponsor Obligations (Other) Often sponsors agree to make somewhat unorthodox contributions in addition to their monetary obligations to the university under sponsored research agreements. These may be in-kind contributions, nonmonetary contributions, or funds to be used for purposes other than those directly related to the defined research agenda. Examples of this type of obligation are the following.

Equipment. Commercial firms in the business of manufacturing and selling specialized types of equipment may contribute to the university equipment to be used by the researcher for the research program that is the subject of the sponsored research agreement. The equipment is generally provided in addition to the monetary contributions that are necessary for the research to be carried out.

Other firms will want to reduce the payments by the cost of the equipment. The issue must be negotiated.

In-kind Goods and Services. Some commercial sponsors may provide in-kind goods or services to the university as part of its legal consideration under the agreement. In order to accept the offer, the researcher must have a use for the product or service so that it will enhance the research project.

Biological Materials. Quite frequently in current biotechnology research, commercial firms may provide the university researcher with some type of biological material, such as a cell line or monoclonal antibody. This type of contribution to the research effort can be an invaluable, perhaps even a mandatory, prerequisite to implementing the research program. Few universities have any hesitation in accepting such biological materials from a sponsor of the research program in which the materials will be used. However, many include provisions in the sponsored research agreement providing for some special obligations regarding the biological material. These are not unlike a short-hand version of a biological materials transfer agreement, which many scientists have become accustomed to seeing in recent years whenever a colleague sends material to him or her to use in research. Most often commercial firms will insist that biological materials that they have produced or acquired by the firm *before* the sponsored research project begins be maintained as confidential (i.e., not disclosed orally or in writing or distributed to third parties, even other university scientists, and used by the university scientist solely for the purposes of the specific research program being carried out under the research agreement). Companies are generally adamant about the need to maintain the confidentiality of these biological materials, because since they were produced *before* the university research project began, the company considers them to be its trade secrets.[107] Companies handle trade secrets—again, the materials developed *before* the university's project began and given to the researcher solely for the research project—differently from biological materials *produced under* or that *arise from* the work under the research agreement (which will be discussed later). In this lone situation, many universities will agree to nondisclosure of the material even to academic colleagues. Again, it is because of the unique trade-secret character of the materials that the university would agree to withhold the materials. As in any agreement dealing with an issue of confidentiality, the contractual clause should provide for a number of exceptions to the university's promise to keep certain defined information confidential. These exceptions also apply a company's trade secrets. Accordingly, there is no obligation by the university researcher to maintain as confidential the defined "materials and information" if (a) the materials and information are in the public domain at the effective date of the agreement or come into the public domain during or after the term of the agreement at no fault of the university; (b) the laws of the federal government, the relevant state government, or a final decision of a court require disclosure; (c) the materials and information are disclosed to the university by a third party not under a duty of nondisclosure to the commercial firm; or (d) the materials and information are already known to the university prior to disclosure of the materials and information by the company. There are many additional exceptions, but these are the ones most routinely used that will provide the broadest freedom to the university researcher.

In most of these situations, the company requests a commitment from the university to return or destroy any of the biological material after the research sponsored by the company is completed. Again, this is the way the company protects the value of its trade secrets.

Most companies will also include a provision, a disclaimer of sorts, stating, for example, that the material is experimental in nature and its characteristics are not fully known. It warns the university not to use it in humans and to handle, store, use, and dispose of it in accordance with applicable state and federal laws and regulations.

Some academic institutions would require that this biological materials transfer-type agreement in the sponsored research agreement be considered part of the research agreement and, as such, be signed by the university and approved by the scientist; other universities would not do so. They would insist, instead, that individual scientists be required to sign personal agreements regarding confidentiality of the biological materials held by the company as trade secrets.[108]

The final and most important point on this issue is that the cardinal principle in the academic community regarding the right and obligation to publish the results and procedures of the scientific research remains tantamount. However, since the materials discussed in this section are trade secrets of the commercial sponsor, there should be no obligation to publish the proprietary characteristics of the material. The scientist should be cautioned to be familiar with the specific university's policies.

Building and Remodeling Costs. In some major sponsored research agreements that will run for a considerable number of years, and that will involve the transferring of significant amounts of funds to the university for a range of research projects, part of the funds may be designated to building a new facility for the university research group.[109] Sometimes the building program is an integral part of the research agreement; on other occasions, it may necessitate a separate legal document with language connecting it to the sponsored research agreement; sometimes the building may be an outright gift. Commercial firms undertake major commitments for "brick and mortar" because it is generally believed that even world class scientists cannot carry out the most creative and productive scientific research without appropriate facilities such as space, equipment, and supplies that can enhance the collaborative and innovative work of a scientific research assemblage. The same point is true for remodeling of university research space with funds provided by the commercial sponsor. Companies would see it as a tragic loss or at least an unwise judgment to fund research activity heavily when the talented scientist in which one has confidence will be made ineffective by outdated facilities.

Even more importantly than some of the contractual provisions dealt with previously in this chapter, commitments by commercial firms to build or remodel research facilities must be handled very carefully by professionals at the university who would handle building and remodeling if done by the university with its own funds.

c. Sponsor Obligations (Technical or Scientific Contributions) Both types of industry research agreements that were identified earlier in this section, sponsored research agreements and collaborative research agreements, may contain provisions for the sharing by the commercial sponsor of tangibles or intangibles, such as scientific information, materials, data, or services. In the case of the sponsored research agreement, the sharing is most often a one-time occasional event; under the collaborative agreement, the sharing of these items is ongoing throughout the term of the agreement and goes from sponsor to university and vice versa.

There are several important points one should remember in working under an agreement wherein valuable scientific components of the defined project are shared. Although final responsibility will reside with the university administration and management of the sponsor, the scientist should generally be aware of the pitfalls—what to look for in order to avoid problems that could interfere with the research function.

1. Who owns the item? Make sure the party turning over the particular article or articles (or service) owns what they are turning over.
2. Read, note, and remember any conditions on the possessing, use, disclosure (to on-campus colleagues and other third parties), and disposal of the product or service.
3. Note any statements regarding scientific publication concerning the information, materials, data, or services. If there is a restriction of any kind, discuss with an administrator at the university to determine how the restriction will work and if it is acceptable.
4. Note if the source of the item requires acknowledgment of it in any publication.

The providing of services by a commercial research sponsor as one of its contribution to the research effort has a few additional concerns the university scientist should note. If the services to be provided are in the form of an individual to work in the campus laboratory, one must make sure the visitor has complied with relevant university policies and procedures. This may involve registering as a visitor with the personnel or human resources department, leaving relevant personal and business information for emergency purposes. It should also entail the addition to the sponsored research agreement of a full explanation of the service to be provided, by whom, the dates when the service will be performed (at least the beginning and ending dates with the actual days to be agreed upon later). The agreement should state clearly that the individual remains the employee of the sponsor, at all times working within the scope of employment of the sponsor, even though in the university laboratory. Consequently, the sponsor should assure the university that its employee is covered by its own workers' compensation and possibly other insurance.

It is exceedingly important that the parties understand that at all times the university laboratory is under the direction and control of the university scientist. The sponsor employee, whether scientist, technician, or computer specialist is in the laboratory as a guest even though he or she may work alongside a university researcher for months. Quite often the purpose of the visitor's on-campus work is to learn a scientific method or process form the academic scientist or to teach the academic scientist a more expedient commercial scientific method. Transfer of certain materials may require explanation and assistance about their use or storage or disposal. In true collaborative research, there is give and take, with the parties' scientists teaching each other what each knows in order to enhance the creative result of the research project.

Another method by which the commercial sponsor can fulfill its obligation of contributing services in addition to funds is to provide the results of services performed by the commercial sponsor's staff at the sponsor's laboratories. Again, it may be for the purpose of providing data or information for a scientist that cannot be obtained from his or her academic laboratory (e.g., synthesis of a compound). Depending on the nature of the research project, this can be an invaluable supplement to the academic research effort.

To reiterate a crucial point: The academic scientist and his or her administration must spell out in detail what he or she needs from the service contribution of the commercial firm. The more concrete and factual the language describing the service obligation, the easier it is to prevent future conflicts over who promised what, when, and where.

d. Assurances of Continuity One of the indispensable obligations of the research sponsor under the sponsored research agreement is to assure the university of continuity of support for the project. For instance, if the agreement allows the commercial firm to withdraw from the agreement at will (or on short notice), the university is left with an incomplete research project and no funds or other contribution to allow the scientist to finish the project. This situation can result in a tragic loss of research continuity, of the scientist's expectations for a significant project, and, perhaps, of research results that could have significantly benefited the public. For these reasons, the university scientist and research administrator must insist upon wholehearted commitment by the commercial firm.

The firm's commitment must run *to the end* of the research project or until one of the target dates[110] (previously agreed to by the parties) is reached, and the parties mutually determine the project should end, or if the sponsor is at that point dissatisfied with the progress of the research and opts out of the agreement. Any opting out by the sponsor must be precisely in accordance with the procedures outlined in the termination clause of the agreement which the parties agreed to during negotiation. It is the university administration's responsibility to see that the commercial firm complies with the provisions of the contract.

Perhaps this is one of the areas where university and industry really have cultural and practical differences. Universities need secure funding; industry makes "quick shifts in research emphasis as a direct result of a change in the industry's marketing interests,"[111] leaving universities at risk for having the bottom drop out of the research project unexpectedly.

Some universities attempt to assure continuity by insisting on provisions in the agreement that require the sponsor to pay already-committed funds on termination if for any reason other than breach of the agreement by the university. For example, if the sponsor wanted to terminate on November 1 of the year and the university had appointed two new faculty members for 1-year terms (to work on the project) when the agreement was initiated on September 1. If the sponsor does not continue to fund these commitments, it would be a dire hardship to faculty and other employees who are dependent on the agreement for their livelihood. Many universities have no way in which they can continue to pay an employee whose salary is funded by a commercially sponsored research agreement, so the issue of assurance of continuity is of extreme importance to the university.

e. Exclusive Sponsor of Research Often, especially in the pharmaceutical field, a sponsor will request from the university the right to be the exclusive sponsor of a broadly defined research project. Most often the sponsor's motivation is to have clear title to a whole body of intellectual property rights covering the university's preexisting research results as well as the results that may arise from the currently funded research. The procedure that can be followed in a routine case of this kind is for the sponsor while considering the research proposal made by the university and scientist to inquire about encumbrances on the particular broad research project based on funding from prior sponsors, commercial, government, or otherwise. Did any other entity fund the research and if they did, what were the strings that were attached? If there were strings attached, have they been severed? Have they been severed in writing? The gist of these inquiries is whether the prior sponsors have been appropriately notified of discoveries under the agreements relevant to them, and whether they had decided to forego rights in the research project, results, and intellectual property. It is imperative for universities to keep clear, current files regarding sponsors (prior and current) or they may lose opportunities for new research funding and technology transfer. After the university has provided the current potential sponsor with satisfactory information regarding obligations to past sponsors, the next significant question is whether the university will agree to decline funding from other sources during the term of the proposed agreement with the current potential sponsor. Some universities will agree and some will not. Other universities try to modify the bare provision by adding to it that the funding it will decline is funding that could adversely affect the current sponsor's rights to the results of the research. The problem with this clause is who will determine what adversely affects the sponsor's rights to the results of the research. It might well be easier to administer the bare promise.

A different, less troublesome problem dealing with the same general issue is where the sponsor and university agree that the sponsor may have exclusive rights to fund the particular research project except for entities listed as exceptions in the sponsored research agreement. Most likely, these exceptions will be grantors such as the National Institutes of Health, other governmental agencies, or private foundations. In these cases, title to preexisting or current intellectual property rights may often be secured by notifying the grantors and otherwise complying with their particular policies. Commercial sponsors generally are agreeable to this solution if the prior sponsors are noncommercial. If the prior sponsor is a commercial firm, the current sponsor will only be satisfied with a legally sufficient release of rights from that prior sponsor.

f. Institutional Performance Obligations Just as commercial research sponsors have obligated themselves to perform in a certain way under the terms of the agreement, such as fund the research in a timely manner, universities also have performance obligations of a binding contractual nature. The essential element that makes the relationship between the parties a legally enforceable contract is that both parties promise to perform in a certain manner or promise to forego performing in a certain manner. The university's obligations may seem obvious: (a) to perform the research; (b) to perform the research within a specified period of time; (c) to report results to the sponsor; (d) to comply with local, state, and federal laws and regulations in carrying out the research; (e) to report to the sponsor discoveries, inventions, or other intangible or tangible intellectual property included in the results of the research; and (f) to provide the basic physical facilities and intellectual environment conducive to research. Certainly the individual researcher also makes commitments to: (a) comply with all of his or her own university's policy and procedures; (b) use his or her best judgment concerning the intellectual merits of a project and how to balance the research activities of acquiring new knowledge and applying existing knowledge; (c) use his or her best efforts when performing the research specified in the sponsored research agreement and not guarantee any particular results (an institutional responsibility also); (d) avoid conflict or apparent conflict of interest situations; and (e) seek to involve graduate students in a sponsored research relationship where possible and appropriate for the educational goals of the students.

The most important commitment the university makes to the sponsor is to perform the research that is the basis of the contractual agreement between them. The relationship between the parties was begun in most cases by the university's sending to the sponsor a proposal that included a statement regarding what research the scientist desired to do; why he or she believes the project has scientific and intellectual merit; and why it might have potential to result in a product or process which can be commercialized. The university and faculty member commit themselves to begin performing the research as described in the proposal and to notify the sponsor if changes in the direction of the research are made. They do not promise or guarantee any particular result, only that their best efforts will be used to conduct the research. Best efforts means to the utmost or the best one can do. It is dependent on the abilities of the individual researcher and requires the researcher to give his or her best try. This is perhaps one of the most important and most talked about issues in the area of commercially sponsored research at universities, primarily because a hint of commercial control of research reaches to the very heart of what scholarly research and freedom of inquiry are all about. It also has a feel of the unscrupulous to think of a for-profit firm telling an academic scientist what research to do, how to do it, or what the result must be. In virtually all of the university policies concerning the performing of industrial sponsored research, the strongest statement in the policy is that the commercial sponsor must not control or direct the research effort. This concern is realistic in that it is surely impossible for a scientist to predict the outcome of his or her research. No doubt many have very good ideas or guesses on how their work will turn out based on their long experience and intelligence. However, that is quite a different matter from guaranteeing that a particular result will be reached. Not only will a university refuse to promise to guarantee any particular results, most insist on including statements that say specifically and emphatically that the university will only use its best efforts in the conducting of the research, and that the university specifically does not guarantee any particular results. Reams (1986), one authority in the field, goes so far as to propose that one of the first provisions that should be written into the agreement is a statement that it will be carried out by the university on a "best efforts" basis, with no financial or other contractual penalty for default by the university except the right of the sponsor to withdraw its support.[112] The proposal that the university should have no penalty, financial or otherwise, for defaulting in its obligations seems to this writer to go somewhat farther than common sense would permit. Perhaps Reams meant that there would be no financial or other contractual penalty for default in terms of the results of the research project not turning out exactly as the sponsor desired. If that is his meaning, it is support for the proposition that no guarantees should be made and, even more, no subtle guarantees should be implied by the researcher to the sponsor. Often scientists seeking research support tend to make guarantees subtly because technician, support staff, or even the faculty member's salary can depend on the receipt of funds from a sponsor. For that and any number of other reasons, including the fact that the university researcher is judged by both his or her peers and supervisors on the quality and quantity of support he or she receives, the faculty member can produce a proposal for research that is quite exaggerated in its expectations. Even if one does this to whet the appetite of the potential commercial sponsor, during the process, the researcher must rewrite the proposal to reflect reality. It must reflect an amount of work, for instance, that is capable of being done within the specified time period, taking into account the other obligations of the researcher. Statements about any commercial possibilities from the results must be realistic and specific, and the commercial field and interests of the sponsor must be recognized so that the obligation to turn results over to the sponsor is specific rather than vaguely implying that the sponsor has rights to everything the researcher does in his or her area of scientific inquiry. The researcher should not describe future research that might be done 5 years in the future and perhaps under the auspices of some other sponsor. Failure to do so could lead to a situation where the sponsor could attempt to come back in 7 years and claim rights to research that was described in the original research proposal (and made a part of the sponsored research agreement) but that was not done until many years later. This may seem farfetched, but it is not, as the writer knows from experience.

To reiterate, the research proposal that is usually attached to the sponsored research agreement as the description of the work to be performed under the agreement must be of the scientist's own making. It must be clear and concise, and it must not describe more research than can reasonably be completed during the term of the agreement with the funds provided by the sponsor.

Another performance obligation of the university is to make periodic scientific progress reports to the sponsor. This obligation is negotiated: The reports can be oral or written, or both, and can be due to the sponsor quarterly, semiannually, or less often. Some sponsors want detailed reports; others want summaries. This is an issue that must be discussed with the sponsor and an agreement reached that is mutually agreeable to both parties.

The university may also have certain obligations for reporting about the financial status of the sponsored research agreement to the sponsor. Universities vary in the frequency or detail they are willing to give to sponsors. This will not be up to the researcher to discuss or negotiate with the sponsor but will have to be negotiated by the appropriate university administrator in accordance with the policies of the particular university.

There are a few miscellaneous points that are made in some university policies on industrially sponsored research agreements that may be helpful for the scientist to know. Many universities provide in their policies that no financial obligations can be assumed by the university except those specifically stated in the agreement, and either the researcher or the university administrator negotiating the agreement must make this clear to the sponsor.[112a] As the research progresses, the sponsor cannot come back to the university and ask the university to allocate more space or employee time unless the sponsor is ready to fund the additions. Other university policies provide that the agreement must clearly state who the sponsor is and where the funds are coming from if they are not regular commercial assets of the named firm. Anonymous sponsorship is not agreeable to academic institutions. Along these same lines, some universities require that the agreements, including the proposal, term, funding sources, and budget be openly available even though the university may not be a government entity subject to one of the open records or freedom of information–type laws. Information regarding any options granted to the sponsor by the university to license results of the research as well as provisions regarding restriction on open communications must be "openly available at some universities."[113]

4. Academic Freedom

Academic freedom is a term with which every research sponsor is familiar. However, there is no doubt that many think (perhaps not aloud) that the concept is diametrically opposed to the commercialization of promising research emanating from the university. On the other hand, for university faculty members it is one of the most highly revered concepts that has evolved in the academic arena. Every university and every faculty member regards the intellectual freedom to teach, to conduct and publicly report scholarly research, to speak out, and to learn as among their most cherished values. The most useful and easy to apply statement on the meaning of academic freedom comes from the 1940 American Association of University Professors' (AAUP) Statement of Principles on Academic Freedom and Tenure:

> (a) The teacher is entitled to full freedom in research and in the publication of the results, subject to adequate performance of his other academic duties; but research for pecuniary return should be based upon an understanding with the authorities of the institution.
> (b) The teacher is entitled to freedom in the classroom in discussing his subject, but he should be careful not to introduce into his teaching, controversial matters which have no relation to his subject. Limitations on academic freedom because of religious or other aims of the institution should be clearly stated in writing at the time of the appointment.
> (c) The college or university teacher is a citizen, a member of a learned profession, and an officer of an educational institutional. When he speaks or writes as a citizen, he should be free from institutional censorship or disci-

> pline, but his special position in the community imposes special obligations. As a man of learning and an educational officer, he should remember that the public may judge his profession and his institution by his utterances. Hence he should at all times be accurate, should exercise appropriate restraint, should show respect for the opinions of others, and should make every effort to indicate that he is not an institutional spokesman.[114]

Traditionally, the concept of academic freedom has specifically included the freedom to conduct and report research; the freedom to teach subject matter without the fear of administrative reprisals; the freedom to publish; and the freedom to speak out privately or publicly on a variety of issues. These freedoms are available to academicians in the United States even more than in many other countries because of the freedoms guaranteed by the Bill of Rights of the U.S. Constitution. The concept of academic freedom set the groundwork for the iron-clad tradition of the university's refusal to allow control or the appearance of control of research by an industrial research sponsor. Unhampered project selections and noninterference with the approach to and procedures of scientific inquiry are nonnegotiable issues with universities. Although industry is apparently coming to terms with this prohibition (as concluded in the Fowler dissertation),[115] it must be raised and emphasized in the initial meeting with a new research sponsor so there is no chance of misunderstanding that could lead to conflict during the performance of the research program.

This issue is as relevant to faculty at private institutions as at public ones. However, at private universities, the faculty member must be certain that his or her employment contract either provides these freedoms or that there is a comprehensive statement in the institution's rules, regulations, policies, or procedures. Often universities adopt other entities statements, such as the one quoted above, the 1940 AAUP Statement.

Part of the principle of academic freedom is freedom to teach; i.e., to work with students with open communication and creative scientific inquiry in order to enhance the student's education in the broadest sense of the word. For these reasons, a faculty member must have the right to bring a student into an industrially sponsored research project with or without the sponsor's concurrence. On the other hand, the faculty has to make sure that the student is not misused in such a project by sponsor-employed scientists who are collaborating. University policies that deal with the subject absolutely prohibit the sponsor's (or faculty's) use of a student as "cheap labor." Also, a student should not be assigned to a commercially sponsored research project that will require undue confidentiality because a student's work must be carefully scrutinized by a number of individuals or the student's education will be detrimentally affected. It is the faculty member's serious obligation to guard against any such occurrence and to see that the student's educational experience is appropriate.

There can be no control by an industrial sponsor over university research done by the researcher under that firm's specific research agreement or over the university researcher not working with the company but doing research. Accordingly, there may be identical or similar research going on in the same or different departments. If a commercial firm funds one researcher, the firm has no control or right to veto the research projects of other researchers. The industrial firm may ask to be informed of such research. It is up to the university and its policies whether to comply with the request. The prime precaution the university must take is to assure that the sponsor does not expect to get access to any research other than that it sponsors and funds.

5. Term of Agreement

It is very important that the ''term'' of the agreement be clearly stated in the agreement. Term is the duration of the agreement from beginning to end. The ''effective date'' (which should be defined in the agreement [e.g., June 1, 1994]) starts the term, and the ''termination'' or ''expiration'' date ends the term. The routine commercially funded agreement is carried out without problems or delays and ''expires'' naturally under that term of the agreement. On the other hand, most agreements provide for termination for default or breach by one party of some provision in the agreement. Generally upon default or breach, such as, failure to make payments by the sponsor or failure to send reports by the university, one party gives the other written notice of the default. If the defaulting party does not cure the breach, such as making the payment or sending the reports, within a certain period of time, the agreement will be terminated. Because the agreement is subject to termination for default, scientists must take compliance with their obligations under the sponsored research agreement very seriously.

When a sponsored research agreement is terminated or allowed to expire, there are a number of provisions or clauses in the agreement that survive and continue to be binding on the parties for years after termination. Examples are:

1. Clauses requiring that identified confidential information owned by the other party be maintained in confidence and not used without the owner's permission. The provision can protect one or both parties depending on which one owns the confidential information.
2. Clauses containing statements about liability and indemnification of each party by the other.
3. Clauses giving the sponsor options to intellectual property arising from the research performed under the agreement.

There are times during the pendency of the sponsored research agreement that the faculty member who is the principal investigator under the commercially sponsored agreement is recruited to another institution. How does this affect the agreement? Although some universities provide in the agreement that it can name another researcher if the principal investigator leaves the institution, it is the rare situation in which this practically works out. In most sponsored research agreements, there is one primary scientist without whom the work could not continue. In that case, if the primary scientist leaves, the sponsor will either terminate the agreement or the parties will mutually agree to terminate it. Most often the company will follow the scientist to his or her new university and fund further research there. However, the sponsor's obligations to the university pertaining to the research done *before* the researcher leaves remain. The result is that if the company still wants to license the results of the research at the original university, it will have to obtain a license from the original university. Ultimately, the company may have to obtain licenses from both universities—the original and current—to have complete legal rights to the product or process it desires to market.

The scientist's laboratory notebooks may be the key to ultimate resolution of conflicts over the research project that is moved by a scientist from one university to another. One must be able to draw a line between work done at each location, although this will not be decisive in determining who owns the resulting inventions (whether patentable or not). It is common for universities to share ownership of inventions (and costs and earnings) when part of the work on a project is carried out at the scientist's original university and part is carried out at the second university. The writer's experience shows that ownership of inventions arising from research results occurring as much as 4 years after the scientist's move can be required to be shared with the prior employer. Again, it must be emphasized that this joint ownership is usually not an impediment to commercializing a result of the research. Agreement must be reached with the sister university, but it usually is not troublesome.

Despite the publicity received by the long-term agreements between university and commercial sponsors, many sponsors prefer a standard 1-year term with an option to continue or renew the agreement if the research has progressed well during the first year. Many sponsored research agreements contain statements specifically so providing if agreement can be reached on the subject matter of the continuing agreement and a new budget can be negotiated. Procedurally, the continuation may be a newly prepared agreement or an amendment to the original agreement. The same issues that were taken into consideration when negotiating the original agreement should be taken into consideration when negotiating a continuation, renewal, or extension of the term.

6. Publication

The freedom for the scientist to publish the results of his or her research is one of the fundamental rights in academic freedom. As a practical matter, it is an absolute requirement for the graduate student to become a university scientist and for the scientist to progress. Although one of the university's missions is to disseminate new knowledge and to disseminate information about new applications for already known information, it is also vital to the individual researcher to be able to publish when he or she thinks appropriate, to control the content of the publication, and to discuss with colleagues the repercussions of the research results for their own and other fields. On the other hand, commercial research sponsors are concerned with having the results of the research they fund publicized too early, so that they may lose patent rights or the competitive edge in the marketplace. These dichotomized positions sound superficially impossible to reconcile, yet most universities and their industrial sponsors have been able to do so in a manner that has made this issue no longer a barrier to university-industry relationships.[116] Fowler (1983) illustrates this statistically in his comprehensive doctoral dissertation.[116a]

Nevertheless, the issue must be understood by the researcher so that it can be explained to the new industrial sponsor, or so the memory of a sponsor who periodically funds scientific research can be refreshed.

a. Delay in Publication It is an anathema to the university scientist to think about delaying the publication of a manuscript, abstract, or book chapter or delay giving an academic seminar, speech, or grand round presentation. If a university research administrator or an industrial sponsor mentions delay in publication, the scientist is usually on his or her feet. One of the factors of which the parties to the negotiation of the sponsored research agreement must be aware is the potential adverse effect of delay of publication for the university faculty member. The scientist knows that if he or she does not get research published in a timely fashion, he or she may be "scooped" by a scientific competitor and as a result lose the opportunity for enhanced reputation (and possible additional research funding). If the discussion can be allowed to develop beyond the obvious meaning of the phrase, it can be proven to the scientist that "true" delay does not have to occur because there is generally a built-in delay for written publications because of the time it takes from submission of the manuscript to publication.

The other view is from the commercial research sponsor who is concerned about the potential adverse effect on patent rights of making the inventions public by written publication or the giving of speeches. The sponsor's concern is derived from several rules of U.S. and international patent law; namely, that publication of an invention or discovery immediately forfeits potential patent rights in virtually all foreign countries; and that under U.S. patent law, the patent application must be filed in the U.S. Patent and Trademark Office within 1 year of the publication or all rights in the United States are lost. For-profit companies are very aware of these rules of law and very conscious of the importance of following them. Trying to market a product without patent protection (which allows the owner to exclude others from making, using, and selling the invention covered by the claims in the issued patent) is very difficult.

In all universities, the scientist is the final decision maker on the content and timing of the publications of his or her research. Some intellectual property policies still give faculty the ultimate right to determine if the invention or discovery is to be made public by publication or patenting.[117] Others state that the faculty member has the right to have input into the decision, but that the university has a right to determine if valuable rights will be lost if the invention or discovery is published and not patented. Regardless of the sponsor's view, most universities will fight to the death to protect the scientist's right to be the final authority on publishing his or her own research findings. In some instances, language of that nature is included in the sponsored research agreement.[118]

The contractual provision that sets forth the agreement of the parties as to delay of publication is one area in which universities differ measurably. Many provide that no research results will be made public until one or other or both parties have the opportunity to prepare and file patent applications covering any invention that is part of the results. Some policies require drafts of manuscripts that describe the results to be sent to sponsors for early review for potential inventions. Most give the sponsor a specific period of time in which to review the manuscript and/or state a specific time period, such as 60 or 90 days, during which the manuscript will not be published. Since it generally takes several months for a journal article to go from draft to a printed publication, agreement to delay for 60 or 90 days does not, in fact, delay the publication. Creative clauses such as these have the ability to satisfy the needs of both parties and are excellent examples of the reconciling of the cultural differences between university and industry.

b. Refraining from Publication There are times when a sponsor may make a special request to the university and scientist in charge of the research under the agreement to refrain from publishing, generally to protect the sponsor's competitive edge in the market. Occasionally, the faculty member may be in no hurry to publish for one reason or the other, or he or she may believe that it is advantageous for the future of his or her research endeavors not to limit the potential for commercialization of the research results. Thus, the faculty member agrees to refrain for a certain time. If the faculty member agreed to refrain from publishing indefinitely to protect the sponsor's rights or marketability of a product or process, his or her university would certainly have something to say about it. Although the institution would take the position that the final decision is the faculty member's, the university would want to assure that the industrial sponsor was not pressuring the scientist, there was no conflict of interest influencing the decision, and the faculty member understood academic freedom and the policies of the university in that regard. Although one would not expect a faculty member to be uninformed about his or her own rights to publish, it is certainly possible, particularly with faculty who may not have gone through some of the past hard-fought battles over academic freedom. At the least, the university research administrator should verify with the faculty member that refraining from publication is not mandated or even condoned by the university. Certainly, in time, the researcher will publish. As long as that decision is solely the researcher's, the university will be satisfied.

c. Editing of Publication Editing of publications has also become a concern of the scientist and university when dealing with commercial firms, and the written sponsored research agreement must speak expressly to it. The issue usually arises under a clause protecting the confidentiality of certain "trade secrets" of the company, which the company shares with the university scientist for the research project to be carried out. Universities leave it up to the faculty member to determine if the trade secret information is necessary for the researcher to carry out the sponsored research agreement funded by the owner of the trade secret. If the company clearly identifies and marks this confidential and proprietary information, as it is often called, confidentiality can be maintained. The faculty member who writes up the procedures of the research project, including the results, may or may not describe the information supplied by the commercial sponsor. However, if it is included, the sponsor will insist that the information be edited out. If the commercial sponsor has identified to the researcher the information that is a trade secret before the researcher made use of the information, then the researcher should not object to editing out the trade secret information (assuming, of course, that the agreement so provided). The reason this editing is approved in the university setting is that the company's trade secret is only of value to the company if it is kept secret. By definition, a trade secret is secret information that gives the owner advantage over others in the marketplace. For the university researcher to disclose it in a publication would be to destroy valuable property. If the information is edited out of the publication, it can usually be done in a way that does not take away from the credibility of the publication. If the faculty member desires, he or she may let the company assist in editing out the trade secrets. However, the researcher should make sure that the trade secrets meet the above criteria before agreeing to delete them. The faculty member should take this assignment very seriously.

Although under these limited circumstances, a faculty member may be asked to edit out certain material owned by the sponsor, it is absolutely verboten for the faculty member to edit out or alter any of his or her scientific conclusions because of comments or requests from a commercial sponsor. It is the faculty member's responsibility to publish the results and the conclusions drawn from those results as accurately as possible. This is common sense as well as a cardinal principal by which universities operate. The commercially sponsored investigator must be alert for any such request explicitly or subtly made by the sponsor and be prepared to say no. Again, the faculty member is the final authority on editing of his or her publication, seminar talk, or any other manner of presenting results of the research, sponsored or not.

To summarize the significant points in the area of publications, it is most important that the faculty member working under a commercially sponsored research project know that at all times, he or she is the final authority when it comes to any aspect of publications, whether writing, editing, presenting, or being asked to delay or refrain from publishing.

7. Preexisting Rights of Institutions

Among the clauses in a sponsored research agreement is one with which few faculty member's seem to be familiar: preexisting rights. It refers to the rights in inventions, discoveries, know-how, and other technology that were owned by the university prior to the effective date of the sponsored research agreement. One standard university sponsored research agreement defines preexisting rights as: "all those previously uncommitted rights of the Board [of Regents] and [the institution] relating to the subject matter of the Research Program which existed before the Effective Date [of this agreement] . . . including, without limitation: _____"[119] This is just one definition, but it explains that a sponsor can be given certain rights, usually an option, to intellectual property rights of the university that exist before the work on the sponsored research agreement begins. In this way, it is possible for the commercial firm to develop a cluster of patents and other technology that may enhance the company's chances to commercialize the technology.

As illustrated in the clause in the example standard sponsored research agreement, it is important to identify the preexisting rights, whether patent applications, issued patents, or other technology.

Patent rights or intellectual property rights clauses will also be found in sponsored research agreements. These are discussed in great detail in Section I. of this chapter and will not be repeated here other than to say that these clauses basically set forth the agreement of the parties as to the ownership and disposition of any intellectual property rights. For example, the university may grant the sponsor of the research an option to license certain technology that arises from the project. Sometimes preexisting rights are included in this clause so that the option applies to the rights that arose before the research under the agreement began as well as the intellectual property right which arose from the performance of the research agreement.

Before negotiating the sponsored research agreement, and particularly the intellectual property clause, the university research administrator and the researcher should identify any preexisting rights and discuss whether they should be included in the intellectual property clause. If the university and researcher choose not to give the sponsor an opportunity for access to the preexisting rights, then they can be omitted from the agreement and not discussed with the sponsor at the time of negotiation of the research agreement. If the sponsor needs them to make, use, and sell a product based on the *results of the research* project, the issue can be dealt with at the time the license is negotiated.

8. Rights to Intellectual Property—Inventions, Know-How, Copyrightable Works, Computer Software, and Tangible Biological Materials

One of the prime issues to be dealt with in negotiating a sponsored research agreement is determining how the rights to all facets of the intellectual property that results from the project will be handled. Will the university retain ownership of them or assign[120] them to the sponsor? Will the university want to retain ownership and merely license rights to the sponsor? What rights will the university retain in either situation? These issues must be sorted out according to the policies of the institution in which the researcher is employed.

a. Definitions The definitions of the various types of intellectual property[121] are:

> A [P]atent is a grant by the United States to an inventor of the right to exclude others for a limited time from making, using, or selling his [or her] invention in this country. It is a printed document in which the invention is fully disclosed and the rights of the inventor are defined. When an inventor secures a patent he [or she] has the opportunity to profit by manufacture, sale, or use of the invention in a protected market or by charging others for making and using it.[122]

Invention means "something invented" in the most simplified Webster form.[123] To invent is to create something for the first time.[124]

According to Mayers in *Drafting Patent License Agreements,* 2nd Ed. (1984), Know-how means

> factual knowledge not capable of precise, separate description, but which used in an accumulated form, after being acquired as the result of trial and error, gives to the one acquiring it an ability to produce something which he otherwise would not have known how to produce with the same accuracy and precision found necessary for commercial success.[124a]

Copyrightable work. A copyrightable work must be an original work set down in a tangible or fixed form. It can be protected under the U.S. copyright laws. It may include traditional written works and other forms such as videotapes, film, music, and paintings. Computer software may also be considered a copyrightable work.[125]

Computer software. Computer software is defined as:

> Computer programs in the broadest sense, user's manuals and other explanatory material which accompany computer programs, and computerized data bases. Computer software also includes: microcode, subroutines, operating systems, high level languages, application programs in whatever form expressed (machine or assembly language, source or object code) or embodied (such as chip architecture, ROM, disk or tape storage, program listings).[126]

Tangible research property (TRP). These materials are defined as:

> tangible (or corporeal) items produced in the course of research projects supported by [the university] or by external sponsors. TRP includes such items as: Biological materials[,] Computer software[,] Computer databases[,] Circuit diagrams[,] Engineering drawings[,] Integrated circuit chips[,] Prototype devices[, and] Equipment.[127]

b. What Intellectual Property the Sponsor Gets Rights to The sponsor will desire to have access to one or more of these categories of intellectual property when funding research in the academic setting. In order to determine during the negotiation period what rights the sponsor might receive, it is important to identify certain characteristics of the rights that might arise, as well as to consider the possible contract provisions that grant rights to the sponsor and the kind that will not.

As a general rule, the sponsor will want some degree of legal rights to any intellectual property encompassed or represented in the results of the research it funds. Many universities provide that the sponsor may have rights to intellectual property that is conceived of or reduced to practice:

1. During the time the researcher is carrying out the defined research program
2. During the term (or duration) of the sponsored research project
3. Using funds supplied by sponsor in the budget (which is part of the sponsored research agreement)

If these three circumstances coincide and intellectual property results, the sponsor will have an excellent claim to the intellectual property *unless* the sponsored research agreement provides otherwise. Some agreements drafted by other universities treat the three circumstances addressed above as conjunctive. If the faculty member makes a discovery in the process of performing the defined research program *or* during the term of the agreement *or* using the sponsor's funds, the sponsor has a valid claim. Quite frankly, this is negotiable. What one university believes to be important, another will not and vice versa. The researcher must make sure that he or she knows the university's policies and that the repercussions are well considered.

Another category of rights that the sponsor of research may desire access to is what the writer will call "improvements" to the intellectual property that arose under the sponsored research agreement. By definition, these improvements arise *after* the conclusion of the sponsored research agreement. This type of improvement can be defined similarly to the way Mayers and Brunsvold (1983) define improvements to licensed technology as: anything that performs the same function as the specific results, of the defined sponsored research program, in a better or more economical way; or any beneficial modification of a component (or material) which may be useful to a potential invention described in the research results, or anything that performs functions similar to those of an invention described in the research results and infringes the claims of the patent application(s) filed on the invention.[128]

One might say, and the writer would agree, that this type of improvement has no place in a sponsored research agreement because the sponsor has bargained to receive some level of rights to (and only to) any invention or discovery encompassed in the results of the agreed-upon research program. As a consequence, most often rights to *improvements to the research results* are left silent in the sponsored research agreement precisely because it is routine (and equitable) for the sponsor *not* to be granted any rights to them. If desired by the sponsor, improvements should generally be the subject of a separate second-generation sponsored research agreement. However, it is important that the sponsor understand clearly what the university intends to happen with improvements so there will be no misunderstanding about it later. Some commercial firms take the common sense approach that they have not bargained for improvements to the research results but only to the results; other firms take it for granted that these improvements will be legally obligated to them and, hence, fall under any option they may have. Which assumption has been informally adopted by a sponsor is hard to determine without posing the explicit question (which is a good idea). One sponsor may maintain a substantial program of continuing research and development, and for this sponsor, the value of future improvements may be as important or more important than the initial research program. That firm must be told of the need to sponsor further research to gain rights to improvements. Another sponsor, usually of more applied rather than basic research, may believe that trying to gain rights to improvements is unimportant because the results of the current research are likely to be a product or process that is ready to market and that will need developmental work only by commercial firm itself and not the academic researcher.[129] Even he or she should be informed on the subject.

One of the past *and* present reasons for the assumption that the sponsor will get some nature of rights to the improvements is that a standard definition of patent rights used in many sponsored research agreements and license agreements includes any continuations-in-part of the relevant patent application. This legal term may well mean in practical laymen's terms, "improvements." Certainly in patent law, continuation-in-part means the addition of new matter to the patent application that is usually improvements to the underlying invention. If the academic researcher and administrator are not careful, he or she can legally commit and obligate those future "improvements," sometimes the subject of a continuation-in-part, to the research sponsor even though the sponsor did not fund the research and the improvement arose *after* and not during the research agreement. Again, a common term used in the agreements to define patent rights is as follows:

> Patent rights shall mean any patent application or patent covering an invention, made during the course of the research program, including any continuations, continuations-in-part, divisionals, reissues, reexaminations, substitutions, extensions or additions thereto, and any corresponding foreign patent application or patent based on such application or patent.[130]

As an aside, it should be made clear to the reader that many sponsored research agreements will not use or define a term like *patent rights* above. Some are simple and basic and do not go into great detail regarding what intellectual property a commercial sponsor may get rights to. Other more complex and detailed sponsored research agreements may use such a defined term as *patent rights* either to describe the option (e.g., university grants to sponsor an option to license any invention, discovery, patent right or technology right that was conceived of or reduced to practice during work under the research program), or to describe the defined terms to be used in the future license, if any. Some universities attach to a sponsored research agreement a model license or basic license definitions and terms so that, should the licensing stage of the project be reached, some of the issues will have already been agreed to.

In some cases, the pertinent clauses in the agreement purposely omit any promise to sponsor of improvements. However, if the university is not careful, some right to the improvements can be inadvertently granted through other clauses such as the definition above. There are several potential answers to this problem, including at a minimum, adding language after the term *continuation-in-part* in the definition of patent rights. That language is the phrase: the new matter which arose from research funded by sponsor, or the shorter phrase: funded by the sponsor.

It would be even better to add a clause stating something along the lines of: It is not the intent of the university to grant to sponsor any rights to (including but not limited to an option to license) improvements to any intellectual property encompassed in the results of the defined research program which are conceived or are reduced to practice *after* the termination of the sponsored research agreement.

The most important point from this section for the researcher to remember is to discuss with the industrial sponsor its expectation regarding this type of "improvements" clearly defining the terminology so there will be no misunderstanding.

Another category of rights that the sponsor may desire to be granted is preexisting rights. Discussed in the previous Section II.B.7, the only comment that will be made here is that this type of right is one that most sponsors will definitely be interested in obtaining access to. These rights arose *before* the effective date of the sponsored research agreement and are owned by the university. Unless the university grants to sponsor an option to license the preexisting rights, the sponsor acquires no rights (options or otherwise) to them.

However, if the company desires a cluster of patents and technology that were developed both before and during the sponsored research agreement, it will be to the university's advantage to include these rights in the option grant to sponsor.

A side issue that is nevertheless of importance to the researcher is whether there are prior sponsors of the specific research project or its ''ancestors'' that have continuing rights to the research so that clear title to the intellectual property would not be available to the current sponsor. Commercial firms are familiar with federal agencies being sponsors. However, they may well ask for proof that the proper notice of election to exploit the invention has been filed with the agency. A larger concern to them will be former commercial sponsors who may have continuing rights, such as an option that has not ended to license technology. It is the obligation of the university to obtain from prior commercial sponsors a formal and legal release of any rights that the earlier sponsor might still have. This procedure is analogous to, but much more simple than, clearing title to a tract of real property or even an automobile. Before an individual will agree to buy, he or she wants proof that no one else has continuing rights to it because of a prior lease or purchase. Most former commercial sponsors will provide the university with a release of any remaining rights to a particular current technology, preexisting rights, or inventions that might arise from a sponsored research program. On the other hand, the former sponsor may not want to give up the last 6 months of its period to exercise an option to a license even though it might keep the researcher from beginning to work with a new sponsor. Sincere appeals from the university or researcher based on the need for further funding, the university's belief that the public will suffer if the research is delayed, and the scientist's right under the principle of academic freedom to continue to make scientific inquiries into the area of the scientist's choice may be successful. These appeals, which in reality are quite practical, are often successful in unencumbering the subject of the research project. New research with a new sponsor can then proceed.

c. Nature of Right That Can Be Committed to Sponsor In the standard university-industry relationship providing for funds for a specific sponsored research project, as we have been discussing, there are several different kinds of legal rights to intellectual property that the commercial firm can acquire. A discussion of them follows.

Universities can grant to the industrial sponsor an assignment of ownership in the inventions and other technology that arise from the work under the sponsored research agreement as long as the university's policies are complied with and no right of a prior sponsor is involved and attempted to be assigned. By standard definition, an assignment is "a transfer . . . to another of the whole of any property, real or personal, or of any estate or right therein."[131] Assignments are actually used for transfer to the sponsor of ownership of patents, patent applications, and other types of intellectual property. It is distinct from a license that passes no title and merely grants to the licensee "an assurance of immunity from suit in respect to acts done within the scope of the license."[132] At one time, it was quite routine for universities to assign rights to intellectual property to corporate sponsors of research. However, in the recent decade, more and more university policies have provided only for licenses to be granted to sponsors. Ownership must be retained by the university. For instance, Yale University provides that

> The agreement should explicitly state Yale's ownership of . . . patents . . . that result from the research activities of the faculty . . . The agreement should also include the conditions under which licenses will be slanted [sic].[133]

However, other universities such as Duke state in their Industrial Research Policy that

> In general, the university retains ownership and control of the patents [resulting from research sponsored by industry]; however, if required by the sponsor, the university will assign and convey to the sponsor under the contractual arrangements rights to all patentable discoveries and inventions resulting from the sponsored research.[134]

Another example of how university policies cover this issue is Stanford University. Its policy provides

> Some arguments [for externally-funded research] permit the university to retain title and grant license rights to the sponsor, some provide for the reverse [i.e., assignment of title to the sponsor] or defer allocation of rights.[135]

The scientist should always be aware of his or her own university's policy in regard to assignment because it is often one of the first questions a potential industrial sponsor asks. Although the particular sponsor may want an assignment, if the scientist's university does not allow assignment, it will be helpful to explain that even though the university does not allow assignment of inventions to sponsors, it does permit an exclusive license to be granted. Most industrial firms will realize that an exclusive license for the life of the patent is legally tantamount to an assignment and will be satisfied without an assignment itself.

One of the other categories of intellectual property rights that can be granted to the commercial sponsor of research is an "option to license" inventions and discoveries (usually whether patentable or not) that arise (i.e., are conceived of or reduced to practice) during the performance of the defined research program during the term of the sponsored research agreement. The option is granted to the sponsor at the time the agreement is entered into (sometimes for additional cash consideration, such as, $5000.00). The term or duration of the option is generally the term of the research agreement plus a specified period of time after the research agreement has expired. This latter time period is strictly a negotiable issue. The sponsor usually wants a longer time period and the university usually wants a shorter one. During the full period of time (i.e., the option period), the sponsor has the opportunity to decide whether to enter into negotiations for a license to the inventions, discoveries, and/or patents. Sometimes sponsors decide early what their commercial intentions are; others ask for a few months' extension of the option period in order to complete the commercial or market evaluation of the invention. It is at the university's discretion to grant the extension or not.

If sponsored research agreements grant an option to license, the agreement should set out clearly the procedure the sponsor will need to follow in order to exercise the option and initiate negotiations for the license. If the sponsor definitely decides not to exercise the option before the end of the option term, it is very important for the university to know it and to have the opportunity to request that the sponsor "release" the option rights so that another sponsor or licensee can be located. Commercial firms have varying policies on this issue which will be negotiated by the parties.[136]

Some universities grant to research sponsors rights of first refusal to license inventions arising from research instead of option rights.[137] The right of first refusal is the sponsor's right to accept the first offer to license the inventions that the university may choose to make. The right may also be activated by the university's receipt of an offer from another firm to license. In that case, the sponsor has the right to meet the offer or to refuse.[138] Despite the fact that legally a right of first refusal and option are very different, the writer's experience is that universities handle rights of first refusal to inventions arising under commercially sponsored research agreements in the same manner as the option. Pursuant to the option, the sponsor has a negotiated period of time during which it must make a decision on whether to enter into negotiations with the university for a license to the inventions. If the sponsor does not act within that time, its option expires and the university is free to seek another licensee. It would serve the scientist well to question the university's administrator regarding the nature of the right granted to the sponsor and precisely how it works.

It is also possible for a university to grant a commercial sponsor outright licenses to the appropriate inventions instead of using an option or right of first refusal. Apparently Massachusetts General Hospital made that type of grant to Hoechst AG in section 6.3 of their 1980 contract.[139] In this case, and others like it, the commercial firm's monetary contribution was so substantial and long term that it could assure Massachusetts General Hospital of its interest in bypassing the discretionary steps and moving forward with commercialization under an immediate license.

Clauses in sponsored research agreements dealing with intellectual property or patents generally make some provision for (a) preparing (including searching), filing, and prosecuting[140] patent applications through the U.S. and foreign patent offices; and (b) protecting, to the extent possible, other forms of intellectual property, such as, copyrightable works, software, videotapes, know-how, and so forth. The provisions often cover who is to do the work; who will bear costs, and who will have authority to review and comment.

If the university is to own (i.e., retain title to) the inventions arising from the sponsored research agreement, it will usually want its own patent attorney to prepare, file, and prosecute the patent applications. In the situation where the university will assign the invention to a commercial sponsor, the sponsor would generally want to control the patent application and would use its own attorney. However, there are more complex situations where the parties will mutually agree to do other than these two general rules. For instance, if the subject matter of a sponsored research agreement is a product or process patented already by a sponsor and the goal of the research is to make it work better, it may be logical for the sponsor to carry out preparation and prosecution of the patent application even though the university retains ownership. Each of these situations requires the cooperation and time of both parties regardless of who owns the invention, prepares, or prosecutes the patent application.

Most universities provide that the sponsor pays all costs of patenting the inventions regardless of who holds title. If the sponsor refuses to pay, the sponsor may well have lost rights to the invention, whether option, right of first refusal, or out-right license. The researcher and university research administration must be cautious to make sure these issues are explicitly and clearly set out in the agreement.

Delay of publication has been discussed in a previous section of this chapter and will not be repeated here except to say that in order to have time to get a patent application filed and prepared, the sponsor will often ask for delay in publication of a journal article, book, or abstract or the giving of an oral seminar. Most universities will grant a short delay, usually 30, 60, or 90 days; some will allow short extensions under extenuating circumstances.[141] However, since written publications take so long to actually be distributed, the requested delay poses no problem. A request for a delay longer than 90 days should be granted by the scientist or university only under the most unusual circumstances.

9. Confidentiality Obligations

In attempting to decipher industry-sponsored research agreements, there are several types of confidentiality obligations that must be recognized and understood:

1. The confidential and proprietary (i.e., trade secret) information owned by the sponsor and transferred to the university scientist
2. The university's confidential and proprietary information developed *prior to* the beginning of the research program defined in the sponsored research agreement
3. The university's confidential and proprietary information developed *during* the research program defined in the sponsored research agreement

In some sponsored research agreements, it is necessary for the commercial sponsor to release to the university certain confidential information owned by the commercial firm. This confidential information is usually of such value to the firm that it can accurately be called a trade secret,[142] and it must be maintained confidentially by the university in order to retain its commercial value. For that reason, the university needs to analyze the problem to determine what its policy will be on keeping confidential such information. The editing of publications discussed earlier in this chapter presents some of the arguments and concern over this issue because the sponsor's trade secrets are often the only information some universities (but not all) will allow to be edited from the scientist's publication. However, the confidentiality obligation the sponsor would like to impose on the university is broader than just editing its trade secrets from publications. Maintaining information in confidence also involves: (a) not disclosing to any third party or group of third parties (e.g., seminar) and, (b) *not using* the information for any purpose unless approved by the sponsor; and not disclosing the information in a written publication, formal or informal. The key is for the researcher *not* to accept the sponsor's trade secret information unless (a) the university's policy allows the maintaining of the sponsor's trade secret in confidence,[143] (b) the trade secret is absolutely vital to the defined-research program, and (c) the scientist is certain that he or she can adequately protect the information from inadvertent or intentional disclosure by other employees and staff. If it is decided that the sponsor's trade secrets will be accepted by the researcher, the sponsor should agree to clearly mark any information or trade secret with the legend "trade secret" or "confidential and proprietary" before it is released to the university researcher. The university researcher would be wise to lock up these trade secrets at all times when not in use in the laboratory or office and to allow no one in his or her laboratory or office to see them unless there is a vital need so that the research can be performed under the agreement. When the research program under the agreement is completed, the trade secret information should be returned to the sponsor with the sponsor acknowledging receipt in writing.

As identified above, there are two other types of confidential and proprietary information relevant to the sponsored research agreement. One is that which is developed *prior to* the beginning of the defined research program; the other is that which is developed *during* the defined research program. Both are owned by the university and are subject to negotiations between the parties depending on the nature of the sponsor's interest in licensing of products. The sponsor's first and foremost concern will be potential licensing of the results of the research that the sponsor is funding. If there is confidential and proprietary information in those results, it enhances the likelihood of the sponsor's obtaining a good competitive position. Accordingly, the sponsor will likely want to keep the information confidential or file a patent application on it. However, what if the scientist wants to publish the results, including the proprietary information in a scholarly journal? The same question is pertinent to the proprietary information developed prior to the current research program. The answer to both questions is that the scientist *can* publish it, but one would hope that the scientist would understand that in order to protect the valuable information, a patent application needs to be filed on the subject of the proprietary information, and that if necessary, publication may be delayed for up to 90 days. Generally there is no real delay because it takes so long for a manuscript to reach the point of an actual printed volume.

What if the scientist does not want to publish the proprietary information? The company can take it via exercising options and licensing and use it to develop and commercialize a product (still holding both types of confidential and proprietary information in confidence).

10. Indemnification and Hold Harmless Clauses

In this litigious day and time, virtually all sponsored research agreements between the university and a commercial firm contain indemnification and hold harmless language. Simplistically, indemnification means that one party agrees by contract to accept certain risks for the other party in the event of monetary losses arising from specific acts or omissions of the first party. In other words, it states which party, as between the university and sponsor, will be financially responsible for damages arising under the agreement. If I suffer a loss because of your action, you contractually agree to bear that loss (i.e., pay a judgment or claim). You bear the risk for me, or as the language is used in most sponsored research agreements, you must indemnify me.

Many of the agreements will use not only the term *indemnify* but also the term *hold harmless*, which is merely a special form of indemnification. It is used when the indemnification is an incidental obligation to services or products, for example.[144] Having both indemnification and hold harmless in the clause is added protection for the indemnitee or the party being indemnified.

Seen most often in patent licenses, indemnifications are an important part of the sponsored research agreement.

After reading many university policies on sponsored research, the writer has found no provision that addresses the indemnification of a commercial research sponsor *by* the university. However, in reviewing a few "form" standard sponsored research agreement, several example clauses were located that show in reality what a university will agree to after negotiating. For instance, one agreement provides

> [University] shall, to the extent authorized under the [State] Constitution and laws hold sponsor harmless from liability resulting from [University's] negligent acts or omissions within the terms of this agreement; provided, however, [University] shall not hold sponsor harmless from any claims, demands, or causes of action arising in favor of a person or entity, growing out of, incident to, or resulting directly or indirectly from negligence (whether sole, joint, concurring or otherwise) of sponsor, its officers, agents, representatives, or employees, or any person or entity not subject to [University's] supervision or control.[145]

Another says

> University indemnifies company against claims by university, university employees, and third parties resulting from negligence on the part of university or its personnel in the performance of this agreement.[146]

The former example contains language required because of the university's status as a state institution, that limits by law how much it can indemnify and hold harmless (i.e., accept the risk for) the commercial sponsor. Some state universities are subject to these types of legal limitations on paying a monetary judgment to a third party based on claims of injury for which the sponsor or the university may be liable. Private universities would not have these statutory protections and could be seen as "the deep pocket" by a plaintiff who is suing the university over acts of the sponsor.

Acts for which indemnification provisions provide protection are generally defined as acts or omissions of the indemnifying party. For universities that are state agencies, it is often only *negligent* acts or omissions of the university for which the university can indemnify the sponsor. This limitation is based on vestiges of the sovereign immunity doctrine. Accordingly, the university should not be expected to offer liability protection to the sponsor. Not only is it considered important for the researcher to know this, the sponsor should be told so that there is no confusion later.

With few exceptions, commercially sponsored research agreements will and should contain language by which the commercial sponsor indemnifies the university for any loss the university may suffer based on the acts or omissions of the sponsor in carrying out the agreement or using the results of the agreement. In other words, if the university is assessed monetary damages based on the sponsor's acts under the agreement, the sponsor pays the judgment for the university. The university suffers no loss of funds. This provision and its wording are extremely important because many universities have no funds set aside to pay a judgment of this nature and its endowment or other funds could be put at risk without the indemnification. In addition, the potential risks to the university increase tremendously when the sponsored research agreement pertains to medical research that may ultimately result in a medical product or process that will be used for human health care. With risks of this magnitude, the sponsor's indemnification of the university must be solid.

If the sponsor is a small company that might not have the funds to make an indemnification good should the situation arise, the university would be wise to insist upon adequate insurance, some naming the university as beneficiary. In addition, the university must insist upon the sponsor's sending evidence of insurance to the university periodically, as well as having the sponsor agree to notify the university should the insurance be discontinued for any reason.

11. Use of Institution's or Scientist's Name

Most universities prohibit the use of their name by a commercial sponsor of research if the use is for commercial or promotional purposes, such as advertising a product, raising investment capital, or going public.[147] Generally the university policies also state that the name of the university may be used by the sponsor for these and other purposes so long as prior written consent is given.[148] A variation of this is Texas A&M University System's policy that states

> If a sponsor desires to use the name of the [Texas A&M] System or a part of the System in connection with the expected program results, provisions for such use will be included in the written agreement. The agreement will specify that advertising referring to the program results will not imply endorsement of the sponsor or the sponsor's product or services by the System or its personnel, and that the data or statements attributed to System personnel will be used in proper context.[149]

These rules may also apply to logos or seals or other trademarks identified with the university. The concern is that an institution of higher education and academic research may lose credibility as well as its reputation for intellectual and objective analysis by appearing to endorse a commercial product. In addition, the sponsor should "prohibit industry from unfairly profiting in the use of [the university's] name."[150] The same may apply to the names of the university scientist and other employees. The university must state these prohibitions in clear language in the agreement.

12. Dispute Resolution

There are today many options when it comes to dispute resolution without one of the parties filing a lawsuit in an attempt to resolve a conflict. Arbitration is a familiar one that may or may not be appropriate in the academic setting because the common (and perhaps inaccurate) belief is the arbitrators try to "split the baby."

Some policies provide for specific types of dispute resolution. For example, Brigham and Women's Hospital provides: "Any disputes or problem of interpretation that arise in connection with this policy shall be resolved by the President with the advice of the Research Committee."[151]

Regardless of whether dispute resolution is an option the parties want to leave open to themselves should the sponsored research agreement not work out satisfactorily, it is imperative that the sponsored research agreement provide that the parties can both opt out of it in an equitable way if there is a failure of a working relationship.

13. Consulting Agreements

Consulting agreements are presented at length in Section III of this chapter, so the discussion here will be brief and related solely to the sponsored research relationship between the university, the researcher, and the commercial sponsor.

By use of the term *consulting* the writer is referring to the activity of the university scientist by which he or she uses expertise to advise, assist, and make recommendations to a commercial firm regarding scientific fields of commercial interest to the sponsor and receives a fee.

Some universities prohibit a faculty member from consulting with a sponsor of his or her own research at the university because of the potential for confusion of roles for the faculty member and, thus, confusion about which agreement (i.e., the consulting agreement or sponsored research agreement) an invention arose under. Ownership becomes an issue and confusion may abound.

Other universities allow what they call "extra-mural" consulting agreements between the university scientist and the commercial sponsor if the university will be deemed to have a financial interest in "any commercialization resulting from the *combined* sponsored research and extramural consulting activities of the faculty member," and potential conflict of interest issues are solved by the university administrator.[152]

Still others seek a mid-point by allowing the scientist to perform consulting services for the commercial firm funding some of the scientist's research but only under these conditions: (a) the subject matter of the consulting arrangement must be clearly stated and must be different from the defined research program in the sponsored research agreement; (b) the consulting provision is written into the sponsored research agreement so there is no doubt that both parties have had the relationship disclosed to them; (c) the payment for consulting comes to the university to be used for research by the faculty member (i.e., funds do not go to the faculty member personally); (d) both faculty member and the sponsor have agreed to keep their two relationships separate; and (e) all inventions that arise under the consulting provision are to be owned by the university.[153]

Without exception and regardless of the type of policies the universities have pertaining to the relationship between consulting agreements and sponsored research agreements, virtually all universities prohibit any consulting that would be a substantial conflict of interest and refer their faculty to that policy. Accordingly, the researcher must keep in the forefront of his or her mind that the university's intent is to fulfill its missions and avoid conflicts of interest while allowing faculty to follow their research interests. In addition, the researcher must know its institution's policies on industrially sponsored research, consulting, and conflicts of interest. If a specific research project will combine two or more of these policies and the result is unclear to the scientist, he or she must contact the appropriate office within the university for clarification.

14. Conflicts of Interest-Commitment

Conflict of interest is also dealt with in Section IV of this chapter, to which the reader is directed. However, a few comments on the subject of conflicts of interest as it relates solely to sponsored research agreements will be made here.

There are many definitions of conflict of interest, several of which have been used in this chapter already. Yet another one is as follows:

> [A] conflict of interest can be said to exist when a member of the university community has a relationship with an outside organization such that his activities within the university could be biased by that relationship in a direction which would ultimately provide direct financial benefit to the individual or a close family member. Conflict of commitment . . . exist[s] when a member of the university community has a relationship which requires a commitment of time or effort to non-university activities such that an individual . . . cannot meet the usual obligations to the university.[154]

The focus of conflict of interest is most often the financial interest that the university scientist holds in the commercial sponsor of the scientist's research. Some universities, very often in unwritten policies, prohibit a faculty member from being the principal investigator of a research agreement sponsored by a commercial firm in which the faculty member or his or her family member holds a substantial financial interest. Others would allow this type of relationship to exist if it was a de minimus financial interest and was fully disclosed. Others would require the relationship to go through a review process on campus either by a committee or administrative hierarchy.

Often universities fear that the situation wherein the faculty member has a financial interest in the firm sponsoring his or her own research can cause adverse effects on the university, such as (a) jeopardizing the scientist's control of the scientific approach to the research project. (There is concern that the sponsor might pressure the scientist into directing the research in a specific commercial direction and that the scientist might bend to that pressure because he or she has money to be made or lost by the results of the research.); (b) the faculty member refusing, directly or subtly, to freely interact with the graduate students, fellows, and other trainees on campus. (The scientist who has a financial interest in the company may feel strongly about keeping the commercial sponsor's scientific information and the results of the scientist's research [which is obligated to the sponsor via contact] secret within the university department. This may or may not be an educational loss to the graduate student, but it is one of great concern to the university and for which they should have a policy.); Harvard University's policy for medical faculty specifically states that: "Students and trainees should not ordinarily participate in research that involves [a commercial firm's] confidential information or otherwise constrains their right to publish or communicate freely"[155]; (3) the faculty member's being influenced by the sponsor on the direction of the scientist's future research. (The fear is that instead of following his or her own intellectual priorities, the faculty member who has a financial interest in his or her own sponsor will orient the future research to the commercial desires of the industrial sponsor.).

The best advice for any faculty member who has an inclination to have a financial interest in an industrial firm that may fund the scientist's research is to be constantly sensitive to the potential for real or apparent conflicts of interest and conflicts of commitment. If a question arises about a situation that the scientist thinks may be potentially inappropriate, he or she should consult the university's policy and contact the appropriate person on campus in order to resolve the issues.

One characteristic of an industrial research sponsor that may make a substantial difference in how a conflict of interest is handled within the institution is the size of the company. Having an equity interest in a large multinational company is quite different from having an equity interest in a very small start-up company. Simply looking at it from a mathematical standpoint would illustrate the difference. Most universities take this into consideration in analyzing potential conflicts of interest and commitment in the sponsored research relationship.

One of the concerns raised by the authorities in this area is that the employer may actually invade the privacy of the individual faculty member in an attempt to gather the information needed to make a scientist's conflict of interest acceptable to the university-employer.[156] The pendulum would have swung too far in the opposite direction. This undoubtedly will be one of the arguments made by faculty members when required by their own university or a federal agency to disclose private financial information. In fact, there will probably be a lawsuit that will have to settle the issue on how deep into the faculty member's finances a university can go (through requesting disclosure) without violating the faculty member's right to maintain his or her personal financial information private and confidential. Although unsettled legally, it is a concern of which the university and faculty member should be aware.

15. Miscellaneous Provisions in Sponsored Research Agreements

There are many, many clauses that can be written into commercially sponsored research agreements to provide varying levels of comfort to the two parties—the university and the commercial firm—on a wide variety of issues unique to particular factual situations. A few of these are:

1. The commercially sponsored research agreement may "in no way compromise the University's policy of non-discrimination."[157]
2. The agreement must "be carried out in a manner consistent with existing governmental regulations regarding such matters as human and animal subjects, recombinant DNA, and the use of hazardous and radioactive materials."[158]
3. Some sponsors request clauses in the sponsored research agreement that give them a right to receive notice by the university of its intent to enter into another sponsored research agreement to do research that is similar to the research sponsored by that first sponsor. One university will advise the faculty member not to accept the second research agreement "only if there is a reasonable possibility that the work done for the second sponsor will infringe on the proprietary rights of the first sponsor under the preexisting sponsored research agreement.[159]
4. Sponsored research agreements almost always state that title to equipment obtained during the term of the agreement (for the purpose of the research project) with the sponsor's funds will be owned by the university.[160]
5. The sponsored research agreement must adhere to the university's policy prohibiting secrecy in research and maintaining the university's open teaching and research environment philosophy.[161]
6. The faculty member is obligated "to notify all government or other current sponsors (if they require) of any other research which is funded by an industrial organization and which requires the use of the same space or facilities."[162]
7. The agreement must explicitly "permit faculty members and students to withdraw from the research program."[163]

8. The sponsored research "agreement may not restrict the research activities of faculty members and students subsequent to their withdrawal from the [sponsored research agreement] or the expiration date of the agreement."[164]
9. "[T]he university must ensure that it does not promise . . . exclusive intellectual property rights in the same research results to more than one sponsor (except with the knowledge and formal consent of all affected sponsors)."[165]
10. University researchers "should be informed of the existence of industrially-sponsored agreements in their [specific] department. . . . "[166]

III. EXTERNAL CONSULTING AGREEMENT

A. Introduction

It is worth stating once again, as the universities do repeatedly, that their mission is teaching, research (and patient care in the case of academic medical centers), and discovery and dissemination of knowledge. Without exception, the full-time university scientist owes his or her overriding obligations to the university. Once this principle has been recognized and become a routine way of life for the scientist, other scientific relationships external to the university may be undertaken with care. Virtually every university declares the importance of, and even encourages, its faculty participation in the nonuniversity scientific arena; each of the same universities remind the scientist that his or her first and foremost responsibility is to the university. Accordingly, the faculty member must be careful not to undertake any nonuniversity or external professional relationship unless it is consistent with the mission and policies of his or her university.

B. Definition and Purpose

Consulting has been defined as

> external (international and domestic) professional activities for pay and includes any activity that: (a) is performed on an individual contractual basis for any individuals(s), firm or agency other than [t]he [u]niversity . . . ; (b) is based upon the professional knowledge, experience and abilities of the faculty or staff member; (c) is undertaken for personal compensation beyond the payment of a nominal honorarium and/or reimbursement for expenses; and, (d) clearly enhances the professional development and expertise of the faculty member.[167]

There are many types of consulting (such as advisory consulting for government organizations), but this definition (and the following discussion) is tailored precisely to the university faculty member and remunerative, professional activities external to the university.

The consulting agreement is just one example of one type of agreement between a university-employed scientist and a commercial firm. Its uniqueness lies in the fact that the parties to the agreement are the individual scientist (even though employed by a university) and a commercial firm instead of the university and the company. The primary value within the consulting arrangement for which the company is willing to pay is the expertise of the individual scientist, based on his or her experience, and reputation within the academic and industrial scientific communities. For that reason, the scientist must understand that the agreement is a legal contract that is personal to him or her whether or not the university countersigns or approves before or after the activity. The agreement will have personal legal repercussions for the individual, so it must not be entered into lightly.

It is instructive to note that traditionally commercial firms have employed university faculty members as consultants in order to gain access to their general expertise. More recently, these private firms have been attracted to faculty because of the firms' specific interests in the results of the faculty members' university-based research. In pursuing this interest, commercial firms have blurred the historic lines between external consulting and university-based sponsored research. Many university scientists have been alert to this change, whereas others have not and have signed any agreement presented by a firm with consulting as the title, failing to determine the true character of the written document.

In the most common form of consulting arrangement, the company pays a specific dollar amount per hour or day, generally for a specific number of hours or days per year, for a term of one or more years, in exchange for a firm commitment by the scientist to consult in a well-defined scientific area in which the company has a commercial interest. The key term that must be defined with exactness is the *scientist's obligation to consult.* For instance, the scientist's obligation to consult may be delineated as the scientist's making himself or herself available by telephone to talk to representatives of the company and/or traveling to the company's laboratories to give general scientific seminars to the company's employees. There are an infinite number of obligations that can be inferred from the commitment to consult. Accordingly, there is no substitute for a clear definition of the scientist's obligations that both parties must expressly accept and make part of the contract.

The scientist must also make certain that the term or time period of the consulting arrangement is spelled out clearly. The number of days or occasions such as visits to the firm should be precisely stated as should the consulting fee (e.g., per day, per visit) and travel expenses.

Scientists must also take care to assure that the defined scientific area that will be the subject of the consulting is an area that is not in some manner already encumbered by or obligated to another commercial firm (or sponsor of other kind). For instance, if a scientist has been working under a sponsored research agreement between the university and a company on Technology A, that initial sponsor will generally have some rights to the results of the research on Technology A, such as an option to license. An option to license generally may be exercised during the term of the agreement or during a period of time following the agreement; the length of which is negotiated by the parties. During the time the option is still available to the sponsor on Technology A, the university and/or faculty member cannot *contract* or deal with another commercial firm for consulting or any other services wherein Technology A would be disclosed, used, or commercialized. In addition, because confidential and proprietary information owned by one or both of the original parties (and existing before the start date of the agreement) is routinely exchanged under a sponsored research agreement, the scientist must take great care not to violate the agreement by disclosing this confidential and proprietary information. However, if the firm desiring the consulting arrangement wants to be assisted in a different subject matter area that can be defined, for instance, as Technology B, there is no conflict or potential breach of the agreement relating to rights to results of Technology A or the confidential and proprietary information disclosed by one of the parties in order for the work on Technology A to progress.

Applying these simplistic principles to the consulting situation, the lessons to be learned by the scientist are (a) do not consult in the specific defined subject matter area in which you have another ongoing agreement, such as a sponsored research agreement, unless the two areas of work can be specifically distinguished one from another; and (b) be careful not to disclose confidential information that has been received by you from one commercial firm (under an obligation of confidentiality) to a firm with which you are consulting and vice versa. The exception to the first point in this advice would be, for instance, a consulting agreement entered into by an epidemiologist that required him or her to be available to the commercial firm to answer questions regarding and give seminars on *general issues* instead of a specific issue in a research project of the firm's in an epidemiological area. This type of general consulting agreement may well not infringe on the researcher's obligations under a sponsored research agreement to another commercial firm. However, the scientist must always remember to give careful consideration to these issues so there is no inadvertent breach of one agreement while seeking to carry out another agreement. Above all, care must always be taken not to disclose confidential information that has been released to him or her under a confidentiality or secrecy agreement.

C. Institutional Policies Regarding External Consulting Agreements

1. Nature and Requirements of Policies

University policies regarding external remunerative consulting agreements of faculty members are unlike the other policies discussed in this chapter because consulting has traditionally been an unquestioned privilege of academic scientists. Many universities had no written policy on the subject until problems began to arise, such as those outlined in Section IV on conflict of interest. As the following discussion will illustrate, there is a continuum for the policies, some at one end reflecting almost a "hands off" philosophy; at the other end of the continuum are policies that go into great detail on requirements, obligations, and prohibitions. Virtually all the policies at any point of the continuum place responsibility on the faculty member for ensuring compliance with all institutional policies (i.e., intellectual property, conflict of interest, outside employment, use of institutional facilities, external consulting, and so forth).

It is important for the faculty member to make specific disclosures to the company with which the faculty member will consult. Among the factual information the company needs to know in order to enter into a lawful agreement that will serve the interests of the parties and take nothing away from the university (the interested nonparty in the agreement) is (a) the nature of the relationship between the university and the faculty member (e.g., employer-employee); (b) any overriding obligations to third parties, such as, the commercial sponsor of some of the scientist's research projects; and (c) the university's intellectual property policy. If the company has questions when the faculty member supplies this information, the appropriate university administrator should be called in for advice even if the particular university does not require disclosure or approval of consulting agreements.

Several universities provide valuable advice to faculty members regarding consulting in addition to their written policies. For instance, Harvard Medical School advises faculty to undertake "exclusive consulting relationships . . . only after careful consideration." Faculty members must anticipate the consequences of agreeing not to work with other organizations. This is especially true for agreements that require the consultant to either get company approval of all other consulting arrangements or not compete in the agreed scientific area of consultation for a specified number of years.[168] Harvard Medical School also provides its faculty with a sample agreement and make available their Office of Technology Licensing and Industry Sponsored Research for advising the faculty member.[169] Most universities, including The University of Illinois at Urbana-Champaign, state in their institutional policies that faculty engaged in

> external consulting work . . . are responsible for ensuring the agreements emanating from such work are not in conflict with university policy. . . .

> Such employees should make their university obligations known to others with whom they make such agreements and should provide other parties to such agreements with a statement of university policies on inventions and discoveries.[170]

The University of Illinois also requires its faculty to report consulting to the university administration.[171] Under the University of Georgia (College of Agriculture) policy, the faculty member is required to submit a request for prior approval to the department chair.[172] Other universities do not require prior notification or approval by the university administration of consulting,[173] but they obligate the scientist to document "if called upon to do by a responsible university officer" that the employee has met all of the conditions of the university's policy.[174] The University of Utah makes it extremely clear that external consulting cannot interfere with the scientist's overriding obligation to the university, and universities require that the consulting should have a relation to the professional interests of the faculty member.[175]

Often universities set quantitative limits on the amount of time that may be devoted to consulting, such as the common 1 day per week. Others state that the limit is 20% of the faculty member's full-time obligation to the university or perhaps just an amount of time that does not interfere with the discharge of the faculty member's obligations to the university. Other policies allow occasional professional consulting engagements but not regular practice.[176]

Other important provisions seen in policies frequently are:

1. Prohibitions or limitations on use of university facilities, equipment, and materials.[177] One allows use of the library and the personal office of the faculty member[178]; another requires reimbursement if the university facilities or equipment are used.[179]
2. Prohibit or use of the name of the university or letterheads.[180]
3. Prohibition of acceptance by scientists of gratuities, gifts, or favors.[181]
4. Consulting activities should be consistent with the university's goals and "can make important direct and indirect contributions to the strength and vitality of the [u]niversity."[182]

Fewer universities include regulations in their policies that:

1. Require some limitation on the amount of fees which can be charged by the scientist[183]
2. Prohibit outside regular consulting which is in direct competition with the university[184]
3. Dispute resolution or disciplinary actions for failure to comply[185]
4. Refer to a conflict of interest policy, either the AAUP/ACE statement[186] or the university's own[187]

5. Require some statement regarding ownership of inventions (patentable or nonpatentable)[188]

Item 5 above, which concerns ownership of inventions, is of major importance to both the university (as the employer of the faculty member) and the company (as the party contracting with the faculty member/consultant). Most, but not all, universities require that scientists employed full-time assign inventions and discoveries to the university. On the other hand, companies require consultants to assign to the company any inventions and discoveries developed by the consultant in carrying out the services required under the consulting agreement. In order to prevent a serious controversy, the faculty member/consultant must be sure to keep the two obligations separate. If the faculty member has followed the institutional policies and the invention arising from the external consulting was, in fact, (a) developed without use of university funds and facilities, (b) is not obligated to a research sponsor, and (c) is able to be distinguished from the scientist's research at the university, it will be the personal contractual obligation of the faculty member to assign rights to the invention to the company. The faculty member would be wise to be extremely cautious so that he or she will not mistakenly agree to turn over rights to intellectual property to the wrong party.

It is possible that instead of money, faculty members can be offered equity or shares of stock by a commercial firm desiring the scientist's consulting services. This type of offer is often made by small start-up firms that have very little cash on hand but do have stock that has not been designated to others. Scientists must be cautious in accepting stock, and they should make certain that they know exactly what their university's policies on equity participation and conflict of interest are. Some universities do not allow acceptance of stock at all, whereas some do not approve it but do not want to know about the stock because consulting is seen as the scientist's personal business. Other universities allow equity participation in some situations and not in others. The faculty member should always advise the company seeking consulting services of the university's policy and try to withstand any pressure by the firm in order to make the decision one for which the faculty member will be responsible.

If the university's policy allows equity participation by the faculty member for consulting, the faculty member must still take care to seek assistance of a tax advisor or attorney in order to understand the actual documentation for the transfer of shares and the personal tax consequences.

D. Common Clauses

The following is a list of the common clauses in an external consulting agreement. Certainly, some creative companies will decide not to use one or to add another one. However, this list can still serve as a guide to issues to be raised in discussing or negotiating an agreement.

1. Names of Parties
2. Term or Period of Agreement
3. Scientific Area for Consulting
    a. Precise and Accurate Description
    b. Areas at be Excluded
        (1) All Others
        (2) Specifically Named Areas
4. Days or Occasions for Consulting
5. Consulting Fee
6. Expenses (travel, and so forth)
7. Exclusive or Nonexclusive Relationship
8. Intellectual Property Rights in Inventions Conceived of or Reduced to Practice During Term of Consulting
9. Confidentiality of Company's Proprietary Information
10. Confidentiality of University's Proprietary Information
11. Obligations to Prior Commercial Firms, including Confidentiality of Their Information
12. Publication
13. Reports (periodic)
14. Scientist's Representations and Warranties
15. Independent Contractor
16. State of Contracting Law
17. Signature of Both Parties

E. Potential Conflicts of Interest

There are numerous opportunities for conflicts of interest when a faculty member enters into an external consulting arrangement with a commercial firm. The most obvious conflict arises in the situation where the faculty member has a private consulting arrangement with the same firm that sponsors the faculty member's on-campus research through a sponsored research agreement with the university. This situation is a fertile ground for conflict and confusion regarding who has rights to various research results. It would take a very astute and organized faculty member with an equally organized staff to keep these assignments properly separate. Although more minor, a situation such as that just described and virtually all other consulting agreements present a problem to the faculty member in terms of avoidance of use of university facilities, equipment, personnel, and supplies. Virtually no university policy allows use of its facilities for the benefit of a private firm with which the faculty member has a private consulting agreement. Reports for the company should not be typed by university personnel on university equipment. Experiments should not be done in the university laboratory. If these activities seem to the faculty member to be appropriate to be done at the university, it probably means the arrangement should be a sponsored research agreement with the university, and not a consulting agreement.

F. Recommendations to Scientist

Recommendations to scientists on the issue of private consulting external to their university positions are rather simple: Know your institution's policy and guidelines (if any) and their interpretations, analyze proposals from companies to determine if the facts support entering into a private consulting agreement or a sponsored research agreement, and then abide by the agreement with great care. If a faculty member is in doubt about whether a particular consulting arrangement complies with the policies, ask the appropriate official on campus. Clarifying before the consulting services are performed will be much easier than sorting things out afterwards. If equity participation is offered, seek advice of your university and your personal tax and/or legal advisor so that you will be ensured of proper transfer and fully understand the personal tax consequences.

IV. CONFLICT OF INTEREST

A. Introduction

Everyone recalls the Biblical verse, "No man can serve two masters. . . ."[189] That quote tersely states the concept of conflict of interest. Despite the claim of comprehension of the abstract term conflict of interest by university faculty members, there was, for many years, no recognition of how the idea was reflected in day-to-day situations in the academic setting. This discussion of conflict of interest will present the issues so that scientists can apply them to workaday situations.

B. Definitions

A conflict of interest is "a conflict between the private interests and official responsibilities of a person in a position of trust."[190] Facts that must exist before a real or potential conflict of interest can arise are that the scientist: must be subject to a "professional code or generally accepted standards of practice,"[191] or a statute that encompasses the same; and must have a relationship with an entity outside the university.

Many people use the term *conflict of interest* to imply wrongdoing in any situation—but one can see this usage daily in the newspapers—this usage is incorrect. It is also incorrect to use *conflict of interest* synonymously with the term *scientific misconduct*, which denotes falsification, fabrication, or plagiarism of data as well as other activities that compromise the integrity of one's research.

There are also so-called conflicts of interest that are unavoidable, and, as such, are not dealt with in this discussion. For instance, a physician might be said to have a conflict of interest among his desires to treat patients professionally, practice economically, make money, avoid malpractice, and be with his family.[192]

Further, some describe as trivial, and so not real conflicts of interest, token gifts such as meals, calendars, or other novelties given to scientists by industrial research sponsors such as pharmaceutical companies. However, some scholars believe even these small items create a conflict of interest problem.

> The key here is not the monetary value of the gift, but whether it is instrumental in fostering a personal relationship [with the company]. . . [T]he subtle threat to the individual['s] character, ('feed[ing] one's human tendencies toward self-interest')[193] and [threat to his or her] duty to his or her patient is minimized by avoiding a personal relationship [with a company] and its attendant obligation.[194]

One of the prerequisites for a conflict of interest is that the scientist must be obligated to a "professional code or generally accepted standards of practice,"[195] or a statute that encompasses the same, such as Utah has and applies to faculty at the University of Utah and other state institutions.[196] That is, the scientist must have a particular set of standards that he or she must obey. These standards may be based on professional licensing body regulations, federal regulations, contractual obligations (to industrial research sponsors or the university-employer), or state laws.[197]

Generally a professional is accustomed to making a commitment to a professional code that serves as a perceptible reference point for analyzing a situation with a potential conflict of interest. The application of this feature of conflict of interest is universally recognized to apply to other university employees, such as a purchasing agent or a trust officer. No researcher would fail to see the conflict in the purchasing agent's owning stock in a company that is making large sales, including a profit to the purchasing agent's employer, the university. However, many scientists have not seen themselves as susceptible to conflicts of interests, especially since the university's rules and regulations have historically been unwritten.

C. Purpose of Institutional Policies

After World War II, federal government grants to researchers began to dominate the arena of academic research. As these federal funding sources began to dry up, scientists looked for untapped resources for research funds. Simultaneously, commercial firms began to see the value in having access to academic research, and to recognize the growth in the competitiveness of Japan. Generally recognized as surging in the late 1970s and 1980s, industrial or commercial interests began funding the research of university scientists. The time for recognition of conflicts of interest had come about as this, the second prerequisite to a conflict of interest developed—a relationship with an entity not part of the university. This entity is most often a commercial for-profit firm (which most of this discussion will stress) but can be nonprofit foundations, associations, and private donors. In reaction to the now-unavoidable issue of conflicts of interest, universities and other organizations interested in academia (including Congress) began surveying universities, writing policies and statutes, and making recommendations.[198] The recommendations without exception focus on the need for universities to govern conflicts of interest through written policies supported fully by the university's governing board.[199] This need for institutional policies on conflict of interest becomes obvious when the reasons for such policies are listed.

1. *Increase the public confidence in universities and scientists.*

> [C]onflict of interest in our research universities can only lead to greater distrust of both the university and the industrial sectors, weakening [the university's] credibility and increasing public suspicion. Whether conflicts of interest allegations have substance or are only appearances of such, they harm [the university's] reputation when publicly reported.[200]

For this reason, universities are adopting conflict of interest policies to show the public at large that the university is capable of dealing with newly created problems so as to ensure the integrity of the scientific enterprise and judgment of researchers.[201]

2. *Enhance awareness of conflict of interest issues of researchers, dispel their confusion over what types of activities involve conflicts of interest that are acceptable or unacceptable, and promote the general norms of ethical scientific behavior*. Many researchers will apparently welcome policies that clarify what may be perceived by some as a "free-for-all."[202] Specific factual situations will need to be reviewed on a one-to-one basis, but general principles can be enunciated in the policy with categories of clearly acceptable and clearly unacceptable conduct spelled out. In addition, as cases are scrutinized ad hoc and decisions made, general fact situations (without names) can be made known periodically to the faculty to aid in the scientist's understanding of the policy and to aid him or her in the structuring of future relationships with commercial firms.

3. *Encourage retention of talented scientists by clarifying the extent to which they can be openly involved with commercial interests without damaging their reputations for ethical behavior, and continue to maintain their primary commitment to teaching and academic research*. There are actual situations in which scientists are compelled because of family illness or other such tragedies to leave academia for higher salaries in industries despite their desire to

remain at the university. Were those individuals allowed to supplement their incomes with relationships with industry that were not conflicts of interest, the university would be able to retain them and the public would benefit from the talents of the scientist.

4. *Educate students (and residents and other trainees) on how to deal with potential conflicts of interest.* Some universities consider a function of their conflict of interest policy to be to teach students about relationships between universities and commercial firms and scientists and commercial firms (and thereby enhance the students' awareness of potential conflict). This early awareness training can lead to generations of scientists who are better able to identify, comprehend, and make difficult decisions regarding potential conflicts of interest without forfeiting career commitments to education, academic research, patient care, or academic administration.

5. *Preserve the researchers' academic research priorities.* Individual researchers can be pressured, subtly or not so subtly, into altering their academic research plans to meet the commercial needs of the industry funding the research. One of the university's primary goals is the production and dissemination of new knowledge, based on its faculty's decisions on what research to pursue. A scientist may need the outward support of the institution and administration to withstand pressure to change his or her academic research goals to coincide with the industrial sponsor's commercial goals that are directed toward profitability (and not furtherance of the knowledge base).

D. Categories of Conflicts of Interest

The two categories of conflicts of interest can be described briefly as the collision of a researcher's dedication, as part of the university community, to advancing and disseminating scientific knowledge, with the researcher's desire or opportunity to increase his or her personal income, and hence, conflict of interest[203]; and the collision of a researcher's full-time commitment of loyalty, time, and energy to his or her teaching, research, and administrative responsibilities with the researcher's desire to perform external professional services, and hence conflict of commitment.

Many university conflict policies make this distinction between conflicts of interest and conflicts of commitment.[204] The following portion of this chapter will discuss these two types of conflicts and the characteristics of each.

1. Conflicts of Commitment

Conflicts of commitment are based on an individual's appointment to a full-time position at the university and a prior or subsequent relationship with a commercial firm that will violate a code of conduct—either the university's written or unwritten policy or the individual scientist's professional or personal code of ethics. Most university-affiliated individuals agree that scientists must "recognize that the primary and overriding obligation of every faculty member, in terms of his or her commitment of time, attention and intellectual energy, is to the university, that is, to the students, colleagues, and general mission of the university."[205] Most would say that there is little room for deviation from this commitment. However, many would also agree that it is possible to arrange external professional activities so as not to interfere with this primary obligation. Many universities, by written policy, approve of outside work to the extent of 20% or 1 day per week of a scientist's total professional output.[206] The policies often provide that this outside work must be considered to be beneficial to the university and the researcher and within the scientist's area of expertise.

Although more unusual, there are also conflicts of commitment that may develop from volunteer activities rather than income-producing activities of the scientist. If a researcher involved in volunteer religious or political or charitable activities disregards his or her primary obligations to the university, he or she may be violative of the university's policy just as if he or she were paid for the services.

It can also be a sign of potential conflict of commitment if a scientist's activities for a commercial firm are too closely related to the scientist's academic research. If there is no line that can be drawn between the two research programs, the scientist probably cannot determine where his or her primary obligation to the university ends and the secondary obligation to the commercial firm begins. The key obligation of the scientist is to further the basic missions of the university; if he or she is not doing so—either by quantity or quality of scientific work—he or she has a conflict of commitment that must be dealt with.

2. Conflicts of Interest

The essential characteristic of conflict of interest as opposed to a conflict of commitment is external personal financial gain by an employed scientist using the employer's resources in violation of the employer's rules. Virtually all universities prohibit this activity either in written policy or by custom. Those who have not had written policies are now finding themselves at a disadvantage in trying to discover for themselves exactly what their faculty are doing externally as well as responding to the concerns of government officials and professional organizations. It is a fair assumption to make that most universities will have written policies formulated in the next few years whether or not such policies are mandated by the federal government for grantees of federal funds.

In looking at personal financial gain of a researcher in order to determine if there is an unacceptable conflict of interest, many policies first define conflict of interest or the synonymous term they have chosen to use.

Many universities, for instance, denominate it, financial interest. Harvard Medical School defines the term *financial interest* in its 1990 policy as follows:

> A "Financial Interest" is an interest in a Business consisting of: (1) any stock, stock option, or similar ownership interest in such Business, but excluding any interest arising solely by reason of investment in such Business by a mutual, pension, or other institutional investment fund over which the Faculty Member does not exercise control, or (2) receipt of, or the right or expectation to receive, any income from such Business, whether in the form of a fee (e.g. consulting), salary, allowance, forbearance, forgiveness, interest in real or personal property, dividend, royalty derived from the licensing of Technology, rent, capital gain, real or personal property, or any other form of compensation, or any combination thereof.[207]

Somewhat differently, the University of Utah Conflicts of Interest Policy defines the terms *substantial interest* and *compensation* instead of the term *financial interest*.

> "Substantial interest" means (1) the ownership, either legally or equitably, by an individual, his/her spouse or his/her minor children, of at least ten percent of the outstanding capital stock of a corporation or a ten percent interest in any other business entity; or (2) the holding of a position in a business entity as an officer, director, or employee.
>
> "Compensation" means anything of economic value, however designated, which is paid, loaned, granted, given, donated, or transferred to any person or business entity for or in consideration of personal services, materials, property, or any other thing whatsoever.[208]

Other university policies, such as that of the University of Chicago, set down the rule providing that "financial interest in . . . professional or commercial activities outside of the University should not compromise the . . . faculty member's obligations to the university,"[209] but do not separately define the term *financial interest*. The common sense usage of the this term along with interpretations made through the numerous detailed guidelines in the policy, give the University of Chicago term *financial interest* meaning even without a separate definition.

It is clear from these few examples that the meaning of the term, whether called *financial interest* or otherwise, often can depend on the context in which it is used.

Under a research contract or grant that is sponsored or funded by a commercial firm, a financial interest of the researcher may be ownership (i.e., stock or equity) in the firm serving as sponsor of his or her research. This financial interest could (and at many universities would) be a conflict of interest. This same factual situation could be expanded to illustrate other personal financial gains by researchers and potential conflicts of interest. If the company gave the researcher a $1,000.00 cash gift or a week's trip to an exotic location in exchange for the conducting of a seminar, a conflict would appear to arise. Likewise, should the commercial sponsor who is funding direct and indirect costs of a research grant or contract attempt, through the researcher or an administrator, to influence the university's position or decision on an issue to the detriment of the university, there is a conflict of interest. An example of this situation would be a commercial firm's meeting with university officials to encourage opposition to the proposed amendments to the federal law governing orphan drugs at the same time the firm is supporting with substantial funds the research of a department chairman. If the university could be convinced to oppose a bill, the passage of which is in the best interests of the university, a conflict would exist. Commercial firms must not be allowed to influence university policies or positions by funding research or making any other type of financial commitment. In addition to those stated above, there are any number of other potential conflicts of interest involving *financial interest* or *private financial gain*. They will be discussed in the section below on acts prohibited by scientists.

E. Differences in State and Private Institutions

As has already been mentioned in passing, state universities, governed by unique state statutes, can be quite different from private independent institutions who have their own privately written codes of conduct with which their scientists must comply. Utah, Texas, and Georgia are examples of states that control state university conflicts of interest through state laws applicable to all public employees.[210] Although these universities may have policies supplementing the statutes, the policies must by carefully drafted to be consistent.

Private institutions and state-supported institutions are, perhaps, more alike than different when it comes to potential conflicts of interest because for both often the vast preponderance of support for research comes from tax-supported public agencies. In fact, some see any type of nonprofit educational or research institution as basically a "public" institution using "public" funds because one of its purposes is to benefit the public and it often solicits funds from the public. For that reason, private institutions must be as diligent as the state-supported institutions in managing conflicts of interest.

F. Applicability to Certain Individuals

Who are the conflict of interest policies[211] applicable to? Certainly full-time university employees (and under a few policies, part-time employees),[212] whether permanent, temporary, or visiting, research or clinical, tenure or nontenure track, are subject to conflict of interest policies. Some authorities recommend that full-time postdoctoral research fellows, residents, and trainees also be included.[213] The University of Minnesota is an example of an institution whose policy does so.[214] Interestingly, the University of California, governed by its unique state law, applies its policy to an extensive list (which is statutorily mandated) of university officials, both faculty and administrative.[215] Principal investigators of privately sponsored research projects are distinguished from other employees by the state law and are required to file rather extensive documents called Principal Investigator's Statement of Economic Interests.[216] Apparently other faculty are not subject to the same requirements.

Some universities specifically exclude categories of individuals from their conflict of interest policy. For instance, the University of Utah makes it clear that its policy "does not include members of . . . any . . . advisory commission, board, or committee serving on a part-time basis."[217] Otherwise their policy applies to permanent employees, full or part-time.

Application of the conflict of interest policy to academic research faculty and clinical faculty is basically the same. However, because the clinical faculty have opportunities for relationships with pharmaceutical firms that a general academic researcher generally would not have, it is important that they be discussed. Clinical trials of medical products or devices are often carried out at medical schools with clinical faculty as the principal investigators. Commercial firms routinely pay the university by the number of subjects who enroll. The amount must be reasonable and not exorbitant or an amount that could sway the investigator's clinical or research judgment and thus result in a conflict of interest. Potential conflicts noted by the federal government are a clinical faculty member's opportunities to refer a patient to a laboratory or other provider in which the faculty member has a financial interest. Clinical faculty members must be careful to be aware of all restrictions that apply to their conduct whether institutional, state, or federal laws.

Most policies extend their applicability to certain family members of the scientist, and the rules that apply to the employed-scientist also apply to the designated relatives. Family members generally include spouse, minor children, others living in one's household, and any trust, estate, or the equivalent that the scientist controls.[218] Some universities, such as Brown University, include parents and siblings of the researchers in their Policy on Duality of Interest or Conflict of Interest.[219]

Most universities apply their conflict of interest policies to administrative staff either by explicit language[220] or by recognizing that many of them will have faculty appointments; thus, drawing them into the policy's coverage. It is recognized that academic administrators, such as deans, associate deans, chairs of departments, and comparable positions at affiliated hospitals have significant authority over others. They make employment decisions and salary, tenure, and promotion decisions, as well as budget overview authority, space and equipment allocation, and are perceived as having tremendous influence over other faculty members. This opportunity for special influence requires that the administrators take special care to avoid a conflict of interest or commitment.[221] Not only could these types of mistakes by the administrator affect his or her function as a role model, they could also have concrete effects on other faculty members.

G. Relationships with Nonuniversity Entities from Which Conflicts of Interest May Develop

Assuming that we have met the first prerequisite to a conflict of interest discussed previously—a code of conduct to which one is expected to adhere—the discussion can now turn to the other prerequisite: a relationship with a commercial entity or, for that matter, even a noncommercial entity if it is governed by forces outside of the university (e.g., association, private foundation, or donor). The types of relationships will be delineated below in order to demonstrate how conflicts of interest may arise for the scientist.

It is now well known that the influx of industry into the university campuses intensified in the late 1970s as astute individuals began to realize that federal funding of research would be declining. The relationships that the commercial firms began to form with universities were at first somewhat simplistic—a check would arrive from a pharmaceutical company for an investigator performing clinical trials using the company's drug, or a faculty member would go to a company to put on a seminar several times a year. In many cases, no contracts would exist and no one would be concerned about the sharing of scientific data with profit-seeking industry. After all, the university's missions emphasized the dissemination of new knowledge and performing services for the benefit of the public. However, as time passed, the relationships became more complex, both from the standpoint of what the commercial firm wanted from the university and what the university wanted from the commercial firm which was to protect its mission and receive needed funds. There emerged a spectrum of relationships that can be categorized, in part, as follows:

1. Sponsored research agreements
2. Collaborative research agreements
3. Materials (or tangible research products) transfer agreements
5. Clinical trial agreements
6. Research consortia agreements
7. Technology licenses
8. Patent licenses
9. Gifts
10. Equity agreements for spin-off companies

Despite the many cultural differences that are widely recognized to exist between the university community and commercial firms ("[T]he academic imperative [is] to seek knowledge objectively and to state it openly and freely, while the industrial imperative [is] to garner a profit, which creates the incentive to treat knowledge as private property.''),[222] all of the types of relationships listed above have been agreed to and put into a binding contractual form agreeable to both parties. This shows, in part, the attempt by many universities and many commercial firms to be flexible in their demands. Universities have developed and adopted intellectual property policies, produced standard agreements, and gathered in groups to find better ways to negotiate. Industry has, to its credit, done the same. The outcome has been mutually beneficial relationships between the two. Universities do not have to forsake their values and missions; commercial firms can still look toward a profit.

Of course, this characterization of the relationships is based on a desire by both parties to figure out how to benefit both without violating either party's values. Some might call it naive. It must be recognized that there are commercial firms whose desire for a profit do not include the values of fair dealing, flexibility, and mutuality of any sort. University researchers exist who are overreaching; interested in receiving funds from commercial firms without fulfilling their contractual obligations. Those entities have to be dealt with in the best way one can, and, frankly, after attempting to fairly negotiate with them, being willing (and courageous enough) to say no. Most often, their reputations precede them and one will have fair warning. The comments that follow will be a brief description of each of the relationships listed above, with a comment or example of potential conflict of interest.

1. Sponsored Research Agreement

An agreement or contract by which a single commercial firm funds the university a specified amount for a named scientist to carry out a specified research project (whether basic or applied research) is generally called a sponsored research agreement. Virtually all universities prohibit the performance of research as a cover for an undisclosed sponsor.

2. Collaborative Research Agreement

This type of agreement is a contract very similar to the sponsored research agreement except that both scientists for the commercial sponsor and university carry out specified scientific research obligations under the particular research project and exchange the results with the other. A conflict of interest could arise if the university faculty member allowed himself or herself to be explicitly or subtly pressured into changing the original research project to one more product and profit oriented after partial results are received from the commercial scientist.

3. Materials (or Tangible Research Products) Transfer Agreements

Agreements between the university and an external entity providing for transfer of specifically identified materials that are the products or by-products of university research. The transfer may be for research purposes only at no cost or for commercial use with a provision for payment to the university. Universities vary in what they consider tangible research materials or products, so the scientist must know his or her institution's policies. An example of a potential conflict of interest would be the transfer to a small biotechnology company (in which the university scientist holds equity) of materials developed at the university without getting university approval or providing for financial return to the university.

4. Clinical Trial Agreements

When a pharmaceutical or biotechnology company desires to have a product tested for Food and Drug Administration (FDA) or other purposes, it often comes to a university scientist who has a prior interest in the type of compound or drug. The agreement between the company and the university providing for the carrying out of the company's protocol is generally called a clinical trial agreement. Many of the provisions and protections are similar to sponsored research agreements.

5. Research Consortia Agreements

This agreement involves more than one commercial firm contracting with the university for access (the type of access to be negotiated) to a body of ongoing research conducted by a group or even department of scientists at the university. Obviously there are many issues to be negotiated. The greatest potential for conflict of interest in consortia agreements is that the group of sponsors will attempt to influence the direction of future research in related areas of inquiry. It takes a strong-willed faculty member to withstand this type of pressure by one, let alone multiple, commercial sponsors.

6. Technology and Patent Licenses

Licenses for technology or patents are contracts by which the university grants to a commercial firm, called the licensee, the right to make, use, or sell products based on technology or patents originating at the university as a result of academic research. A potential conflict of interest would arise if the university faculty member became so involved with the commercialization of the technology that he or she stopped having time out of class to confer with students. This would be an obvious conflict of commitment.

7. Gifts

Gifts, whether trivial gifts given directly to the scientist, or multimillion dollar donations to the university for research, are voluntary transfers of property without expectation or promise of return. However, gifts are a fertile field for conflicts of interest because they may well encourage a subtle type of return to the donor by the university scientist. The scientist may purchase through the university goods and services from the donor company or withhold disappointing research results in order not to offend the donor who may make future donations.

8. Equity Agreement for Spin-off Companies

In recent years, many small for-profit companies have been formed based on valuable technology developed at the university. In many cases, the inventor of the technology to be transferred and commercialized is offered equity or shares of stock in the company. The agreement by which the equity is promised to the faculty member may be between the company and the faculty member or the company and the university. This arrangement can also be a hotbed for conflicts of interest. Hence, the faculty member cannot be too careful. New businesses require concentration, energy, and almost constant attention. Depending on the level of involvement of the faculty member that accompanies the equity interest, a conflict of commitment can arise. If the faculty member considers himself or herself to be a so-called founding scientist with no duties, the conflict may be nonexistent. However, if the founding scientist role includes management duties, it may be impossible for the faculty member to fulfill his or her obligations to the university to teach, research, serve on committees, and so forth. A conflict of interest also may occur if the faculty member suppresses discussion of new scientific findings with laboratory or university colleagues because they have commercial applications that must be quickly pursued.

H. Prohibited Acts

This section will delineate the acts prohibited by selected conflict of interest policies. General prohibitions will be discussed first; specific prohibitions will follow. Although many potential conflicts of interest can be cured of their ethical problems by disclosure or approval by the university, there are some that rarely can be cured. These specific prohibitions fall into that category. Generally, the policies of state universities that are dependent on state laws[223] for their rules regarding conflicts of interest restate the philosophy of the law as their general prohibition. Private institutions tend to stress the individual university's historic philosophy.

1. "[T]o promote the public interest and strengthen public confidence in the integrity of the university, [the university must] establish standards of conduct [prohibiting] . . . actual or potential conflicts of interest between their duties to the university and their private interests."[224]
2. "[A]cquisition of private economic . . . interests . . . [must] not interfere with the . . . discharge of public duties."[225]
3. Financial interests and outside activities should be arranged so as not to interfere with the researcher's primary commitment to the university.
4. "[G]uidelines [must] be applied in accordance with the spirit of the mission of [the university] in education, research, and patient care . . . and the interpretation and application by a . . . committee. . . ."[226]

5. "[A] university [must not] allow for unregulated exploitation of faculty business operations."[227]
6. "[Faculty members] are obligated to avoid conflicts of interest where possible. [They] must protect the public trust accorded them."[228]

There are far too many specific prohibitions to list. However, a selection will give the reader an impression of the types of prohibitions many universities believe to be important.

1. In directing students, faculty members are obliged not to make judgments based on concerns of commercial sponsors but in the educational and scholarly best interests of the students.[229]
2. The use for personal profit by a faculty member of unpublished information arising from research of university colleagues or assisting an external entity by giving it exclusive access to such information is prohibited.[230]
3. Direct negotiations by a faculty member with a potential commercial research sponsor over issues of university concern, such as intellectual property rights, publication, or university direct and indirect costs are prohibited.[231]
4. It is prohibited for the faculty member to agree to conduct a substantial amount of research that could be carried on within the university at an off-campus location to the detriment of the university.[232]
5. Consulting arrangements that are inconsistent with the university's intellectual property policy or with a commercial research sponsor are prohibited.[233]
6. The influence by a faculty member owning equity in a commercial firm over negotiations between the university and a commercial firm for another faculty member's research.[234]
7. If a faculty member owns, operates, or is a major investor in a company that does business with the university.[235]
8. It is generally prohibited for a faculty member to have a consulting agreement (by which the faculty member receives funds for personal use) with one commercial firm and a sponsored research agreement (of which he or she is the principal investigator) with a different commercial firm on the same scientific subject matter.

Many potential conflicts of interest can be averted by disclosure or by approval of the university (with or without conditions). However, many of of those listed above are of the nature that they cannot be cured.

I. Institutional Conflicts of Interest

Despite what many think, the university or institution can find itself in a position of conflict of interest. If the university holds an equity interest (i.e., owns shares of stock) in a company in which one or more faculty members also hold equity positions, it is said by some that the university has placed itself in a potential conflict of interest[236]; namely, its desire for financial gain could unduly influence its professional judgments,[237] as well as destroy the traditional academic values on which the university is based and lower morale of faculty, both involved and uninvolved in the venture.

Harvard Medical School's 1980 decision not to participate as an equity owner in the genetic engineering company based on its own scientists' technology is, to a nonparticipant, an example of a potential institutional conflict of interest avoided. Harvard would have owned approximately 10–15% of the new company, although the company would not have been permitted to use the university's name or be located on the campus. Originally, some of the university administrators had thought the company could provide a source of funds for education and research "while hastening translation of basic knowledge into useful products and devices."[238] Ultimately the university president stated that the most important thing was to preserve academic values, and apparently owning equity in the company would create "potential conflicts with these values."[239]

Opposition to the venture was voiced by faculty members who worried about the effect of the company on their freedom to teach and do research. Many felt the company's necessary secrecy could not be reconciled with the free and open exchange of ideas that is fundamental to academic research and teaching. Faculty also raised issues of how compatible the roles of employer and business partner would be for the university. Would the university's judgment as an employer be influenced by a scientist's participation or nonparticipation in the new company? Would students be treated as cheap labor by industry?[240] Would the basic research enterprise be contaminated?[241] Ultimately the matter ended there—no company and no university involvement and, therefore, conflict of interest. Harvard lost a substantial opportunity to have its technology marketed as well as a large amount of funding.

In order to handle its potential institutional conflicts of interest, universities must develop guidelines that will facilitate analysis of moving forward with relationships that contain potential conflicts and at the results of rejecting the relationship. Analyzing, studying, and comparing one's basic missions to an opportunity with potential conflicts will help the university face the future partnerships with industry without fear of surprise even though it remains risky.

J. Prohibited Acts by Commercial Firms

Often the relationships with a commercial firm that raise the conflict of interest issues for a scientist are: (a) funding of the scientist's research under a sponsored research agreement or contract, or grant, or donation to the university; and (b) licensing by the university of technology developed by the scientist as well as future improvements made by that faculty member. It is prohibited in written policies (and unwritten as well) for commercial sponsors and commercial licensees, both of whom hope eventually to be marketing the university's faculty-developed technology, to attempt to control the faculty member's scientific approach to the research program being funded currently or to the future improvements on the licensed technology. It is completely unacceptable for a commercial firm to try to pressure the scientist into producing products instead of having a product arise form the academic research plan. This type of pressure is detrimental to both the individual scientist and the university.

Also, it is important that the scientist (and the university) ensure that the commercial firm receive only the research results for which the firm contracted.[242] This can be monitored by the office of research administration (who will verify with the scientist the correctness of the scope of the research proposal and the results) or the manager of technology transfer in the licensing of future improvements.

It is also prohibited behavior for a commercial sponsor or licensee to use the institution's resources for any purpose without making provision for such use in the budgets of the agreement with the university. Sending a commercial employee to do research in a scientist's laboratory on campus (without adequate payment of costs and assumption of responsibility) is a use of public resources for private gain and should be clearly prohibited. This is but one example of this proposition. There are many others that should also be prohibited specifically or in a general way that is clear to all parties, especially commercial sponsors and faculty members.

K. Institutions' Requirements of Scientists

Since many, many institutions have policies on potential conflicts of interest, and many of those that do not have them are searching for the best policy for their institution, the following discussion will describe what institutions often require of scientists in the potential conflict of interest situation. The distinctions among the various policies gathered by the writer were less dramatic than one might expect (as it was when the Association of American Universities made its survey in 1985).[243] This discussion will include provisions common to many universities.

Many universities introduce their conflicts of interest policies with statements regarding the individual scientist's accountability and integrity. For example, in the University of Minnesota's policy on Disclosure of Financial Arrangements with Industry, it states:

> It has been, and continues to be, assumed that all individuals will be alert to the possible effects of outside activities on the objectivity of their decisions, their obligations to the university, and the university's responsibility to others.[244]

The University of Pittsburgh provides in their policy simply that, ''Responsibility for respecting and meeting the obligation due to the university rests with the individual faculty member and administrator.''[245] Harvard University's Faculty of Arts and Sciences Statement of Policy on Conflicts of Interest provides that, ''[T]he university has . . . provided its members with guidelines on conflicts of interest and commitment that leave as much as possible to the individual discretion.''[246] Perhaps the strongest statement on individual accountability and integrity comes from the policy of the Harvard University Faculty of Medicine, which states:

> [There are] two assumptions underlying the explicit rules and implicit norms governing faculty behavior at the Harvard Medical School: that the vast majority of scientists are honest and conduct their research with the highest standards and integrity; and that for the vast majority of cases, self-regulating structures and processes in science are effective.[247]

The principal obligation of faculty members is to embrace teaching and research and participation in the broader university community by taking part in governance and administrative duties as well as community service.[248]

Most universities that have conflicts of interest policies require some type of written disclosure of potential conflicts of interest by their faculty. Interestingly, of the 46 institutions responding favorably to the Association of American Universities' (AAU) question: ''Do you have a conflict of interest policy?'', 45 replied that they did and that they required disclosure of potential faculty conflicts of interest. The question was part of AAU's 1985 survey of university policies on conflicts of interest.

The disclosure requirement may be no more than a one-time obligation to provide full information to and obtain approval of a designated academic administrator for any arrangement in which a conflict of interest is implicit or potential.[249] On the other hand, the disclosure may entail a more complex procedure such as that adopted by Harvard University's Faculty of Medicine.[250] It requires numerous steps by each faculty member:

1. Complete and submit a disclosure form on an annual basis.
2. If changes arise during the year which the faculty member believes may either (a) give rise to a conflict of interest, or (b) eliminate a conflict previously disclosed, updated forms must be submitted.
3. Disclosure forms should be returned to the designated office, with a copy to one's department chair for review.
4. Faculty members are encouraged to seek assistance in the event of questions or special circumstances.
5. Disclosure forms will be considered confidential.[251]

In addition to the procedure for disclosing conflicts, Harvard's Faculty of Medicine also included in their policy a lengthy section on how the disclosure forms are reviewed.[252]

Yet another mode of disclosure of a real or potential conflict is mandated in Cornell University's policy. It requires disclosure to the dean in sufficiently detailed form to permit an objective evaluation. The disclosure must be made prior to the faculty member's making the commitment or initiating the activity which the potential conflict. If the disclosure reveals a real or apparent conflict, the dean will attempt to resolve the matter informally with the faculty member. Should no resolution result, the dean will refer it to an advisory committee that will recommend to the provost a resolution. The provost may accept, reject, or modify the recommendations. If the faculty member still disagrees, he or she may request further consideration. After reconsideration, the provost's decision is final.[253]

By far most of the policies require that the disclosure include members of his or her immediate family,[254] and the disclosure format is the same for the applicable family member as it is for the faculty member. Others require disclosure by immediate family (specifically defined as spouse and minor children,[255] close members of your family,[256] spouse and/or dependent children),[257] immediate family (spouse, brothers, sisters, parents, or children),[258] and spouse, minor children, and other persons living in the same household.[259]

Many of the university policies require that the disclosures go to university committees. Some universities, who expect only a few disclosures, have an in-house committee review all disclosures. Others use committees as a last resort if the conflicts cannot be worked out informally with a university administrator. Although the committees are intended to make determinations in specific cases, they also serve as advisory bodies for policy decisions. In few cases is the committee the final authority. Generally the committee makes recommendations and final decisions are made by the chief executive officer or even the Board of Regents.

Only a few of the universities expressly provide for action if a faculty member fails to comply with the conflict of interest policy either by ignoring it altogether or simply not keeping disclosures up to date. The University of Pittsburgh simply says a faculty member will be subject to discipline up to termination for failure to comply.[260] The Harvard University Faculty of Medicine policy is also specific about the sanctions for the faculty member who knowingly fails to comply. The sanctions will be those under the university's applicable disciplinary policies and procedures, including, but not limited to (a) formal admonition; (b) the inclusion in the faculty member's file of a letter from the dean indicating that the individual's good standing as a member of the faculty has been called into question; (c) ineligibility of the faculty member for grant applications, Institutional Review Board approval, or supervision of graduate students; (d) nonrenewal of appointment; or (e) dismissal from the faculty of medicine.[261] Even though other universities may not explicitly state in their policy the sanctions for failure to comply with the conflicts of interest policy, if the policy has been properly adopted by the university as part of its official rules and regulations, the university is still within its rights to take disciplinary action against the faculty member under the standard faculty discipline policy.

Universities, mentioned previously, whose conflict of interest policy is actually a part of a state statute, may have legal sanctions against the noncompliant faculty member. The statute governing the University of California, for instance, speaks of civil and criminal liability, including fines, in addition to the university's own discipline policy.[262]

Up to this time, we have discussed codes of conduct or obligations of faculty pertaining to conflicts of interests being based on contracts, employment agreements, or letters of appointment, professional codes, and state statutes. However, the future may find faculty learning to live with these local codes of conduct as well as overriding laws and/or regulations imposed by the Congress or one of the federal agencies. The source of these laws and/or regulations will be the continuing concern by many entities about conflicts of interests of federally funded scientists. The most recent example was the effort of the National Institutes of Health (NIH) and Alcohol, Drug Abuse & Mental Health Administration (ADAMHA) to develop conflict of interest guidelines. Published in the *NIH Guide for Grants and Contracts* (September 15, 1989), an NIH official stated that we were in "a climate that requires [NIH] to [develop guidelines]."[263] The guidelines provided that NIH and ADAMHA would write and distribute conflict of interest guidelines, review each university's conflict of interest policy, routinely review the waivers and exceptions universities grant, and verify the universities' compliance before making a future grant of research funds. Requiring disclosure of financial interests and outside professional activities of the scientist, spouse, dependent children, and other dependents (at the time of proposals are submitted), the guidelines also required retention of disclosures and actions taken by universities for at least 3 years after termination of a given award. When a university identified a faculty member as being in a conflict of interest situation, NIH would be notified immediately and the university would take steps to safeguard federal funds. The guidelines further prohibited the scientist from having personal equity holdings or options in a company that would be affected by the result of his or her research or that produced a product or device being evaluated in the research project.

The university community and other interested parties were outraged over the guidelines, which were seen by many as too broad and too simplistic, whereas at the same time lacking definition of the conflicts in which they were interested. The guidelines also appeared contrary to current government policy encouraging commercialization of university inventions. An AAU representative made the excellent point that, "Conflicts of interest is essentially inescapable [and the scientific community needs] more thoughtful principles that spell out which conflicts of interest society will tolerate and which are unacceptable."[264] Other academic administrators disagreed with the need for more paperwork—an additional administrative burden.

At the time this book is going to press, the next set of NIH regulations are rumored to be coming out soon. Apparently it is expected that universities and their faculty members will find the regulations much more palatable, giving primary responsibility for the conflict of interest issues to the university administration.

L. Recommendations to Scientists

The best recommendation one can make to a scientist is contained in the white paper entitled: "Conflicts of Interest in Academic Health Centers" produced by the Association of Academic Health Centers in 1989. It quotes Mark Twain on page iii: "Always do what is right: this will gratify some people—and astonish the rest." Perhaps all academics still wish that the ethical issues faced on campuses were that simplistic. We have found that the issues are becoming more and more complex. For a scientist to achieve the traditional goals of the academic—successful science and teaching leading to promotion, publication, tenure, and recognition as increasing scientific knowledge for the benefit of mankind—one must recommend thoughtful care in all of these activities. No longer can one assume the exactness or integrity of a colleague's data, the motivation for a token gift, or the effects on one's own research from a financial interest. Faculty should tread lightly, making certain to:

1. Obtain a copy of all relevant policies of your university (such as Conflict of Interest, Outside Employment, Consulting, Intellectual Property). If your university is a component of a larger system, be sure to have the system rules as well as your campus' policies. Make sure your copy of the policy is current.
2. Study the policies until you can understand them. If any portion is unclear, locate the office administering the policy so that you can ask questions about the meaning of the policy and the interpretations made by the administration.
3. If there is a particular policy that you may be planning to bring into effect for yourself, ask if the university has kept records of past situations involving the same policy. If not confidential, ask to review the records.

4. Disclose to your supervisor and chairman your plans if the university's conflict of interest, outside employment, or consulting policy will possibly be applicable.
5. If the university's policy has procedures to follow, do so as soon as appropriate and with great care. Be completely honest and show your professional demeanor.
6. Take extra care if your actions under any of the policies, especially the conflict of interest policy or the sponsored research agreements, will have influence over a student's educational experience.
7. Comply to the letter with the university's financial policies, including assuring that after a grant award or an award under a sponsored research agreement that the expenditure of funds is in accordance with the university and the sponsor's policies.
8. Comply with federal regulations applicable to your research, such as human subjects, animal welfare, and biohazards.

NOTES

1. Harvard University Faculty of Medicine, *Guide to Protecting and Managing Intellectual Property*, 28 (1985) (patent policy adopted in 1934); D. Blumenthal, S. Epstein and J. Maxwell, Commercializing university research, 314 N. Engl. J. Med. 1621 (1986) (Wisconsin Alumni Research Foundation was established in 1925 as a nonprofit independent foundation to administer inventions of faculty at the University of Wisconsin.)
2. Northwestern University, *Guide to Intellectual Property* (draft), 3 (1990); Yale University, Report on the Committee on Cooperative Research, Patents, and Licensing, 10 (1984).
3. Brown University, Brown University Patent and Invention Policy, 1 (1981).

3a. Harvard Medical School, *Guide to Protecting and Managing Intellectual Property*, 24 (1985).

4. The University of Chicago, Policy on Patents and Software, 1 (1989).
5. Harvard University Faculty of Medicine, *Guide to Protecting and Managing Intellectual Property*, 24 (1985).
6. The University of Illinois at Urbana-Champaign, The General Rules Concerning University Organization and Procedure, Article III, Intellectual Property, 7 (1988).
7. Brown University, Brown University Patent and Invention Policy, 1 (1981); Auburn University, Auburn University Patent Policy, 1 (1974).
8. Yale University, Yale University Patent Policy, 1 (1989).
9. Vanderbilt University, *Faculty Manual*, Chapter Two, Inventions, Discoveries, Patents, 56 (1989–1990).
10. Brown University, Brown University Patent and Invention Policy, 2 (1981).
11. Duke University, Policy on Inventions, Patents and Technology Transfer, 4 (undated).

12. Auburn University, Auburn University Patent Policy, 3 (1974).
13. Florida State University, University Sponsored Educational Materials, §6C2-4.63, 379 (undated).
14. Stanford University, *Stanford Research Policy Handbook, Policy on Tangible Research Property*, 101 (1989). (Tangible research product is defined as biological materials, computer software, computer databases, circuit diagrams, engineering drawings, integrated circuit chips, prototype devices, or equipment.)
15. The University of Texas System, *Regents' Rules* and *Regulations, Patent Policy* Part 2, Ch. V. (1982).
16. The University of Illinois at Urbana-Champaign, *The General Rules Concerning University Organization and Procedure*, 7 n.1 (1988).
17. Stanford University, *Stanford Research Policy Handbook*, 91 (1989).
18. Massachusetts Institute of Technology, *Guide to the Ownership, Distribution, and Commercial Development of MIT Technology, Policies and Procedures*, 5 (1981).
19. The University of Pennsylvania, *Research Investigators' Handbook*, Copyright, 36–37 (1990).
20. Florida State University, *University Sponsored Educational Materials*, §6c2-4.63, 379 (undated).
21. Stanford University, *Stanford Research Policy Handbook* (1989); The University of Pennsylvania, *Research Investigators' Handbook* (1990); Northwestern University, *Guide to Intellectual Property*, (draft) (1990).
22. Northwestern University, *Guide to Intellectual Property*, 3 (draft) (1990).
23. The University of Illinois at Urbana-Champaign, *The General Rules Concerning University Organization and Procedure*, 7n.1 (1988).
24. Stanford University, *Stanford Research Policy Handbook*, 101 (1989).
25. Northwestern University, *Guide to Intellectual Property*, 8 (draft) (1990).
26. *Idem*.
27. University of Washington, *Patents and Inventions Procedures and General Information, Transfer of Nonpatented Technology*, 5 (1983).
28. Harvard Medical School, *Guide to Protecting and Managing Intellectual Property*, 15–16 (1985).
29. *Idem*.
30. Duke University, *Policy on Inventions, Patents and Technology Transfer*, 4 (undated).
31. University of Pennsylvania, *Research Investigators' Handbook*, Patent Policy, §I., 32(1990).
32. University of Illinois at Urbana-Champaign, *The General Rules Concerning University Organization and Procedure*, Article III Intellectual Property, 9 (1988).
33. The University of Georgia, Patent Policy, 4–5 (1982).
34. The University of Nebraska Lincoln, *Handbook on University of Nebraska Patent Procedures*, Form 2 (undated).
35. Brigham and Women's Hospital, *Policies on Hospital-Industry Relationships*, V-2 (1989).
36. Stanford University, *Stanford Research Policy Handbook*, 86 (1989).
37. *Idem*.

38. University of Illinois at Urbana-Champaign, Sponsored Research and Rights in Intellectual Property, 1 (1988).
39. Harvard Medical School, *Guide to Protecting and Managing Intellectual Property*, 6 (1985).
40. University of Pennsylvania, *Research Investigators' Handbook*, 35 (1990).
41. The University of Utah, *Policy and Procedure Manual, Patents and Inventions* (1984).
42. Stanford University, *Stanford Research Policy Handbook*, (1989).
43. Auburn University, Auburn University Patent Policy, 6 (1974).
44. University of Nebraska Lincoln, *Handbook on University of Nebraska Patent Procedures*, 15 (undated); University of Washington, Patent and Invention Policy, 2 (1983).
45. Stanford University, *Stanford Research Policy Handbook*, 88 (1989).
46. The University of Pennsylvania, *Research Investigators' Handbook*, 32 (1990).
47. Yale University, Yale University Patent Policy, 1 (1989).
48. Auburn University, Auburn University Patent Policy, 1, (1974).
49. Harvard Medical School, *Guide to Protecting and Managing Intellectual Property*, 5 (1985).
50. The University of Pennsylvania, *Research Investigators' Handbook*, 32 (1990).
51. Auburn University, Auburn University Patent Policy, 6 (1974).
52. The University of Utah, *Policy and Procedure Manual, Patents and Inventions*, 10 (1984).
53. Northwestern University, Patent and Invention Policy for Northwestern University, 1 (1974).
54. The University of Illinois at Urbana-Champaign, University Policy on Inventions and Patents, 2 (1988).
55. The University of Virginia, *Financial and Administrative Policies Manual, Patent Policy*, 15.5.2.6 (1986).
56. The University of Pennsylvania, *Research Investigators' Handbook*, 31 (1990).
57. H. Mayers and B. Brunsvold, *Drafting Patent License Agreements*, 2nd Ed., 177–178 (1984).
58. Harvard Medical School, *Guide to Protecting and Managing Intellectual Property*, 16 (1985).
59. University of Pennsylvania, *Research Investigators' Handbook, Policy for Commercial Sponsors*, 42–42 (1990); Yale University, *Yale University Faculty Handbook, Other University Policies Affecting Faculty Employment, Sponsored Research*, 96–100 (1986).
60. Stanford University, *Research Policy Handbook*, 82 (1989).
61. Note that there are a few universities who assign rights to commercial research sponsors.
62. Yale University, *Yale University Patent Policy*, 1–2 (1989).
63. University of Illinois at Urbana-Champaign, *Sponsored Research and Rights in Intellectual Property*, 2 (1988).
64. 34 U.S.C. ch. 18 §§200–212 (1990).
65. Sanford University, *Research Policy Handbook*, (1989).

66. Northwestern University, *Patent and Invention Policy for Northwestern University*, 2 (1974).
67. Yale University, *Yale University Patent Policy*, 2 (1989).
68. Harvard Medical School, *Guide to Protecting and Managing Intellectual Property*, 19 (1985).
69. Brown University, *Brown University Patent and Invention Policy*, 6 (1981).
70. University of Illinois at Urbana-Champaign, *The General Rules Concerning University Organization and Procedure*, 8 (1988); The University of Texas System, *Regents' Rules and Regulations, Intellectual Property Policy*, Part 2, Ch. V, (1990); University of Pennsylvania, *Research Investigators' Handbook, Patent Policy*, 33 (1990).
71. Harvard Medical School, *Guide to Protecting and Managing Intellectual Property*, 40 (1985); University of Illinois at Urbana-Champaign, *The General Rules Concerning University Organization and Procedure*, 8 (1988).
72. University of California, *A Patent Primer for Faculty and Staff*, 16 (1988); University of Delaware, *University Policy on Patents*, 5 (1987).
73. Harvard Medical School, *Guide to Protecting and Managing Intellectual Property*, 40 (1985).
74. Vanderbilt University, *Faculty Manual, Income*, 58 (1989–1990).
75. University of Utah, *Policy and Procedure Manual, Patents and Inventions*, 4–5 (1984).
76. University of Illinois at Urbana-Champaign, *The General Rules Concerning University Organization and Procedure*, 8 (1988).
77. The University of Utah, *Policy and Procedure Manual, Patents and Inventions*, 4–5 (1984).
78. University of California, *A Patent Primer for Faculty and Staff*, 16 (1988); The University of Texas System, *Regents' Rules and Regulations, Intellectual Property Policy*, Part 2, Ch. V, (1988).
79. Harvard Medical School, *Guide to Protecting and Managing Intellectual Property*; Harvard University *Memorandum on Patent Procedures and Royalty Distribution*, 40 (1985).
80. Northwestern University, *Patent and Invention Policy for Northwestern University*, 2 (1974).
81. Brigham and Women's Hospital, *Policies in Hospital-Industry Relationships*, V-3 (1989).
82. *Idem*.
83. Florida State University, *University Sponsored Educational Materials*, §6C2-4.63, 381 (undated).
84. Carnegie-Mellon University, *Intellectual Property Policy, Organizational Announcement* No. 307, 9 (1985).
85. University of Delaware, *University Policy on Inventions*, 5 (1987).
86. D. R. Fowler, University industry research relationships: The research agreement, 9 J. Coll. Univ. Law 516 (1982–1983).
87. The University of Pennsylvania, *Commercial Sponsors*, 42 (1983).
88. The University of Minnesota, *Guidelines on Interactions With Industry*, 1 (undated).

89. *Idem.*
90. Harvard Medical School, *Guide to Protecting and Managing Intellectual Property*, 35 (1985).
91. Association of American Universities, *University Policies on Conflict of Interest and Delay of Publication*, 3–4 (1985).
92. D. R. Fowler, *A Study of the Need for and the Impediments to Improved and Novel University-Industry Research Relationships*, 105 (1983) (Ph.D. dissertation, Claremont Graduate School).
93. Consulting agreements are discussed in Section III of this chapter.
94. D. R. Fowler, University-industry research relationship: The research agreement, 9 J. Coll. Univ. Law 515–532 (1982–1983).
95. Execution means the signing of the agreement by authorized representative of the parties.
96. D. R. Fowler is the author of numerous excellent publications, including University-industry research relationships, 27 Research Management 35 (January-February 1984); University-industry research relationships: The research agreement, 9 J. Coll. Univ. Law 515 (1982–1983). He also wrote a doctoral dissertation that would be very helpful to anyone interested in this chapter's subject. See Fowler, *A Study of the Need for and the Impediments to Improved and Novel University-Industry Research Relationships* (Ph.D. dissertation, Claremont Graduate School) (1983).
97. B. D. Reams, Jr. is author of an excellent and comprehensive volume: *University-Industry Research Partnerships: The Major Legal Issues in Research and Development Agreements*, Quorum Books (1986).
98. D. R. Fowler, University-Industry Research Relationships: The Research Agreement, 9 J. Coll. Univ. Law 515, 518 (1982–1983).
99. B. D. Reams, Jr., *University-Industry Research Partnerships: The Major Legal Issues in Research and Development Agreements* (1986).
100. *Id.* at 51.
101. D. R. Fowler, University-Industry Research Relationships: The Research Agreement, 9 J. Coll. Univ. Law 515 (1982–1983).
102. The University of Minnesota, *Guidelines on Interactions With Faculty*, 11 (undated).
103. For example, $300,000 total funding amount plus 15% overhead ($45,000) equals $345,000. Three equal payments of $115,000 each are to be made on December 31, 1991; December 31, 1992 and December 31, 1993.
104. B. D. Reams, Jr., *University-Industry Research Partnerships: The Major Legal Issues in Research and Development Agreements*, Quorum Books 180 (1986).
105. *Idem.*
106. *Idem*, at 272–274.
107. Trade secrets are commonly defined as secrets of any character that are owned by, unique, and important to a business and give the business an advantage over its competitors.
108. B. D. Reams, Jr., *University-Industry Research Partnerships: The Major Legal Issues in Research and Development Agreements*, Quorum Books 320 (1986). A formal, written policy of Washington University has not been located, but this po-

sition is represented in a negotiated agreement, Washington University-Anheuser-Busch Companies, Inc., Micromixing Research Agreement of 1983. However, many universities take this position.

109. Research groups can be defined as groups within a department who regularly collaborate; groups who make up a center of specialty in a particular area of scientific interest, or even an entire department.
110. D. R. Fowler, University-industry research relationships: The research agreement, 9 J. Coll. Univ. Law 515 (1982–1983).
111. B. D. Reams, Jr., *University-Industry Research Partnerships: The Major Legal Issues in Research and Development Agreements*, Quorum Books 157 (1986). See also J. C. Williams, industry-university interactions: Finding the balance, Engineering Education (March, 1986) 320–325.
112. B. D. Reams, Jr., University-Industry Research Partnerships: The Major Legal Issues in Research and Development Agreements, Quorum Books (1986).
112a. The University of Pennsylvania, *Policy for Commercial Sponsors*, 3 (1983).
113. Harvard Medical School, *Guide to Protecting and Managing Intellectual Property*, 36 (1985).
114. American Association of University Professors, 1940 Statement of Principles on Academic Freedom and Tenure (reprinted in 3 EDUC. L. §§. 11.01[2] [a] (1989).
115. D. R. Fowler, *A Study of the Need for and the Impediments to Improved and Novel University-Industry Research Relationships* (Ph.D. dissertation, Claremont Graduate School) (1983). But see B. T. Watkins, Colleges tie with business see threatening academic freedom, Chronicle of Higher Education, 14 (June 2, 1982).
116. *Idem.*, at 105.
116a. *Idem.*
117. The University of Texas System, *Regents' Rules and Regulations, Intellectual Property Policy*, Part 2, Ch. V., V-9 (1990).
118. The University of Texas Southwestern Medical Center at Dallas, *Sponsored Research Agreement* (undated).
119. *Idem.*
120. To assign means to give to another party title or ownership to property.
121. Intellectual Property is intangible property such as patents, trade secrets and copyright.
122. U.S. Department of Commerce, Patent and Trademark Office, Patents & Inventions: An Information Aid for Inventors, 1 (1980).
123. Saliwanchik, Legal protection for microbiological and genetic engineering inventions, 1 (1982).
124. *Idem.*
124a. H. Mayers and B. Brunsvold, *Drafting Patent License Agreements*, 2nd Ed. (1984).
125. Harvard Medical School, *Guide to Protecting and Managing Intellectual Property*, 9 (1985).
126. Harvard Medical School, *Guide to Protecting and Managing Intellectual Property*, 11 (1985).

127. Stanford University, *Stanford Research Policy Handbook, Policy on Inventions, Patents and Licensing* (1989).
128. H. Mayers, and B. Brunsvold, *Drafting Patent License Agreements*, 2nd Ed., 68–70 (1984) (see same definitions and concerns regarding licenses).
129. *Idem.*
130. The University of Texas Southwestern Medical Center at Dallas, *Sponsored Research Agreement*, 8 (undated).
131. *Black's Law Dictionary*, 4th Ed. *West* (1951).
132. H. Mayers and B. Brunsvold, *Drafting Patent License Agreements*, 2nd Ed., 32 (1984).
133. Yale University, *Yale Faculty Handbook*, 98 (1986); See also Northwestern University, *Patent and Invention Policy for Northwestern University*, §VIII Inventions or Discoveries Arising From Sponsored Programs, 2 (1974).
134. Auburn University, *Auburn University Patent Policy*, §1.4.3 Industrial Research, 4 (1974).
135. Stanford University, *Stanford Research Policy Handbook, Policy on Inventory, Patents and Licensing*, 86 (1989).
136. See Section II.B.5 of this chapter entitled Termination of Rights—Rights Surviving After Termination.
137. Washington University–Mallinckrodt, Inc., Hybridoma Research Agreement, §8.11 (1981), reprinted in R. D. Reams, Jr., *University-Industry Research Partnerships*, Appendix B, Quorum Books, 235 (1986).
138. 1A Corbin on Contracts, §§260–261 (Supp. 1990).
139. Reprinted in R. D. Reams, *University-Industry Research Partnerships*, Appendix A, Quorum Books, 193–194 (1986).
140. Prosecution in this context means the process of the Patent & Trademark Office's (PTO) reviewing a patent application and the applicant's responding until the patent issues or is finally rejected.
141. The University of Minnesota, *Guidelines on Interactions With Industry*, 5 (undated).
142. "A trade secret may consist of any formula, pattern, device or compilation of information which is used in one's business, and which gives him an opportunity to obtain an advantage over competitors who do not know or use it. . . . " Restatement of the Law of Torts, §757, Comment (b), reprinted in H. Mayers and B. Brunsvold, Drafting Patent License Agreements, 2 ed., American Law Institute 177 (1984).
143. Yale University, Research Agreement, 4 (undated).
144. See generally R. D. Reams, Jr., *University-Industry Research Partnerships: The Major Legal Issues in Research and Development Agreements*, Quorum Books, 61-2 (1986).
145. The University of Texas Southwestern Medical Center at Dallas, Sponsored Research Agreement, 12 (undated).
146. The University of Delaware, Research Agreement Between Company and the University of Delaware, 9 (1990).
147. University of Pennsylvania, *Research Investigators' Handbook, Policy for Commercial Sponsors*, 42 (1990).

148. University of Delaware, Research Agreement Between Company and the University of Delaware, 8 (1990).
149. Texas A&M University System, *Administrative Policy and Reporting Manual*, §C.10.2., 11–12 (1983).
150. R. D. Reams, Jr., *University-Industry Research Partnerships: The Major Legal Issues in Research and Development Agreements*, Quorum Books, 90–91 (1986).
151. Brigham and Women's Hospital, *Policies in Hospital-Industry Relationships*, V-3 (1989).
152. The University of Pennsylvania, *Research Investigators' Handbook*, 43 (1990).
153. Interview with G. M. Mussey, Director of Contracts Management, The University of Texas Southwestern Medical Center at Dallas, Texas (July 30, 1990).
154. Duke University, *Faculty Handbook*, *University-Industry Guidelines*, Appendix Q, 13 (1987).
155. Harvard University Faculty of Medicine, *Policy on Conflicts of Interest and Commitment*, 4 (1990).
156. R. D. Reams, Jr., *University-Industry Research Partnerships: The Major Legal Issues in Research and Development Agreements*, Quorum Books, 71 (1986).
157. The University of Pennsylvania, *Research Investigators' Handbook*, 42 (1990).
158. *Idem.*
159. Duke University, *Faculty Handbook*, *University-Industry Guidelines*, Appendix Q, 12 (1990).
160. University of Delaware, *Research Agreement Between Company and the University of Delaware*, 4 (1990).
161. University of Minnesota, *Guidelines for Interaction With Industry*, 5 (1982).
162. Yale University, *Faculty Handbook*, 98 (1986).
163. *Idem.*
164. *Idem.*
165. The University of Illinois at Urbana-Champaign, *Research Administration Bulletin, Sponsored Research and Rights in Intellectual Property*, 1 (1988).
166. Harvard University Faculty of Medicine, *Protecting and Managing Intellectual Property, Statement on Research Sponsored by Industry*, 36 (1990).
167. The University of Georgia, College of Agriculture, *Faculty Consulting Policy and Reporting Procedure*, Policy No. 5, 1 (1987).
168. Harvard Medical School, *Guide to Protecting and Managing Intellectual Property*, 20 (1985).
169. *Idem.*
170. The University of Illinois, *The General Rules Concerning University Organization and Procedure*, 7 (1988).
171. *Idem.*
172. The University of Georgia (College of Agriculture), Faculty Consulting Policy and Reporting Procedure Policy No. 5, 1 (1987).
173. The University of Utah, *Remunerative Consultation and Other Employment Activities, Policy and Procedures Manual* No. 2-26, 3–4 (1980).
174. *Idem.*

175. *Idem.*; The University of Georgia, College of Agriculture, *Faculty Consulting Policy and Reporting Procedure*, Policy No. 5, 1 (1987).
176. Colorado State University, *Scholarly and Creative Activities of the Faculty, Consulting Policy*, §D.8.6 (1985).
177. Stanford University, *Research Policy Handbook, Policy on Outside Consulting (Faculty)* 72 (1989).
178. Auburn University, *Consulting and Extra Compensation Policy*, 2 (1984).
179. The University of Delaware, *Outside Employment* (1990).
180. Auburn University, *Consulting and Extra Compensation Policy*, 2 (1984).
181. The University of Pittsburgh, *Handbook for Faculty, Conflict of Interest*, 3 (1980).
182. The University of Chicago, 205. *Outside Professional and Commercial Interests of Faculty*, 205.1. (1989).
183. The University of Pittsburgh, *Handbook for Faculty, Outside Professional Service*, 41 (1980).
184. Colorado State University, *Scholarly and Creative Activities of the Faculty, Consulting Policy*, §D.8.6 (1985).
185. The University of Minnesota, *Policy on Outside Consulting Service Activities, and Other Outside Work*, 9 (1983); The Texas A&M University System, *Administrative Policy and Reporting Manual*, 4 (1981).
186. Duke University, *Faculty Handbook, Consulting Activities and Conflicts of Interest*, 5/7 (1987).
187. Stanford University, *Research Policy Handbook, Policy on Outside Consulting (Faculty)* 73 (1989).
188. Auburn University, Patent Policy, §1.4.5 (1974); University of Chicago, *Outside Professional and Commercial Interests of Faculty*, 206.3 (1989; K. E. Dillon and K. L. Bane, Consulting and conflict of interest, Educational Record 52–72 (1980). (This is an excellent, although dated analysis of consulting policies of one hundred universities.)
189. Matthew 6:24 (King James).
190. *Webster's Third New International Dictionary* (1961), reprinted in Council on Scientific Affairs and Council on Ethical and Judicial Affairs of the American Medical Association, Conflicts of Interest in Medical Center/Industry Research Relationships, 263 JAMA 2790 (May 23/30, 1990).
191. Association of Academic Health Centers, *Conflicts of Interest in Academic Health Centers* 19, (1989).
192. *Idem.*, at 20.
193. M. Chren, C. S. Landefeld, and T. H. Murray, Doctors, drug companies, and gifts, 262 JAMA 3448, 3451 (December 22/29, 1989).
194. *Idem.*, at 3451.
195. Association of Academic Health Centers, *Conflicts of Interest in Academic Health Centers*, 19 (1989).
196. UTAH CODE ANN. §§67-16-1 to -14 (1953).
197. State statutes or laws will apply to faculty at state-supported institutions, such as The University of Utah; UTAH CODE ANN. §§67-16-1 to -14 (1953).

198. See generally American Association of University Professors and the American Council on Education, Preventing Conflict of Interest in Government-Sponsored Research at Universities, (December, 1964) (hereinafter *AAUP/ACE Statement*), reprinted in Association of American Universities, University Policies on Conflict of Interest and Delay of Publication, 10 (February, 1985) (hereinafter *AAU Report*) (The AAUP/ACE Statement "provides a detailed discussion of conflict of interest and encourages individual universities to establish procedures to address it."); Pajaro Dunes Conference: Draft Statement, 9 J. Col. & Univ. L. 533–539, (1982-3) (an agenda of pertinent issues) CAL. GOV'T. CODE §§81000-91014 (1989) (requiring public employees to disclose potential conflicts of interest; original enactment in 1974).
199. *Idem.*
200. R. Ditzel, The Commercialization of University Technology in the 1990's, 3 Albany Law School Annual Conference on Intellectual Property 8-1, 8-25 (1990).
201. Harvard University Faculty of Medicine, *Policy on Conflicts of Interest and Commitment* (1990).
202. M. Chren, C. S. Landefeld, and T. H. Murray, Doctors, drug companies, and gifts, 262 JAMA 3448, 3451 (December 22/29, 1989).
203. Council on Scientific Affairs and Council on Ethical and Judicial Affairs of the American Medical Association, Conflicts of Interest in Medical Center/Industry Research Relationships 263 JAMA 2790 (May 23/30, 1990).
204. University of Pennsylvania, *Conflict of Interest Policy for Faculty Members, Research Investigators' Handbook* (1990); Cornell University, *Cornell University Conflicts Policy*, (1986); University of Minnesota, *Policy on Outside Consulting, Service Activities and Other Outside Work* (1982); University of Minnesota, *Disclosure of Financial Arrangements With Industry* (1988).
205. A. B. Giamatti, The university, industry and cooperative research, 218 Science 1278 (December 24, 1982).
206. Cornell University, *Faculty Handbook*, 118 (undated).
207. Harvard University Faculty of Medicine, *Policy on Conflicts of Interest and Commitment*, A-1 (1990).
208. The University of Utah, *Policy and Procedures Manual* No. 2-30, *Conflict of Interest* (1981).
209. The University of Chicago, *Outside Professional and Commercial Interests of Faculty*, 205.1 (1989).
210. UTAH CODE ANN. §§67-16-1 to -14 (1953); TEX. CIV. PRAC. & REM. CODE ANN. §6252-9b (Supp. 1990); GA. CODE ANN. §§45-10-1 to -70 (Supp. 1989).
211. For the balance of the chapter the term *conflict of interest* will be used to include both conflict of interest and conflict of commitment.
212. The University of Utah, *Policy and Procedures Manual* No. 2-30, *Conflict of Interest* (1981).
213. Association of Academic Health Centers, *Conflicts of Interest in Academic Health Centers*, 70 (1989).
214. The University of Minnesota, *Disclosure of Financial Arrangements With Industry*, (1982).

215. University of California, *Conflict of Interest Code*, Appendix A, (1988).
216. *Idem.*
217. The University of Utah, *Policy and Procedure Manual* No. 2-30, *Conflict of Interest*, 1 (1981).
218. Association of Academic Health Centers, *Conflicts of Interest in Academic Health Centers*, 72 (1989).
219. Brown University, *Policy on Duality of Interest or Conflict of Interest* (1978); Cornell University, *Cornell University Conflicts Policy*, 2 (1986).
220. Harvard University Faculty of Medicine, *Policy on Conflicts of Interest and Commitment*, A-1 (1990).
221. *Idem.*
222. A. B. Giamatti, The university, industry and cooperative research, 218 Science 1278, (December 24, 1982).
223. UTAH CODE ANN. §§67-16-1 to -14 (1953); TEX. CIV. PRAC. & REM. CODE ANN. §6252-9b (Supp. 1990); GA. CODE ANN. §§45-10-1 to -70 (Supp. 1989).
224. The University of Utah, *Policy and Procedure Manual* No. 2-30, *Conflict of Interest*, (1981).
225. *Idem.*
226. Harvard University Faculty of Medicine, *Policy on Conflicts of Interest and Commitment*, A-1,4 (1981).
227. R. Killoven, Institutional conflict of interest, 3 Research Management Review 1, 7 (1989).
228. Colorado State University, *Conflict of Interest*, §D.8.7, D-8 (1985).
229. The University of Pittsburgh, *Policy Governing Conflicts of Interest and Conflicts of Commitment*, 5 (1986).
230. *Idem.*, at 11.
231. *Idem.*, at 12.
232. *Idem.*, at 13.
233. Harvard University, *Faculty of Arts and Sciences*, *Statement of Policy on Conflicts of Interest* (1982).
234. Cornell University, *Cornell University Conflicts Policy*, 3 (1986).
235. *Idem.*, at 5.
236. B. J. Culliton, Pajaro Dunes: The search for consensus, 216 *Science*, 155, 156 (April 19, 1982) (University investments in proprietary companies in which faculty are principals "are ill advised unless there are sufficient safeguards to avoid adverse effects on the morale of the institution.")
237. C. K. Gunsalus and J. Rowan, Conflict of interest in the university setting: I Know it when I see it, 3 *Research Management Review* 13 (1989).
238. M. Knight, Harvard rules out role for now in genetic engineering company, 18 The New York Times 1, Col. 5, (November 17, 1980).
239. *Idem.*
240. *Commercialization of Academic Biomedical Research: Hearings Before The Subcommittee on Investigations and Oversight and The Subcommittee on Science, Research & Technology of the Committee on Science and Technology*. 97th Cong., 1st Sess. 46 (1981).

241. *Idem.*
242. C.K. Gunsalus and J. Rowan, Conflict of interest in the university setting: I know it when I see it, 3 *Research Management Review* 13, 19 (1989).
243. Association of American Universities, *University Policies on Conflict of Interest and Delay of Publication* (1985).
244. University of Minnesota, *Policy on Outside Consulting, Service Activities and Other Outside Work*, 2 (1982).
245. University of Pittsburgh, *Policy Governing Conflicts of Interest and Conflicts of Commitment*, 3 (1986)
246. Harvard University's Faculty of Arts and Sciences, *Statement of Policy on Conflicts of Interest*, 1 (1982).
247. Harvard University Faculty of Medicine, *Policy on Conflicts of Interest and Commitment*, 1 (1990).
248. The University of Pittsburgh, *Policy Governing Conflicts of Interest and Conflicts of Commitment*, 3-4 (1986).
249. Yale University, *Faculty Handbook*, 103 (1986).
250. Harvard University Faculty of Medicine, *Policy on Conflicts of Interest and Commitment*, B-2 (1990).
251. *Idem.*
252. *Idem.*
253. Cornell University, Conflicts Policy, 5–9 (1986).
254. The Texas A&M University System, *Administrative Policy and Reporting Manual*, Faculty Consulting, Outside Professional Employment, and Conflicts of Interest, B.4.12, 5 (1989).
255. Duke University, *Faculty Handbook*, Appendix Q, 14 (1987).
256. Vanderbilt University, *Conflict of Interest Policy* 36, (undated).
257. The University of Minnesota, *Policy on Outside Consulting, Service Activities and Other Outside Work* (1982).
258. Brown University, *Policy on Duality of Interest or Conflict of Interest* (1978).
259. Harvard University Faculty of Medicine, *Policy on Conflicts of Interest and Commitment* (1990).
260. The University of Pittsburgh, *Policy on Governing Conflicts of Interest and Conflicts of Commitment*, 25 (1986).
261. Harvard University Faculty of Medicine, *Policy on Conflicts of Interest and Commitment* (1990).
262. CAL. ANN. GOV'T. CODE §81000-91014 (1989).
263. J. Palco, Conflict over conflict of interest, 245 Science 1440 (1989).
264. *Id.*
265. Association of Academic Health Centers, *Conflicts of Interest in Academic Health Centers, iii* (1989).

APPENDIX

Intellectual Property Policies

AUBURN UNIVERSITY PATENT POLICY

Introduction

This document constitutes the patent policy of Auburn University and applies to Auburn University faculty and other employees and students in the areas of discoveries, inventions and patents for the benefit of the public. The patent policy embodies both the procedural and substantive aspects concerning discoveries, inventions and patents as set out in Sections 1.0 and 2.0 and as summarized in the next paragraph. The policy has been developed to insure that prospective inventors are aware of the assistance available to them and the steps necessary to insure that a patentable idea reaches fruition. It was developed under the philosophy that by patenting an idea it not only offers an opportunity for personal reward but helps to promote the use of the patentable idea. Many unprotected ideas will not be utilized by private industry because of the high risk in capital investment. Thus, many ideas are lost to the economy.

Auburn University faculty, other employees, and students shall report to the University through the Secretary-Director of the Auburn Research Foundation, all discoveries or inventions deemed patentable arising from research or other activities conducted within the field of their official duties and responsibilities or programmed or substantially supported by the University. Such potentially patentable discoveries and inventions should be reported using the disclosure of invention form available from the Secretary-Director. Inventors are cautioned that a patent application must be filed within a year after open publication of the disclosure of an invention or the idea becomes part of the public domain and unpatentable.

The University's equity, if any, in inventions or discoveries of faculty members, other employees and students made outside the scope of their official duties and responsibilities but in which there is some use of University facilities will be determined by the University. The University's equity in inventions arising from projects conducted by students and faculty or employees as a part of formal instruction or thesis projects will also be determined by the University. The ownership and control of patent rights to patentable discoveries and inventions resulting from research sponsored by industrial or governmental agencies will be determined on the basis of the contractual agreement between the sponsor and the University. The University's equity in discoveries and inventions not covered by such contractual agreements will be determined by the University. Inventions and discoveries made or developed solely in the course of consulting work performed for outside agencies with the concurrence or approval of the appropriate Vice President shall not be considered having been made or developed in the course of University employment unless otherwise provided by the approving authority or unless a significant use of University facilities is involved. The equity in inventions arising from consulting engagements by faculty and other employees not having prior concurrence or approval by the appropriate Vice President will be determined by the University.

The policy is intended to encourage the patenting of discoveries and inventions and the development and utilization of patents for the benefit of the public and the inventors for all patents in which the University has an equity. Furthermore, the policy offers a mechanism for patenting discoveries and inventions of the faculty, employees and students in which the University has no equity but may acquire equitable interest in order to assist the faculty, other employees or students. The policy is general and reflects the tenor of Auburn University in its activities to assist prospective patentees develop their discoveries and inventions for patenting for the benefit of the public and the inventors.

1.0 Substantive Policy

1.1 *Parties*

This policy shall exist between Auburn University and all faculty, other employees and students of the University. This relation between the parties shall arise as a part of the employment contract between the University and its faculty and employees and shall constitute a condition of enrollment for all students and, prior to the use of any University facilities which might result in an invention or discovery, the individuals will execute a signed and witnessed agreement to this policy.

1.2 *Duration of Policy*

The policy shall be effective between Auburn University and each faculty member or other employee commencing with his employment and continuing for a period of one year subsequent to his termination of employment. Similarly, the policy shall be effective between the University and each student commencing with his enrollment and continuing for a period of one year subsequent to the termination of his last period of enrollment.

The policy is subject to change and a copy of the current policy will be kept on file in the office of the Secretary-Director of the Auburn Research Foundation.

1.3 *Class of Intellectual Property Rights Affected*

This policy covers patentable subject matter and patents but does not extend to copyrightable or copyrighted material. Patentable subject matter involving the act of invention covers:

1. Processes, machines, compositions of matter and other articles of manufacture. ''Process'' includes improvement of a known process or an entirely new method for making a composition of matter or article of manufacture.
2. Design.
3. Plants - Asexually and sexually reproduced plants, with some exceptions.

Invention involves: a

A. Conception, a mental act;
B. Reduction to practice.

Unpatentable subject matter covers:

1. Discovery of natural phenomenon;
2. Product of nature (except antibiotics, vitamins);
3. Mere ideas or concepts;
4. Inventions relating to atomic weapons.

1.4 *Equities of the Parties*

1.4.1 *Internally Sponsored Projects*

The patent rights to inventions and developments resulting from research or other activities conducted by faculty, other employees or students within the field of their official duties and responsibilities or programmed or substantially supported by the University are retained by the University. Disposition of patent rights is handled by the University. However, the faculty members, other employees and/or students, who are the inventors of record may participate in the income from discoveries and inventions on which patents are obtained in accordance with Section 1.5.

In the event Auburn University does not elect within six months after submission of a disclosure to pursue a patent or patents on any discovery or invention, the inventor or inventors may request the release of all rights to the invention or discovery and they may dispose of and handle them as they desire.

The University or its agent will bear all costs of pursuing the obtaining of patents for all inventions and discoveries retained by the University, but these costs will be deducted from the proceeds prior to any payments to the inventor or inventors.

1.4.2 *Personal Research*

In order to provide incentive and encouragement to faculty members, other employees and students of the University in pursuing their personal research interests outside the scope of their official duties and responsibilities and not programmed or substantially supported by the University although slight use may be made of University facilities, any resulting invention or development which may be patentable may be submitted to the University for acceptance or rejection, or may be disposed of and handled by the inventors as they desire. However, all personal discoveries or inventions will be reported to the University, regardless of submission for acceptance or rejection by the University for a determination in accordance with this section. Generally, projects conducted by students and faculty or employees as a part of formal instruction or thesis projects are considered personal research and outside the scope of official duties and responsibilities and programming unless such thesis projects are conducted as a part of a contractual research effort or substantially supported by the University.

Faculty, other employees and students submitting patentable ideas and discoveries resulting from personal research to the University accepted for patenting will participate in the income from these patents in accordance with Section 1.5. The University or its agent will bear the costs of pursuing the obtaining of patents for all accepted inventions and discoveries, but these costs will be deducted from the proceeds prior to any payments to the inventor or inventors.

1.4.3 *Industrial Research*

The ownership and control of patent rights to patentable discoveries and inventions resulting from research sponsored by industry at the University are determined by the contractual arrangement between the sponsor and University. In general, the University retains ownership and control of the patents; however, if required by the sponsor, the University will assign and convey to the sponsor under the contractual arrangements rights to all patentable discoveries and inventions resulting from the sponsored research. Where the University retains ownership and control of patents resulting from sponsored research, the University treats these patents the same as though resulting from its own research and Section 1.4.1 is applicable.

1.4.4 *Government Research*

Auburn University complies with the government patent policies as set forth by the various Executive Departments and Agencies of the Federal government. The question of ownership and control of patents resulting from Federally or State supported research is determined in each specific contract. Contracts with the State of Alabama shall be governed by Section 1.4.1 unless otherwise provided.

1.4.5 *Consulting*

Faculty and other employees desiring to perform consulting work for outside organizations are required to obtain the prior concurrence in accordance with current University policy.

Inventions and discoveries made or developed solely in the course of consulting work performed with prior concurrence in accordance with current University policy shall not be considered as having been made or developed in the course of University employment unless otherwise provided in the concurrence or unless a significant use of University facilities is involved, in which case the University will determine whether or not Section 1.4.2 or other section applies.

Inventions arising from consulting engagements by faculty and other employees not having prior concurrence or approval by the appropriate Vice President and involving substantial use of University facilities will be governed by Section 1.4.1, notwithstanding any agreement between the consultant and his client. Students working with faculty on consulting projects are governed by this same policy.

1.5 *Division of Equities*

Faculty, other employees and/or students as inventor or co-inventors may receive jointly percentages of the net proceeds as indicated below:

% of Net Proceeds	Net Proceeds
30	Up to $10,000
25	Next $10,000
20	Next $10,000
15	All over $30,000

The net proceeds is defined as the proceeds minus all costs of obtaining the patent and of defending and/or marketing the patent. The percentage of net proceeds to the inventor or co-inventors will be distributed annually.

1.6 *Agreement as to Division of Equities*

Auburn University shall, if it determines that more than one individual is entitled to an equity in an invention or discovery, determine the manner in which the equity awarded to the group shall be divided among the individuals constituting the group, provided, however, that if such individuals have agreed among themselves, such agreement shall be conclusive.

1.7 *Assignment of Equities*
Notwithstanding any determination of Auburn University, or any other provision of this policy, faculty, employees and students shall have no equities or rights whatsoever in inventions and discoveries belonging to the University unless and until a written agreement has been executed by the University and University faculty, other employees and/or students consistent with the determination of the University guided by this policy.

The inventor or inventors are required to execute such as assignments and other documents that the University determines to be necessary in obtaining the patent.

1.8 *Effective Date of this Policy*
This policy is effective as of the date of its signing by the President of Auburn University. Nothing in the Auburn University Patent Policy shall be deemed to change any agreements made under previous policies.

2.0 Procedural

2.1 *Inventors*
All University faculty, other employees and students shall report to Auburn University through the Secretary-Director of the Auburn Research Foundation any ideas or discoveries deemed patentable arising from situations described in Section 1.4 using a disclosure of invention form, available from the Secretary-Director. Inventors are cautioned that a patent application must be filed within a year after open publication of the disclosure of an invention or the idea becomes part of the public domain and unpatentable.

2.2 *Auburn University*
Auburn University shall consider promptly all reports of inventions and discoveries and shall make the determinations required herein within a reasonable time and in no case longer than six months. The University personnel or students involved shall be entitled to appear before the Secretary-Director of the Auburn Research Foundation and present evidence to the report. The University's determination shall be made in writing and shall contain a statement of findings and grounds for decision. In all cases where the University declines to pursue a patentable idea or potential patent the ownership of the rights to such idea or patent are released in toto to the inventor/s unless prior restraints exist between the university and a sponsoring agency. The University shall maintain the confidentiality of all patentable ideas so as not to jeopardize the patentability of such ideas.

The University shall have the right to determine how and for what consideration any patent rights assigned to the University will be used.

2.3 *Dissemination of this Policy*
This policy shall be published in the Faculty Handbook or other appropriate document distributed to the faculty and employees and the Tiger Cub, the official publication

for students of Auburn University. The policy shall be referenced in initial employment papers of faculty and other employees of Auburn University, Administrative Digests and other appropriate documents for faculty, other employees and students.

APPROVED THIS 15th DAY OF March, 1974

Harry M. Philpott
President, Auburn University

UNIVERSITY OF CALIFORNIA PATENT POLICY

I. PREAMBLE

It is the intent of the President of the University of California, in administering intellectual property rights for the public benefit, to encourage and assist members of the faculty, staff, and others associated with the University in the use of the patent system with respect to their discoveries and inventions in a manner that is equitable to all parties involved.

The University recognizes the need for and desirability of encouraging the broad utilization of the results of University research, not only by scholars but also in practical application for the general public benefit, and acknowledges the importance of the patent system in bringing innovative research findings to practical application.

Within the University, innovative research findings often give rise to patentable inventions as fortuitous by-products, even though the research was conducted for the primary purpose of gaining new knowledge.

To encourage the practical application of University research for the broad public benefit, to appraise and determine relative rights and equities of all parties concerned, to facilitate patent applications, licensing, equitable distribution of royalties, if any, to assist in obtaining funds for research, to provide for the use of invention-related income for the further support of research and education, and to provide a uniform procedure in patent matters when the University has a right or equity, the following University of California Patent Policy is adopted.

II. STATEMENT OF POLICY

A. An agreement to assign inventions and patents to the University, except those resulting from permissible consulting activities without use of University facilities, shall be mandatory of all employees, for persons not employed by the University but who use University research facilities, and for those who receive gift, grant, or contract funds through the University. Exemptions from such agreements to assign may be authorized in those circumstances when the mission of the University is better served by such action, provided that overriding obligations to other parties are met and such exemptions are not inconsistent with other University policies.

B. Those individuals who have so agreed to assign inventions and patents shall promptly report and fully disclose the conception and/or reduction to practice of potentially patentable inventions to the Director of the Patent, Trademark, and Copyright Office. They shall execute such declarations, assignments, or other documents as may be necessary in the course of invention evaluation, patent prosecution, or protection of patent or analogous property rights, to assure that title in such inventions shall be held by the University or by such other parties designated by the University as may be appropriate under the circumstances. Such circumstances would include, but not be limited to, those situations when there are overriding patent obligations of the University arising from gifts, grants, contracts, or other agreements with outside organizations.

   In the absence of overriding obligations to outside sponsors of research, the University may release patent rights to the inventor in those circumstances when:

(1) the University elects not to file a patent application and the inventor is prepared to do so, or
(2) the equity of the situation clearly indicates such release should be given.

provided in either case that no further research or development to develop that invention will be conducted involving University support or facilities, and provided further that a shop right is granted to the University.

C. Subject to restrictions arising from overriding obligations of the University pursuant to gifts, grants, contracts, or other agreements with outside organizations, the University agrees, for and in consideration of said assignment of patent rights, to pay annually to the named inventor(s), or to the inventor(s)' heirs, successors, or assigns, 50% of the first $100,000 of cumulative net royalties and fees per invention received by the University, 35% of the next $400,000 of cumulative net royalties and fees per invention received by the University, and 20% of all additional cumulative net royalties and fees per invention received by the University. Net royalties are defined as gross royalties and fees, less 15% thereof for administrative costs, and less the costs of patenting, protecting, and perserving patent rights, maintaining patents, the licensing of patent and related property rights, and such other costs, taxes, or reimbursements as may be necessary or required by law.

When there are two or more inventors, each inventor shall share equally in the inventor's share of royalties, unless ass inventors previously have agreed in writing to a different distribution of such share.

Distribution of the inventor's share shall be made annually in February from the amount received during the penultimate calendar year. In the event of any litigation, actual or imminent, or any other action to protect patent rights, the University may withhold distribution and impound royalties until resolution of the matter.

D. In the disposition of any net income accruing to the University from patents, first consideration shall be given to the support of research.

III. PATENT RESPONSIBILITIES AND ADMINISTRATION

A. Pursuant to Standing Order 100.4(gg), the President has responsibility for all matters relating to patents in which the University of California is in any way concerned.

B. The President is advised on such matters by the Intellectual Property Advisory Council (IPAC), which is chaired by the Senior Vice President - Academic Affairs. The membership of IPAC includes representatives from campuses, Agriculture and Natural Resources, the Department of Energy Laboratories, and the Director of the Patent, Trademark, and Copyright Office. IPAC is responsible for:

1. reviewing and proposing University policy on intellectual property matters including patents, copyrights, trademarks, and tangible research products;

2. reviewing proposed exceptions to established policies; and
3. advising the President on related matters as requested.

C. The Senior Vice President - Administration is responsible for implementation of this Policy, including the following:

1. Evaluating inventions and discoveries for patentability, as well as scientific merit and practical application, and requesting the filing and prosecution of patent applications.
2. Evaluating the patent or analogous property rights or equities held by the University in an invention, and negotiating agreements with cooperating organizations, if any, with respect to such rights or equities.
3. Negotiating licenses and license option agreements with other parties concerning patent and/or analogous property rights held by the University.
4. Directing and arranging for the collection and appropriate distribution of royalties and fees.
5. Assisting University officers in negotiating agreements with cooperating organizations concerning prospective rights to patentable inventions or discoveries made as a result of research carried out under grants, contracts, or other agreements to be funded in whole or in part by such cooperating organizations, and negotiating with Federal agencies regarding the disposition of patent rights.
6. Recommending to the President appropriate action on exemptions from the agreement to assign inventions and patents to the University as required by Section II. A., above.

UNIVERSITY OF ILLINOIS

THE GENERAL RULES CONCERNING UNIVERSITY ORGANIZATION AND PROCEDURE

July 14, 1988

ARTICLE III. INTELLECTUAL PROPERTY

SECTION 1. INVENTIONS AND DISCOVERIES

(a) *Objectives*. The University recognizes that inventions and discoveries (hereinafter "inventions"[1]) may result from the activities of University employees in the course of their duties or through the use of University facilities, funds, or resources. It is the University's policy that such inventions shall be administered so as to be brought into practical use for public benefit at the earliest possible time. In the development of its technology, the University's objectives are to encourage and facilitate such development by all means appropriate to a publicly supported institution and to advance the respective interests of the University, its research sponsors, and its inventors.

(b) *Ownership*. Except as otherwise specified by the University in writing, inventions shall belong to the University if conceived or reduced to practice: (1) by a University employee as a result of the employee's duties; or, (2) through the use, by any person, including a University employee, of University resources such as facilities, equipment, funds, or funds under the control of or administered by the University.

In the event there is a question about an invention as to whether the University has an ownership claim, the invention should be disclosed to the University. Such disclosure is without prejudice to the inventor's ownership claim. In determining ownership interest in an invention, a designated official may determine that the University has no property interest in the invention because its conception and reduction to practice was unrelated to the inventor's University duties, involved only insignificant use of University resources (such as office space or libraries normally available to all faculty and staff) or for such other reasons as may be set forth in the guidelines. The inventor will receive a written statement as to the University's property interest. Such decisions shall be reported to the Board of Trustees on a periodic basis, but in no event less than once a year.

The University may also accept an offer to assign ownership in an invention or other technology which does not meet the foregoing conditions provided that such action is determined to be consistent with the University's objectives in the development of technology. Inventions so accepted shall be administered in a manner consistent with the administration of other University-owned inventions.

(c) *Disclosure*. Employees and others who conceive inventions or reduce them to practice as provided in paragraph (b) shall promptly and fully disclose each invention to the University and describe the circumstances under which it was conceived or reduced to practice.

(d) *Disposition*. The University shall have the sole right to determine the disposition of University-owned inventions subject to any prior contractual obligations to external

---

[1]For purposes of this policy, the definition of an "invention" includes "any device, process, material, plant, product, or other innovation."

sponsors. Such disposition shall be in a manner which, in the judgment of the University, is in the best interests of the University, its inventors, its research sponsors, and the public.

After evaluation of an invention and review of applicable contractual commitments, the University may develop the invention through licensing, may release the invention to the sponsor of the research under which it was made, may release the invention to its inventor(s), or may take such other actions as are consistent with its objectives regarding inventions.

(e) *Preferential-consideration-for-sponsors-of-research.* Research agreements with private sponsors shall provide that all inventions which are conceived or reduced to practice by University employees as a part of the sponsored research shall belong to the University. The sponsor may receive an option to license such inventions on terms to be negotiated, said option to be exercised within a specified period following the disclosure of an invention or the filing of a patent application. When the nature of the proposed research allows identification of a specific area of technology or application which is of interest to the sponsor, the University may accept research agreements with terms which entitle the sponsor to specific commercial rights within the defined field of interest. Otherwise, the specific terms of licenses and rights to commercial development shall be based on negotiation between the sponsor and the University at the time of exercise of an option by the sponsor and shall depend on the nature of the invention and its application, the relative contributions of the University and the sponsor to the work, and the conditions deemed most likely to advance the commercial development and acceptance of the invention. In all cases where exclusive licensing is deemed appropriate, such licensing agreements shall require diligent commercial development of the invention by the licensee.

(f) *Distribution of income.* When income is received by the University from a University-owned invention, all payments or obligations directly attributable to patenting, marketing, licensing, protecting or administering the invention shall first be deducted from such income. The income remaining after such deductions is defined as net income.

The inventor shall receive 50% of the first $200,000 of cumulative net income from each invention and 25% of cumulative net income thereafter. The remaining net income shall be allocated in support of the University's research programs.

In cases of multiple inventors, the inventors' share shall be divided among the coinventors as they shall mutually agree at the time of formal assignment of the invention to the University. Should the inventors fail to agree mutually on a division, the University shall determine the division. In cases where several inventions are licensed under a single licensing agreement, the University shall determine and designate the share of net income to be assigned to each invention.

(g) *Relations with inventors.*

(1) The University will inform inventors as to its decisions regarding inventions which they have disclosed. Should the University decide to abandon development of a University-owned invention, ownership will be assigned to the inventor(s), subject to the rights of sponsors and to the retention of a nonexclusive, royalty-free license to police the invention for University purposes.

(2) The University may require inventors to execute an assignment and related documents necessary to further the patenting and development of an invention and to cooperate with the University in such activities. The inventor's obligation to assign rights to the University is subject to the provisions of the Illinois Employee Patent Act, which provides in part:

> "A provision in an employment agreement which provides that an employee shall assign or offer to assign any of the employee's rights in an invention to the employer does not apply to an invention for which no equipment, supplies, facilities or trade secret information of the employer was used and which was developed entirely on the employee's own time unless (a) the invention relates (i) to the business of the employer, or (ii) to the employer's actual or demonstrably anticipated research or development, or (b) the invention results from any work performed by the employee for the employer. Any provision which purports to apply to such an invention is to that extent against the public policy of the State and is to that extent void and unenforceable. The employee shall bear the burden of proof in establishing that his invention qualifies under this subsection."

(3) Employees engaged in external consulting work or business are responsible for ensuring that agreements emanating from such work are not in conflict with University policy or with the University's contractual commitments. Such employees should make their University obligations known to others with whom they make such agreements and should provide other parties to such agreements with a statement of University policies on inventions and discoveries.

SECTION 2. ADMINISTRATION OF INVENTIONS AND DISCOVERIES

(a) *Administrative responsibility*. The President shall designate the University officers and offices responsible for the administration and disposition of inventions and discoveries, with primary responsibility delegated to campus users particularly for such matters as the disposition of inventions (Sections 1.b, 1.d), preferential terms for research sponsors (Section 1.e), and resolving disputes among co-inventors (Section 1.f). Licenses and options for licenses may be granted in the name of the Board of Trustees of the University of Illinois when approved by the President or his designee and executed as are other University contracts.

(b) *Administrative guidelines and procedures*. General guidelines and procedures for the administration of inventions and discoveries shall be established by the President in consultation with the University Intellectual Property Committee and the campuses. Detailed operational guidelines and procedures for the administration of campus-based responsibilities shall be established by each campus.

(c) *University Intellectual Property Committee*. The University Intellectual Property Committee shall be appointed annually by the President. The Committee shall consist of a faculty representative from each campus appointed by the Chancellor with the advice of the Research Board and Graduate College Executive Committee on that campus and, unless otherwise designated by the President, of the Vice-Chancellor for Research on each campus, the Vice-President for Academic Affairs, the Vice-President for Business and Finance, and the University Counsel. The Committee shall make recommendations to the

President regarding procedures, guidelines, and responsibilities for the administration and development of inventions and such other matters as the President shall determine. The Committee shall report periodically to the Board of Trustees and the President on inventions disclosed and the disposition thereof.

(d) *Exceptions to policy*. Recommendations for exceptions to the provisions of Section 1 of this Article shall be made by the Intellectual Property Committee to the President for presentation to the Board of Trustees.

(e) *Appeals*. Subject to rules and procedures developed by the University Intellectual Property Committee, a University inventor may appeal to the Committee to seek resolution of complaints or questions regarding the matters contained in or authorized by Sections 1 and 2.

(f) *Timeliness*. It is the intent of the University that inventions once disclosed by the inventors, will be handled in a timely and expeditious manner by the designated persons on each campus and within the Central Administration. With this end in mind, an inventor with an invention under consideration will be put in contact with an appropriate administrator who will monitor the progress of the invention through the patent process. Appeals (Section 2[e]) on the basis of timeliness may be made by either the inventor or by this administrator.

SECTION 3. COPYRIGHTS

(a) *Introduction*. The policy of the University with regard to copyrightable works is intended to foster the traditional freedoms of the University's faculty, staff, and students in matters of publication, through a fair and reasonable balance of the equities among authors, sponsors, and the University. At the same time, it is intended to ensure that copyrightable materials in which the University has a legitimate interest are utilized in a manner consistent with the public interest.

(b) *Ownership*. Ownership in copyrightable works produced by authors who are University staff members or students shall remain with the authors except in the following situations where such rights of ownership as it may require shall vest with the University:

(1) The terms of a University agreement with an external party require the University to hold or transfer ownership in the copyrightable work, or

(2) Works expressly commissioned in writing by the University, or

(3) Works created as a specific requirement of employment or as an assigned University duty. Such requirements or duties may be contained in a job description or an employment agreement which designates the content of the employee's University work. If such requirements or duties are not so specified, such works will be those for which the topic or content is determined by the author's employment duties and/or which are prepared at the University's instance and expense, that is, when the University is the motivating factor in the preparation of the work.

In a particular employment situation, if an author is uncertain about the ownership of the works referred to in this paragraph (b), before undertaking the assignment the author shall be entitled to request in writing and to receive a clarifying written statement from the author's department head or supervisor.

When the University commissions the preparation of a copyrightable work by an author who is not a University employee, the contract with such author shall specify that the author convey by assignment such rights as are required by the University.

With respect to copyrightable works owned by the University pursuant to this paragraph (b), the authors may be required to execute such documents as are necessary to vest ownership in such works in the University or its designee. Authors who produce works for which ownership vests in the University pursuant to this paragraph (b) shall warrant that such works do not infringe any preexisting copyright.

(c) *Sponsored Works*. A sponsor's right to use or own a work prepared under its sponsorship will be determined by the provisions of its contract with the University.

(d) *Licensing*. As provided in paragraph (b), copyrightable works developed by University staff or students using University resources usually and customarily provided (such as office space, library facilities, or regular salary) are owned by the authors. Such works need not be licensed to the University. Copyrightable works developed with use of University resources over and above those usually and customarily provided shall be owned by the authors but licensed to the University. The minimum terms of such license shall grant the University the right to use the copyrightable work in its internally administered programs of teaching, research, and public service on a royalty-free, non-exclusive basis. All expenses incidental to such licensed use shall be borne by the University.

(e) *Copyright Registration and Notice*. In general, only those University-owned works which have an income potential or which the University intends to develop, use, or publicly circulate should be protected by copyright notice. Such copyright notice should be composed and affixed in accordance with United States Copyright Law.

The responsibility for determining the need for copyright protection and recommending copyright registration of a University-owned work shall rest with the department head or supervisor where the work was prepared. Such recommendation shall be forwarded through the dean or equivalent officer to the office responsible for the registration of copyrights. The registration procedure for the University is a centrally administered responsibility, which may be delegated by the President.

(f) *Student Works*. Unless subject to the provisions of paragraph (b) or provided otherwise by written agreement, copyrightable works prepared by students as part of the requirements for a University degree program are deemed to be the property of the student and are subject to the following provisions:

(1) The original records of an investigation for a graduate thesis or dissertation are the property of the University but may be retained by the student at the discretion of the student's major department.

(2) The University shall have, as a condition of the degree award, the royalty-free right to retain and use a limited number of copies of the thesis, together with the right to require its publication for archival use.

(g) *Administration*.

(1) *University Press Publications*. The University Press shall be responsible for copyright registration of works owned by the University and published by the Press, and for administering contracts with its authors. Such contracts shall define the rights and obligations of the author and the University, and shall be processed as are other University contracts. A copy of the printed agreement currently in use by the Press shall be made available upon request.

(2) *Other Copyrightable Works*. The procedures and responsibilities for maintaining records for copyright registration, royalty collection and distribution, marketing, and

such other actions as are appropriate, including, but not limited to assignment with respect to copyrightable works shall be established by the President. All contracts for the marketing of copyrightable works shall be executed in accordance with Article II.

(h) *Distribution of income*. The University will normally distribute fifty percent (50%) of the net income received from the marketing of University-owned copyrightable works to the author(s) of such works. Other distributions may be justified in unusual circumstances or by written agreement with the author(s). Such income distribution arrangements shall be proposed by the department head or supervisor of the unit where the work is performed, and are subject to review and approval by the Vice-Chancellor for Research, who shall also mediate disagreements arising from such determinations. A record of income sharing agreements shall be reported to the University Intellectual Property Committee by the Vice-Chancellor for Research. Unresolved disagreements concerning such income distribution, or circumstances in which the proposed author's share exceeds fifty percent (50%) of net income, will be reviewed by the Intellectual Property Committee which will make recommendations to the President for final disposition. Net income is gross income received by the University less expenses attributable to copyright registration, marketing, or other requirements for use and sale of materials outside the University. After deduction of the authors' shares, the remaining net income will be allocated as determined by the President, who may seek the advice of the University Intellectual Property Committee.

(i) *Complaince with the Copyright Act*. The Copyright Act and implementing regulations specify numerous conditions on the reproduction and use of works protected by copyright and provide penalties for non-compliance. These regulations address such matters as photocopying, public broadcast programs, and various other practices. University units whose activities involve any such regulated usage are responsible for knowing applicable regulations, monitoring their continuing evolution, and conducting their programs in full compliance with the applicable laws and regulations.

Certain University units provide services to users which involve the transmission, communication, or duplication of materials protected by copyright but where the ultimate use of the materials is not subject to supervision by the service unit. Service supervisors in such circumstances shall inform all service users of laws and regulations applicable to such usage by means of posted notices, bulletins, or other appropriate methods. The responsibility to avoid infringement of the copyright extends to all individuals who use such material.

(j) *The PLATO System*. The PLATO[2] System is a computer-based educational system developed at the University of Illinois and originally covered, wholly or in part, by the claims of U.S. Patent Number 3,405,457. The term "PLATO System" shall also include any successor systems or enhancements thereof developed, conceived, or proposed by the Computer-based Education Research Laboratory of the University. The copyright provisions applicable to lessonware or software using the University PLATO System are as follows.

---

[2]Control Data Corporation has acquired the exclusive commercial rights from the University to use the name PLATO.

Unless otherwise agreed in writing, all PLATO Authors granted access to any University PLATO System, all Participating Institutions,[3] and all Joint Venture Institutions[3] shall grant to the University an irrevocable, nonexclusive, royalty-free license to copy, use and distribute any and all lessonware, software, or related auxiliary materials which are developed as a result of such access without restriction in connection with its internally administered programs of teaching, research, and public service. All expenses incidental to such copy and use shall be borne by the University.

Except for the foregoing non-exclusive license grant requirement, which shall supersede prior Board actions, the University policy with respect to copyrightable PLATO lessonware or software shall be the same as for other copyrightable works. Ownership by the University in lessonware or software developed prior to March 17, 1983, shall remain in effect, except as may be otherwise specified by written agreement between the author and the University.

[3]Participating Institution: any external institution or organization leasing or purchasing at full cost one or more terminals connected to the University of Illinois PLATO System. Joint Venture Institution: as above, except costs are shared between the institution or organization and the University.

POLICIES ON INDUSTRIAL SPONSORED RESEARCH

GUIDELINES FOR RESEARCH PROJECTS UNDERTAKEN IN COOPERATION WITH INDUSTRY†

Preamble

Harvard University welcomes industrially supported research agreements because they can stimulate its investigators, promote technological transfer, and provide the University with valuable support. At the same time, it recognizes the need to avoid arrangements that might compromise, or seem to compromise, its intellectual principles and purposes and the freedom of inquiry the members of its Faculties enjoy. As an institution, the University benefits from public research funds and the public's trust, and it has an obligation to develop its research discoveries with concern for the public's interest.

These guidelines outline some general principles—concerning how and why research is conducted at the University—with which all research agreements with non-governmental external sponsors (referred to, hereafter, as industrially-supported research agreements) should conform.

I. Conduct of Research

The exchange of information and the discussion and interchange of ideas are basic elements of all University research. Agreements to perform secret research in Harvard laboratories are unacceptable.

A. The proscriptions on secrecy in cooperative university-industry research agreements must conform with those that apply to federally-sponsored research. If a research project involves confidential information that would inhibit free and open interactions among scholars, the University should not accept it.

B. It is essential that the research of students and postdoctoral-fellows-in-training contribute, and be perceived to contribute, to their scholarly development. Even as individuals, students should not ordinarily participate in agreements that involve confidential information or otherwise constrain the right to publish or communicate freely. Exceptions consistent with the University's principles should be approved in advance by the student's department or at a higher level. In addition, departments should periodically review the work of students engaged in industrially-sponsored research to see that the educational commitment of the University to its students is maintained.

C. Agreements may permit industrial sponsors to examine completed manuscripts for potential patent applications, but agreements may not restrict the rights of investigators to publish their findings nor to communicate their research results freely in other ways consistent with protecting patentability and with other ethical and professional standards.

   Agreements to treat University-based research as confidential, to withhold publication or delay it significantly, or to permit sponsors to modify materials submitted for publication, are ordinarily unacceptable.

D. The responsibility for research programs and flexibility in directing them must remain with principal investigators. Sponsors may consult on matters of mutual concern but they may not dictate how research shall proceed.

---

†Voted by the President and Fellows of Harvard College October 3, 1983.

E. All research proposals must be approved by appropriate departmental authorities.
F. Faculty members should be informed of the existence of industrially-sponsored research agreements in their departments, and any special provisions in such agreements should be explained to them.
G. General information on the subject, duration, funding sources, and budget of each industrially-sponsored research agreement should be openly available, along with information on whether there are any associated exclusive or nonexclusive patent agreements or other restrictions on open communication.
H. The support of a major portion of a faculty member's research by a single corporate sponsor is generally undesirable; whenever such support is permitted, the research should be periodically reviewed and approved by the appropriate Dean.
I. It is expected that these guidelines on free and open exchange of information will be followed in all instances that involve concepts, processes, products, and other information about natural phenomena. There may, however, be situations (for example, studies involving records of individuals or identifiable industrial organizations, or university-coordinated educational programs of an apprenticeship character) where exceptions to the guidelines are consistent with the University's educational, professional, and scholarly principles. Such exceptions should be granted only after detailed review by the appropriate Dean with advice from appropriate faculty committees, and the reasons should be publicly explained.

II. Motivation for Research (Conflicts of Interest)

Faculty members have a responsibility to maintain the scholarly character of their research. Faculty members who propose to participate in industrially sponsored research must disclose in confidence to the designated individual or committee, the nature of all personal commitments to, and remuneration by, the industrial sponsor. (In this context, personal commitments and remuneration are understood to include equity options and holdings as well as income from consulting or other services.) Questions emerging from such disclosures should be resolved with the Dean of the Faculty and/or the appropriate faculty committees before the University accepts the sponsored research agreement.

III. Quality of Research (Project Review)

The distinction between fundamental intellectual inquiry and commercially targeted development is not always clear and no faculty committee should attempt to define a line of demarcation. Many investigators are committed to studying tools and processes whose primary purpose is to benefit the health or welfare of society in areas that might have commercial value. Nevertheless, the following University principles and policies apply:

A. The primary assurance of the quality of research carried out under University auspices stems from the scholarly qualifications required of individual principal investigators.
B. Since most federally or foundation sponsored research proposals are subject to peer review, prior approval by the Dean of the scientific content of proposals to conduct such research is often perfunctory. In the absence of other review processes, the Dean or the Dean's designee may require a more careful examination of the suitability of an industrially-sponsored research proposal before approval. In particular, each Dean should develop a mechanism for reviewing such proposals when they are

large, extend over several years, and/or entail a substantial commitment of a department's resources. Continuing projects should be subject to periodic review.

C. The responsibility faculty members bear for the intellectual development of their research students and the responsibility of departments and Faculties to oversee graduate student research are especially important when industrially-sponsored research is involved, and when conflicts of interest may appear to exist for faculty members or for the University.

IV. Commercial Applications of Research (Patent-Licensing)

A. License negotiations should be governed by the principle that any invention deriving from university-based research should be developed fully and rapidly in the public interest. In many instances, this goal will best be achieved by granting non-exclusive licenses to every company that wishes to develop the research commercially. Sometimes, exclusive licenses may be more appropriate. When they are, for example, to justify the investment a pharmaceutical company must make in testing before commercial introduction of a new product, the potential exclusive licensee should satisfy Harvard: (1) that commercial products for public use will be developed more effectively through exclusive licenses than through non-exclusive licenses; and (2) that adequate resources can and will be committed to bringing these commercial products promptly into use.

B. The University should retain the right to "march-in" if, after a suitable period, a licensee has not developed and/or commercialized the discovery satisfactorily.

C. The *prospective* granting of exclusive licenses in industrially-sponsored research agreements raises a number of questions. When one does not know what the invention will be, one cannot demonstrate in advance that an exclusive license will be needed nor that the sponsor will be able to develop it effectively. Accordingly, prospective exclusive licensees should be required to give evidence of their willingness and ability to develop and commercialize the kinds of products or processes likely to result from the research.

In cases in which the investigator or the University has a significant private financial relationship with a potential industrial sponsor of research, it should ordinarily be presumed that a prospective exclusive license would create at least the perception of a conflict-of-interest. Prospective exclusive licensing in such situations should only be contemplated after the Dean or his or her designee (or, for agreements not emanating from a single Faculty, a committee of Deans or their designees) has examined the private financial relationships, the suitability of the research, and the licensing terms. The Harvard Corporation should be advised of the background and conditions of any such proposed arrangement.

Control of Material Samples[‡]

In view of industry's growing interest in limiting the use of material samples supplied to faculty members to promote research of mutual interest, the following practice should ordinarily be implemented:

[‡]Voted by the Committee on Research Policy, March 5, 1985.

1. A material sample should be supplied directly to a faculty member for use in research conducted in his or her laboratory. Any arrangements, including personal consulting agreements, constraining the faculty member's right to use or distribute sub-samples of this material to other parties either inside or outside the University should be between the industrial donor and the faculty recipient. The University will not formally endorse such arrangements. No limitation on the right to disclose or publish is acceptable.
2. At the time any agreement of this nature in initiated its terms should be disclosed to the Faculty of Arts and Sciences Committee on Technology Transfer. The faculty member should be contacted on an annual basis to seek assurance that the Faculty's *Guidelines for Research Projects Undertaken in Cooperation with Industry and Policies Relating to Research and Other Professional Activities Within and Outside the University* have been observed.

POLICY INFORMATION FOR POTENTIAL COMMERCIAL SPONSORS OF RESEARCH AT THE UNIVERSITY OF PENNSYLVANIA

Prologue:

The University of Pennsylvania has a long tradition of welcoming support for research projects from commercial sponsors. Such support aids our pursuit of new knowledge and, through the cooperative interactions it spawns, often facilitates the practical applications of such knowledge to the benefit of society.

Modern research universities, such as Pennsylvania, are seen and see themselves as institutions, unique in our community, that engage in sustained inquiries into fundamental areas of the humanities, of science, and of technology without regard to immediate practical consequences, and that promote the free exchange and discussion of the results obtained. As such they are often called upon for technical help and objective advice in meeting the needs of both the public and private sectors. Our policies regarding the conduct of commercially sponsored research agreements, as summarized below, reflect this image of the University. We believe that these policies, in preserving our academic ideals while recognizing the legitimate proprietary needs of the sponsor, provide a sound basis for mutually satisfactory research relationships.

1. General Considerations

The University seeks support for sponsored research conforming to its Guidelines for the Conduct of Sponsored Research which include the following criteria:

a. It is of intellectual interest to the principal investigator, who will usually be a member of the standing faculty, and who will be responsible for the conduct of all or the major portion of the work.

b. It in no way compromises the University's policy of nondiscrimination, nor the freedom of inquiry of faculty members, nor their abilities to disseminate their research findings and conclusions.

c. It is consistent with the University's basic objectives of creating and disseminating new knowledge and is appropriate to the purposes and objectives of a tax-exempt nonprofit educational institution. In addition, sponsored programs that aid the educational purposes of the University will be especially welcome. Such aid might, for example, take the form of financial support for student theses or dissertations.

2. Negotiation and Implementation of Agreements

a. Every sponsored program must include a formal research proposal. Such a proposal requires the written approval of the cognizant deans and departmental chairpersons. In the case of a proposal having faculty from more than one school, the written approval of the cognizant deans, or Vice Provost for Research, as appropriate, is required. Proposals submitted through an institute must also have the director's approval.

b. Responsibility for negotiating any commercially sponsored research agreement rests with the Office of Research Administration under the general policy guidance of the Vice Provost for Research. A detailed outline describing the administrative structure of the program is required as part of an agreement.

c. The University undertakes research on a best efforts basis, but disclaims any guarantee of results.

d. The University expects full reimbursement of the allowable costs of research, including direct and indirect costs, the latter to be charged at no less than the rate that is applied to grants from the Federal Government. Only in unusual circumstances, and with the express approval of the cognizant dean, will deviations from this policy be permitted.

e. It is University policy not to undertake to keep proprietary information provided by the sponsor confidential, because of the negative impact such an obligation has on free communication of research results within the University and elsewhere. Exceptions to this policy will be considered in cases when the confidential information provided is peripheral to the main intellectual thrust of the work. For example, confidential information can sometimes be handled in coded form, so that results can be discussed without divulging proprietary information. The University reserves the right to refuse to accept proprietary information.

f. All research agreements will be carried out in a manner consistent with existing governmental regulations regarding such matters as human and animal subjects, recombinant DNA, and the use of hazardous and radioactive materials.

3. Publications and Publicity

a. The University accepts no limitations on its right to disseminate all findings and conclusions derived from sponsored projects except when the privacy of an individual is concerned or in the limited sense discussed in sections 2e, 3b, and 4c.

b. The University agrees to submit a copy of any proposed disclosure to the sponsor for review prior to its planned submission for publication. The sponsor may then request a delay in publication, not to exceed three months, for patent or trade secret protection (see part 4). In unusual circumstances, and with the approval of the Vice Provost for Research, a further delay may be allowed, which in no case will exceed one year from the date of initial submission to the sponsor. It is University policy to identify fully and openly the sponsor of the research program, either by acknowledgement in a publication or, if such acknowledgement is unwanted, in response to public inquiry. An exception may be made when in the judgment of the Provost such a condition is not harmful to the University nor to the integrity of the research.

c. A sponsor may not use the name of the University or of its employees or agents in any way in advertising or product-promotion activities without the prior written approval of the President of the University.

4. Patents, Trade Secrets and Licenses

a. The University strongly prefers patent to trade secret protection of intellectual property, because patent disclosure is consistent with the University's objective of disseminating new knowledge. Accordingly, patent protection will be sought for a research discovery with potential commercial importance unless both the sponsor and University agree that the discovery is not patentable.

b. The University prefers to retain title to inventions and other intellectual property resulting from sponsored research. This insures that the University will be able to direct vigorous efforts toward transferring the technology for the public benefit. The normal mechanism for this is through royalty-bearing licenses, preferably non-exclusive, though exclusive licensing may be more appropriate in specific instances. Other arrangements for disposition of intellectual property may be negotiated on a case by case basis, provided the University's property rights and the public benefit are protected.

c. When it is determined that trade secret protection is required, that portion of the sponsored agreement requiring such protection will be transferred out of the University as soon as possible, over a period not to exceed sixty days. Such transfer is necessary because the secrecy that trade secret protection requires is contrary to the University goal of openness in research. During the transfer period, University affiliated personnel may serve as extramural consultants to the sponsor. The University will have the right to publish the research developments that occurred prior to the date on which the University agreed that patent protection was unavailable, after a delay of not more than one year to protect the sponsor's interests.

d. In case of trade secret protection, the trade secret will be owned by the sponsor. The legitimate property interests of the University for the work performed prior to transfer out of the University will be protected by an agreed upon formula for royalty payments resulting from commercialization.

5. Potential Conflicts of Interest

a. The University permits extramural consulting agreements between a principal investigator and a sponsor. In such cases the University must be deemed to have a financial interest in any commercialization resulting from the combined sponsored research and extramural consulting activities of the faculty member. To insure a lack of conflict of interest, such agreements will be reviewed by the Vice Provost for Research.

b. The University recognizes the potential problems that could result when research in the same general area is funded by competitors, especially when it occurs within the same department. These problems may include the confidentiality of proprietary information and the desire of a sponsor for exclusive rights to research developments. Such arrangements will therefore be especially carefully considered by the faculty of the department and by the University administration, in order to minimize conflicts of interest.

c. Ownership of a substantial financial interest in a commercial entity by a faculty member whose sponsored research might be supported by funds from that commercial entity raises special concerns about potential conflicts of interest. Such arrangements must therefore be specifically approved by the Provost.

6. Protection Against Liability

a. The University will not be held liable for loss or damage suffered by the sponsor in the course of the research contract association with the University. Furthermore, the sponsor must agree to defend and indemnify the University in the event of any form of liability or claim arising from the sponsored research, except those arising from the University's negligence or willful acts.

b. For research agreements sponsored cooperatively by a consortium of companies the sponsors will bear the burden and expense of meeting any regulatory requirements or obtaining any necessary clearances associated with antitrust issues and will further provide warranties and/or indemnification against any antitrust violations.

7. Additional Information

Five additional documents which are relevant for potential sponsors of research at the University of Pennsylvania are 1) the *Guidelines for the Conduct of Sponsored Research*,

2) the *Patent Policy*, and 3) the *Conflict of Interest Policy for Faculty Members*, 4) *Organization and Implementation for Research Administration*, and 5) *Sources of Research Support*. These documents are available from the Office of Research Administration.

EXTERNAL CONSULTING POLICIES

UNIVERSITY RESEARCH COUNCIL

CARNEGIE MELLON POLICY ON CONSULTING BY FACULTY

Preamble

This policy pertains to consulting by tenure-track and research faculty (as defined in the Policy On Research Faculty Appointments, available in the Faculty Handbook) at Carnegie Mellon.

The University believes that its educational program, and effective teaching in all its aspects, can flourish only when sustained by continuous, active participation of its faculty in research, enriched in many cases by interaction with industry, artistic organizations, business, government, and other activities and institutions of our society.

This interaction is of greatest value when it contributes significantly to the public welfare, offers an opportunity for professional challenge and growth, or otherwise enhances the effectiveness of a faculty member's service to the University.

Interaction can come in many forms, one of which is consulting. Consulting includes those professional activities related to the faculty member's discipline for which remuneration (beyond a nominal honorarium or reimbursement for expenses) is received.

The potential magnitude of outside professional activity, particularly when it entails consulting, is such that orderly procedures must be followed to avoid ethical and legal conflicts of interest and to ensure that such activity does not conflict with the proper discharge of University responsibilities. Essential to the effectiveness of such procedures is complete disclosure of outside professional activities.

Disclosure

Liaison between the head of a department and faculty members is the principal means of communication and disclosure in matters involving outside professional activities. It is the obligation of faculty members to keep their department heads continually informed in adequate detail as to all outside professional activities, service on external committees, and other special assignments.

Standards and Criteria

Personal responsibility, integrity and high ethical standards are the principal factors in avoiding conflicts of interest. The University expects that all members of the faculty will conduct their outside activities in a manner which reflects credit on themselves, their profession, and the University without need for specific criteria or rules of conduct. The principal safeguards against abuse include the standards required by professional colleagues and the rigorous process by which the University evaluates and selects individuals for appointment and promotion.

Full-time Service

The obligation inherent in full-time service is difficult to define since in academic life it means far more than a stated number of hours per week. In a context in which the faculty member has substantial freedom in arranging his or her professional life, it implies an overriding interest, loyalty, and first responsibility to the University. This obligation, therefore, must be defined qualitatively, depending on principle rather than formula.

The University has traditionally granted full-time members of the tenure-track faculty the right to devote up to an average of one day (of University time) per week to outside, paid, professional activities, where that activity is consistent with that person's role as a member of the faculty and where that activity also enhances the contribution of the faculty member to the University. Such activity benefits both the faculty member and the University. This document extends this traditional consulting privilege of tenure-track faculty to research faculty.

The faculty member in participating in such activities must accept the responsibility to arrange his or her work so that during the academic year (and summer, if on salary) the faculty member will devote the appropriate amount of time to activities connected with the primary appointment. In particular, when effort is expended on sponsored research contracts, the faculty person must be aware that if the effort is 100%, the intent of the sponsor is that 100% of effort is equivalent to five days of work per week. Thus, consulting activities that occur while being paid 100% from research contracts would be in addition to five days of work per week. It follows similarly with other levels of effort, e.g., if 80% of effort is on research contracts, the expectation is that four days of work per week will be devoted to those research contracts. It is the responsibility of faculty to ascertain whether their commitment to sponsored research leaves time for consulting. For purposes of accounting, it is to be understood that this standard will be applied, on average, over each three-month period of the academic year and over each month of the summer.

THE UNIVERSITY OF GEORGIA COLLEGE OF AGRICULTURE POLICY NUMBER 5

FACULTY CONSULTING POLICY AND REPORTING PROCEDURE

The University of Georgia and the College of Agriculture recognize that appropriate consulting activities may be mutually beneficial to the faculty and the university. These benefits include: enhancing the knowledge, expertise and experience of the faculty; assisting in maintaining awareness of real would conditions; opening and expanding of contacts and communications with other institutions organizations, industry, governmental agencies and other clients; and fostering of public service activities otherwise unavailable. Thus, appropriate consulting activities constitute enrichment and continuing education activities which enhance the professional development and reputation of faculty. This may result in improved teaching, research and service. Hence, reasonable participation in consulting is encouraged within the parameters in Section 802.1602 A, B & C of the Board of Regents Policy Manual and set forth herein.

Definition of Consulting

Consulting is defined as external (international and domestic) professional activities for pay and includes any activity that: (1) is performed on an individual contractual basis for any individual(s), firm or agency other than The University of Georgia; (2) is based upon the professional knowledge, experience and abilities of the faculty or staff member; (3) is undertaken for personal compensation beyond the payment of a nominal honorarium and/or reimbursement for expenses; and, (4) clearly enhances the professional development and expertise of the faculty member.

Some activities may be qualified as consulting which benefit the institution and the faculty member/consultant that are undertaken without compensation in excess of expenses and honoraria.

Consulting does *not* include private business undertakings. Faculty may participate in certain business activities, provided they do not conflict with their college and university duties. Such business activities must be of such a character as not to damage in any way the prestige or image of the college or university.

Consulting does not include those responsibilities to the College of Agriculture which are a part of a given faculty member's regular professional responsibilities for which that individual is already being compensated.

Guidelines for Consulting

With appropriate prior administrative approval, a faculty member, employed on a full-time basis, may render professional consulting services. Such consulting activities must not preempt, and *must* be scheduled around, the responsibility of a faculty member to meet assigned duties and obligations to teaching, research and extension expected of a full-time faculty member.

To avoid conflict of interest and insure the continued credibility of the College of Agriculture, faculty may engage in consulting only when such activities conform to the principles listed below. These criteria are not designed as a substitute for good judgment by faculty or administrators, but to provide guidelines within which consulting activities can be evaluated.

1. Consulting should not be undertaken if it encroaches upon the time and energy required for the full and complete performance of the duties to which a member is assigned.

2. Faculty members should make clear in writing to those who employ them as consultants that their work is private and unofficial and that the college and/or university is not responsible in any way for the services rendered or results obtained.
3. A faculty member may not consult for an enterprise owned or operated whole or in part by him/herself or a relative.
4. Consulting activities must in no way compromise the position of the college or the university.
5. Duties which college personnel should perform as a public service within the state of Georgia are not to be performed by *any* faculty member acting as a consultant.
6. University resources, including equipment and supplies, are not to be used in consulting. When college or university resources are required, the activity shall be performed on a contractual basis with the college or university rather than as an individual consulting activity.
7. Both the fact and semblance of conflicts of interest must be avoided. If the question arises, the interest and the reputation of the college and the university must come first.
8. Care must be exercised to avoid establishment of consulting activities that infringe upon the activities or units of the university or those which may conflict with services provided by recognized professional consulting services within the state of Georgia.
9. The *maximum* number of consulting days permissible for a member of the faculty on a full-time appointment (without using annual leave) is *two days per month*. Consulting days are *not* cumulative. Any consulting days exceeding two per month must be taken as annual leave.

Consulting Requests Approval and Reporting

All consulting activities, including local, state, national and international, must have prior written approval by the department head of the faculty member initiating the request, the appropriate division chairman, associate deans, and the dean. At the Georgia Station and Coastal Plain Station, approval of the resident director must also be obtained. In extension, approval of the appropriate assistant and associate directors must also be obtained. The department head is to determine and sign-off that the consulting activity does not pose a conflict of interest and is consistent with the role and responsibility of the faculty member and the college. The associate deans and dean will determine if the proposed consulting service is or is not in conflict with services presently being offered by the college.

With such prior written approval, a college faculty member may be engaged as a consultant. On June 30 of each year all existing approvals for consulting must be affirmed based upon appropriate information from the faculty and the department concerned. Consulting fees are the prerogative of the faculty member but, in general, faculty should not engage in consulting activities at a fee below the standard fee for services rendered by an independent consultant.

4/87

Attachment: Consulting Form, 4/87
Board of Regents Policy
Manual Section 802.1602

REQUEST TO ENGAGE IN CONSULTING SERVICES BY UNIVERSITY OF GEORGIA COLLEGE OF AGRICULTURE PERSONNEL*

The Regents' policy on consulting services by faculty to individuals, firms, or agencies outside the University is interpreted as being positive and beneficial to the University and the College. The policy of the College of Agriculture requires prior approval for each consulting commitment by submitting a request for approval to the appropriate department head, division chairman, director(s), program leader, and dean.

Permission to consult covers only the specifically described activity and time period covered in each request. Request renewals must be approved annually on or before June 30. The initiation of the approval procedure is the responsibility of the individual seeking permission to consult.

Activity proposed: (include name of agency, firm, or program and where the activity will be undertaken):

Relationship of the activity to the University's programs (explain how the activity will provide experience and knowledge useful for you in teaching, research, extension, or administration):

Will the activity interfere with your University commitments? Yes__ No__

Will University time other than weekends and annual leave be used? Yes__ No__.
If yes, how much time (hours per month)__________
Dates of service: From__________ To__________
Will University resources, including personnel, facilities, supplies or equipment be used? Yes__ No__. If yes, describe arrangements as to how UGA is to be reimbursed for University resources used.

Applicant	Department	Date

APPROVAL:

Department Head/Division Chairman	Resident Director/Program Leader
Associate Dean/Director	Dean and Coordinator (Date)

*Consulting is defined as the employment in any capacity of a faculty member of the College of Agriculture by an agency, firm or individual other than The University of Georgia.

CONFLICTS OF INTEREST POLICIES

A Statement of Policy for Interaction between The University of Connecticut and The Public and Private Sectors

POLICY ON CONFLICT SITUATIONS

Introduction

Members of University faculties are dedicated to objective treatment of facts, ideas and values. This important role in society is based on credibility. Consequently, this places a special responsibility on University staffs when they as members of the academic community act in any other capacity.

With the development of the Connecticut Technology Park adjacent to the campus which offers ever-increasing opportunities for interaction, University officials determined that it was the appropriate time to establish policies and guidelines for such interaction to ensure that it remain within limits acceptable to the University as principal employer and that the faculty member is protected from inadvertent conflicts of interest. To this end, relevant state statutes, *University Laws and By-Laws*, current administrative policies and procedures, and appropriate literature from other Universities were reviewed. Further, faculty members with experience in these matters were interviewed.

Conflicts and Prohibitions

Conflicts fall into two categories:

1) *Conflicts of Interest* are situations in which a University employee may have an opportunity to influence University administrative, business or academic decisions in ways that could lead to personal gain or give improper advantage to others.

2) *Conflicts of Commitment* are situations in which a University employee's external activities, often valuable in themselves, and even when they result in no personal gain or improper advantage for others, nevertheless interfere improperly with his or her paramount obligations to students, colleagues and the University. (Further discussion found in section on CONSULTING POLICY.)

Faculty members must avoid actual conflict of interest and conflict of commitment situations as well as the appearance of conflict. The potential of real and perceived conflicts that can arise when academic employees are simultaneously involved in more than one organization must be identified by the individuals themselves and made known to the appropriate administrators so that any potential detrimental influence can be avoided.

Should a faculty member be involved in ownership or management of an outside company, care must be taken to assure that such outside involvement does not distort the direction of University research, color the supervision of graduate students, or change a collegial and open environment into one of secrecy and striving for personal financial advantage.

It is a potential conflict of interest for faculty and all other University personnel, acting in a private capacity, to solicit or bid on contracts which might otherwise go to the University. This applies whether the employee is acting as a sole proprietor (consultant), as a principal in a company or as an employee of a company.

Finally, the Code of Ethics for public officials applies to all employees of the State of Connecticut, including faculty and staff of the University of Connecticut. The major prohibition contained in the Code of Ethics is directed towards any activity which is in substantial conflict with the proper discharge of duties or employment in the public interest.

That means that no employee may engage in an activity if he or she has reason to believe that there will be a direct monetary gain or loss as the result of his or her *official* activity. If, however, the gain or loss is no different than that of a group or business or profession, then there is *not* an interest that is in substantial conflict with the discharge of duties or employment in the public interest.

There are additional prohibitions in the Code of Ethics. No state employee may accept other employment which will impair independence or judgment as to official duties or require the disclosure of confidential information acquired in the course of those duties. Further, no employee is permitted to knowingly disclose for financial gain confidential information obtained in the course of official duties. This applies even if the gain is to accrue to a close relative or a business with which the employee is associated.

No employee is permitted to appear for compensation before certain state agencies.[1] This does *not* apply to teaching or research employees in public higher education if the appearance does not violate any other provision of the Code of Ethics.

No employee is permitted to solicit or accept anything of value based upon an understanding that his or her official action would be influenced.

No state employee or business with which he/she is associated may enter into a contract with the state for more than $100 without an open, public bidding process.

There are also prohibitions that affect employees who leave state service.

No former employee may, for one year, represent anyone concerning a matter in which he or she participated while a state employee and in which the state has an interest. The former employee may not, for one year, represent anyone for compensation before his or her former agency if the state has a substantial interest in the matter. There is also a prohibition against employment with businesses subject to regulation by the employee's former agency; and against a former employee who, having substantially participated in a contract negotiation or award, accepts employment with the contracting party within one year of the employee's departure from state service.

CONSULTING POLICY

Traditional Scholarly or Academic Professional Activity

Traditional scholarly or academic professional activities which include participation in conferences, seminars, professional meetings; serving on review panels for government-granting agencies; presenting colloquia, invited lectures, editing or reviewing for scholarly or scientific publications; giving concerts or exhibits are currently encouraged by University policies governing promotion, tenure and reappointment, and are subject to general requirements on germaneness to professional objectives, and the fulfilling of University obligations to teaching, research and service. These activities are not circumscribed by policies or procedures pertaining to consulting, rather they are subject to full

[1]Those agencies are: the Banking Department, the Claims Commissioner, the Commission on Hospitals and Health Care, the Insurance Department, the Department of Liquor Control, the Department of Motor Vehicles, the State Insurance Purchasing Board, the Department of Environmental Protection, the Department of Public Utility Control, the Connecticut Siting Council, the Division of Special Revenue within the Department of Revenue Services or the Connecticut Real Estate Commission.

disclosure in departmental annual reports and the rules and regulations concerning travel and absence from the University. Such activities do not require that permission be requested in advance on consulting activity forms or reported on annual audits of consulting activities. It is important, however to caution that these traditional scholarly activities must not interfere with the required duties of faculty and are, in that respect subject to similar, if not identical, constraints as those placed on consulting. Department heads and deans may need to weigh all such activities together in recommending approval or disapproval for any particular external activity.

External Professional Activity

The external activities which fall under consulting are governed by the *University of Connecticut Laws and By-Laws*, specifically XV.M (p. 26 12th Edition, 1985); the relevant Connecticut Statutes; and the policy and procedural statements issued from the Offices of the Vice President for Academic Affairs and the Vice President of Health Affairs. Adherence to these policies and procedures must be maintained to satisfy the general accountability requirements of the University and to ensure that the University's goals and responsibilities with respect to teaching, research and service are met.

External professional commitments must have prior approval. Whether or not an external professional activity is "for pay," approval depends on its professional appropriateness and time requirements so that conflict of commitment situations can be avoided.

I. University of Connecticut Campuses, Except the Health Center

With respect to the "one day per week" rule for external professional activities, it is evident that the rule is variously interpreted as one day in seven; all the time that remains after four seven-and-one-half-hour days due the University; etc. To make explicit what period of time "one day per week" designates might impinge on the personal time of the employee (where "week" equals seven days) or impinge on the professional time due the University (where "week" is measured in terms of classified staff obligations). Given the requirement that the employee's professional obligations to the University will be met, the designation of allowable time for external professional activities needs to be defined in each case. The "one day per week" rule, therefore, can only be understood as a general guideline that external professional activities must be limited by internal professional requirements. Since the prior approval form requires a statement of the time to be spent on the activity, each case needs to be determined on its merits by the administrators responsible for the employee's activities. Excessive time proposals can either be rejected by these administrators or recommended for released time arrangements.

II. University of Connecticut Health Center

The basic policy of the University of Connecticut Health center with regard to outside income for faculty and managerial exempt staff was presented to the Health Affairs Committee on April 2, 1987 and the whole Board of Trustees on April 10, 1987. The policy is as follows: when income-generating consulting is judged to be consistent with a faculty member's or manager's other responsibilities and appropriate to his or her professional advancement, permission to engage in such consulting may be given with the stipulation that the income be given to the Health Center.

1) Restrictions defined by Health Center Policy
The policy of the Health Center further restricts activities which might have been allowed under the University rules and specifies that income from approved activities cannot ordinarily be retained.

A. *The independent practice of medicine and dentistry* is specifically prohibited by the University By-Laws. Health Center policy requires that all practice must be conducted in association with and billed through University Physicians or University Dentists.

B. *Consulting* may be approved in accordance with University policy. In view of the demanding schedules of Health Center faculty, consulting as frequently as one day per week would ordinarily not be approved.

C. *Income* from approved consulting must be given to the Health Center with the exception of "occasional fees or honoraria." See 2. *Money Which May Be Retained*, below.

2) Money Which May Be Retained
There are three types of outside income which result from a faculty member's professional activities but may be retained. These are as follows:

—income derived from a copyright held by the faculty or staff member
—income derived from the faculty member's legal share of a patent or license
—occasional fees and honoraria

The faculty member's right to retain the first two is defined under state law. The third is a matter of Health Center policy. The Schools of Medicine and Dental Medicine have defined occasional fees and honoraria as those resulting from speaking engagements or similar presentations and not exceeding $4,000 per academic year. This amount will be annually reviewed by the Dean's Advisory Committee. Funds received for services to NIH study sections or advisory committees or the equivalent in other federal, state, or local government agencies may be retained and, to the extent that the amount retained exceeds expenses, will be included in the $4,000 cap. Activities which generate occasional fees and honoraria may be undertaken when they do not interfere with the faculty member's assigned duties.

3) Approval Process
The rules of the Health Center require that all outside consulting, whether paid or unpaid, receive written approval which is reviewed by the Vice President for Health Affairs.

CONFLICT OF INTEREST POLICY FOR FACULTY MEMBERS OF THE UNIVERSITY OF PENNSYLVANIA

The following policy, adopted by the Faculty Senate on November 17, 1982, and approved by the Provost on March 1, 1983, replaces the Policy on Extramural Consultative and Business Activities of Faculty Members which appears on pp. 41–47 of the 1979 Handbook for Faculty and Administrators.

I. Introduction

This policy applies in full to all Standing Faculty, Standing Faculty-Clinician Educators, and all full-time members of both the Associated Faculty and Academic Support Staff, hereinafter simply designated faculty members. Parts of it also apply to those with part-time faculty appointments; these cases are noted in the appropriate sections. The details of this policy derive from the following general obligations:

1. All employees are required to conform to the mores and ethical standards of the University and the rules promulgated to enforce them.
2. Employment as a faculty member presumes a primary commitment of time and intellectual resources to the academic mission of the University and its functioning as a community.

The following sections cite specific types of activity that have commonly been found to conflict with these obligations, and the procedures and regulations that have been devised to identify and resolve such conflicts. They are intended to serve as examples and not as a comprehensive compilation. Situations not covered by them will be judged in the light of the above general obligations.

Examples of actions that run counter to the first general obligation include nepotism, discrimination on the basis of irrelevant characteristics, inappropriate use of the University's name, and exploitation of any aspect of association with the University for unacceptable purposes or private gain. They are proscribed at all times for all faculty members, extending to those in part-time employment as noted in the relevant sections of this document. Excessive commitment of time or mental effort to extramural engagements or other non-University activities during the academic year constitutes a violation of the second general obligation. As used in this policy, the academic year is defined for each faculty member as that portion of the year during which they receive a salary from the University for services.

II. Conflict of Interest in the Allocation of Time and Effort to Extramural Activities

The University recognizes that its faculty members are not employees in the usual sense, and that a precise allocation of academic time and effort is inappropriate. Their pursuit of knowledge in their areas of competence is presumed to be a lifelong commitment. A limited association of faculty members with government, professional agencies, and public or private organizations is appropriate, especially when it may enhance their competence as scholars.

Policy on Extent of Extramural Activities. Forms of extramural activity include part-time engagement for a fee as a technical or professional consultant or practitioner and formation of association with business enterprises or non-profit organizations. In principle, both such associations are approved under the following conditions:

1. Faculty members should not engage in such extramural associations to an extent that detracts significantly from their availability for normal academic duties. These commitments in aggregate should not exceed one day per seven-day week during the academic year. Exceptions to this will be permitted only in unusual circumstances and require the specific approval of the president or provost, the appropriate department chair, and the appropriate academic dean or director.
2. Faculty members shall make known to their department chairs and academic deans or directors the prospect of each *continuing engagement*, including, at least, all engagements expected to extend for a substantial portion of an academic term. Faculty

members should decide to enter a relationship only if, after discussion with their department chairs and academic deans or directors, there is concurrence that the proposed engagement will not conflict with the faculty members' professional obligations to the University, or with University's outstanding or prospective commitments for teaching and research.

3. In addition to the prospective disclosure cited above, all faculty members must report on the extent of their extramural activities of all types as detailed in Section VII below.

III. Conflict of Financial Interest between the University and Extramural Organizations

Members of the faculty or of their immediate families (including parents, children, siblings, spouse) may have significant investments or interests or hold official positions in extramural business organizations, whether or not they have undertaken to perform continuing work or services for them. Such economic or official relationships are of concern if:

1. The organizations are engaged in activities that parallel activities in which the University is currently or prospectively engaged, and in which faculty members play (or might appropriately play) a role in their academic capacity; or
2. The organizations have a present or prospective relationship with the University, e.g., as suppliers of goods or services or as parties to research contracts, and the conduct of those relationships may involve faculty members in their academic capacities; or
3. The engagements undertaken by faculty members under the aegis of extramural business organizations might be suitable and appropriate activities for execution within the University.

A. Policy on Disclosure of Relationships with Organizations that are Suppliers or Potential Competitors of the University. In either of these situations, faculty members shall be required to report the facts and circumstances to their department chairs and academic deans or directors so that appropriate steps may be taken to avoid conflicts of interest, especially ones in which faculty members may benefit from a knowledge of confidential information.

In the foregoing it assumed that those with part-time faculty appointments will not normally participate in University decisions that could engender such conflicts of interest for them. Whenever this condition does not obtain, the policy stated above extends to them. Furthermore, in any circumstances in which part-time faculty members are engaged in externally sponsored research projects contracted with the University, or in which they stand to benefit from a knowledge of confidential information, full disclosure of their relationships with relevant extramural organizations and of the facts pertaining to any potential conflict is required.

B. Policy on Acceptance of Engagements through Extramural Organizations. Faculty members with positions or connections in extramural organizations who wish to undertake engagements through those organizations rather than the University are obliged to offer first to the University each such engagement (grant, contract, client, etc.) in which they would assume one or both of the following relationships to the engagement:

1. Owner, executive or other principal decision-making position responsible for the conduct of that business enterprise; and/or

2. Principal investigator or other substantial responsibilities for the satisfaction of the engagement.

By requiring that each engagement be offered to the University, the following ends are served:

a. The disclosure of the type, scope and extent of extramural activities is achieved, in accord with University policy;
b. The decision as to whether an engagement is appropriately undertaken as a University or extramural activity is shared with the University administration, thereby avoiding possible conflicts of interest, and the appearance of such conflicts.

Faculty members intending to conduct engagements in business enterprises with which they are associated shall disclose in writing to their department chairs and deans:

1. The nature and terms of the proposed enterprise, and
2. The reasons why it should be conducted as an extramural activity.

If the chairs and deans agree that the engagements are not appropriate as a University activity, and if they conclude that the other conditions of the extramural consulting policies of the University will be met, then they will advise the faculty members to proceed. Otherwise, they may require that the engagements be conducted within the University.

IV. Disclosure of University Affiliation in Publications of Extramural Organizations

Faculty members who form or associate with extramural business enterprises or non-profit organizations should exercise particular care that their University affiliation is appropriately cited in publications of such organizations. Problems that can arise from failure to observe this injunction include:

1. Such an organization, by reason of the participation of faculty members, might be considered to have some formal or informal relationship to the University.
2. Faculty members by reason of their positions in such organizations, might be expected to discharge duties and responsibilities for those organizations that would be inconsistent with their primary duty to the University.

A. Disclaming University Relationships. A business enterprise or non-profit organization, with which a faculty member has a connection, may release to the public from time to time publications concerning itself and its activities. In all such publications it may be desirable and, in many cases, required by law that a faculty member's affiliation with the University be disclosed.

The impact of such disclosure will depend on the circumstances. At one extreme a faculty member might serve as a member of the board of directors of an established business or non-profit organization, where there is not even a remote implication that such organization is in any way connected with the University of Pennsylvania. At the other extreme all or a large number of the principals of an organization (officers, directors, promoters and substantial shareholders) may be faculty members. In such cases, there is a strong implication that the organization may be connected with the University of Pennsylvania, even that the University bears some responsibility for its activities and success.

In these cases, an express statement of the form:

The__________has no connection, directly or indirectly, with the University of Pennsylvania.

in prominent type, should be included in all publications released by such organization. The provost shall have the power to require such a statement to be included in all organizational publications that refer to faculty members, when it is in his/her judgment necessary.

The foregoing rules extend to part-time faculty members, when their association with the University is mentioned in an organizational publication.

B. Affirmation of Obligations to the University. A faculty member may have a position of responsibility (continuing or temporary) with an extramural business organization. In such cases it should be made clear in any publications of the organization that the obligations, in terms of both time and responsibility, of the faculty member to the extramural organization are limited by and subject to the policy of the University of Pennsylvania. This alerts both the public and the faculty member's business associates that duties to the extramural organization are thus limited. This is especially necessary in the case of corporate officers who are normally regarded as owning a comprehensive fiduciary duty to the corporation and its shareholders. The suggested format for such a disclosure is:

> J. Smith, a vice president of this corporation, is a member of the faculty of the University of Pennsylvania and as such is subject to limitations by the University on the time that may be devoted to the affairs of this corporation. In any instance where the interest of this corporation may conflict with the interest of the University of Pennsylvania, J. Smith will resolve such conflict in favor of the University of Pennsylvania.

The provost shall have the power to require such a disclosure in any instance where he/she adjudges it necessary.

V. Conflict of Interest in Externally Sponsored Research

Detailed information and regulations concerning sponsored research may be found in the 1977 Research Investigator's Handbook, modified by the "Guidelines for the Conduct of Sponsored Research," published in *Almanac* 4/7/81. Copies of both are available from the Office of Research Administration.

All faculty members, including those in part-time employment, participating in externally sponsored research projects contracted with the University should refrain from any use of their employment that is motivated by, or gives the appearance of being motived by, the desire for private gain for themselves or other persons, including particularly those with whom they have family, business, or financial ties. The potential for private gain is present when any private agencies or enterprises in which they have significant financial interests or consulting, managerial, or other employment relationships stand to benefit. Examples of situations from which conflicts of interest may arise include, but are not limited to, the following:

1. Undertaking or orientation of sponsored research to serve the needs of a private agency or enterprise in which a responsible staff members has an interest.
2. Purchase of major equipment, instruments, materials or other items for externally sponsored research from any agency or enterprise in which a responsible staff member has an interest.

3. Acceptance of any limitations on the free publication of and access to the results of any sponsored research Exception may be granted by the provost for privileged information, but only in the form of a delay in the release of such information. The delay will only on rare occasions exceed three months. Those wishing to engage in research of a kind whose results cannot be so disseminated may only do so as an extramural consulting activity under the conditions described in Section II.
4. Transmission to any private agency or enterprise, use for personal gain, or other unauthorized use of the work product, results, materials, records, or information gathered from sponsored research that is not made generally available through publication or other free access.
5. Acceptance of gratuities or special favors from a private agency or enterprise with which the University conducts business in connection with a sponsored research project.

A. Disclosure to Responsible University Officials. Before participating in any sponsored research project, all faculty members must give written notice of their extramural consulting relationships or other sponsored research projects that may relate in any way to the project to the appropriate department chairs and through them to the deans and vice provost for research. Any significant financial or managerial interests that may relate in any way to the project must be disclosed in writing to the vice provost. Any faculty members engaged in sponsored research projects must disclose in the same manner any change in their outside activities or interests. In the light of such disclosures, the University will take appropriate steps to neutralize or eliminate potential conflicts of interest.

B. Distribution of Effort. The sponsoring agency supporting research must not be misled as to the amount of intellectual effort that faculty members are actually devoting to these research projects. A system of precise time accounting is incompatible with the inherent character of the work of faculty members, because the various functions that they perform are closely interrelated and do not conform to any meaningful division of a standard work week. However, if externally sponsored research agreements provide that faculty members will devote a definite fraction of effort to the projects, or if it is agreed that they will assume specified responsibilities in relation to such research, demonstrable relationships between the stated efforts or responsibilities and the actual extent of their involvement are to be expected. Each faculty member, in such circumstances, shall confirm the fraction of effort devoted to the projects in the effort reports required of all faculty members who are so engaged.

C. Advice and Guidance. Any questions concerning potential conflicts of interest, appropriate distribution of effort, or other problems associated with externally sponsored research, should be addressed to the office of the Vice Provost for Research.

VI. Employment of More than One Family Member

University policy permits the employment of more than one member of a family (including parents, children, siblings, spouse), whether or not the persons concerned are in the same academic or administrative department. The University's sole concern, in such appointments as in all others, is that faculty or staff members are the best candidates with respect to the requisite academic or administrative qualifications for employment. However, the University recognizes that the appointment of two or more family members, especially within the same department, can lead to abuses and generate pressures and prejudice among colleagues. To guard against such conflicts, the following rules must be observed:

1. No faculty member shall participate in any way whatsoever in the decision to employ, promote, reappoint, or terminate a member of his or her family on the Standing Faculty or Associated Faculty.
2. No faculty member shall participate in any other vital decision, including salary determination, affecting a member of his or her family on the Standing Faculty or Associated Faculty.

Furthermore, faculty members should take care to avoid conflicts of interest or the appearance of such conflicts in the employment of a family member in a non-faculty position. All decisions regarding such employment should be conducted in strict conformance with the relevant regulations, contained in the University of Pennsylvania Personnel Policy Manual, available from the Office of Personnel Relations.

The above requirements extend to those with part-time faculty appointments whenever they may exercise decision power over the employment of a family member.

VII. Requirements for Reporting Extramural Activities and Obligations

At the end of each academic year, each faculty member shall submit to his or her department chair and dean a report of his or her extramural activities during that year, containing the following information:

1. Number of days (or hours, if preferred) of extramural activities for fee (include consulting, professional practice, outside teaching commitments, lectures for honoraria, etc.);
2. Names of organizations (government agencies, private firms, partnerships, etc.) for which the extramural activities conducted represented a continuing engagement;
3. Number of days (or hours, if preferred) of extramural activities on behalf of business enterprises in which they have financial interests or official positions.
4. Names of business organizations in which the faculty member is a significant owner, partner, officer, director, or staff member, etc.

The last item shall also be reported by all part-time faculty members for whom any of the following conditions obtain:

The organization is a supplier of the University and the part-time faculty member participates in the decision to engage its services.

The organization supplies goods or services to the University to be used in the performance of externally sponsored research projects in which the part-time faculty member participates.

The part-time faculty member is privy to confidential University information that could be used to the business advantage of the organization.

The affiliation of the part-time faculty member with the University may be mentioned in any publication of the organization.

Forms for the reporting of extramural activity are available from the Office of the Provost.

All faculty members must also report on a continuing and timely basis to the appropriate administrators the relevant circumstances, as noted in the sections cited, whenever any of the following conditions are met:

1. They have or wish to initiate a relationship with an extramural business organization that is or may become a supplier or competitor of the University (Section III.A.)
2. They wish to undertake an engagement (grant, contract, client, etc.) through an extramural organization (Section III.B.)
3. They intend to participate in a sponsored research project that may be related to their other sponsored research projects, to any of their extramural consulting relationships, or to any organization in which they have significant managerial or financial interests (Section (V.A.).

Revision of Conflict of Interest Policy

The following revision of the second paragraph of Section V of the Conflict of Interest Policy for Faculty Members is approved, following earlier publication in Almanac (April 30, 1985), endorsement by the Research Committee, and approval by the University council. It will be incorporated in the Handbook for Faculty and Academic Administrators.

Thomas Ehrlich, Provost
Sheldon Hackney, President

Original Text

All faculty members, including those in part-time employment, participating in externally sponsored research projects contracted with the University should refrain from any use of their employment that is motivated by, or gives the appearance of being motivated by, the desire for private gain for themselves or other persons, including particularly those with whom they have family, business, or financial ties. The potential for private gain is present when any private agencies or enterprises in which they have significant financial interests or consulting, managerial, or other employment relationships stand to benefit.

Revision

The University encourages its faculty members, including those in part-time employment, to participate in externally sponsored research projects whether supported by government agencies, foundations, associations, or other non-profit organizations; or by corporations, partnerships or other for-profit entities. In any sponsored project, faculty members are expected to avoid use of the project for their private financial gain other than in the form of salary support or of royalties resulting from commercialization of intellectual property rights in accordance with University policies. However, there may be unusual circumstances where the interests of the University would be served if a faculty participant in a sponsored project were to assume an entrepreneurial role, as, for example, by direct participation in a private enterprise providing funds in support of the project. Assumption of such a role would not be a violation of these guidelines if approved in advance and reviewed periodically by the relevant Dean and the Vice Provost for Research.

9

Technology Transfer and Commercialization

Joyce Brinton

Harvard University, Cambridge, Massachusetts

I. AN OVERVIEW OF BIOTECHNOLOGY LICENSING

A. A Historical Perspective

It may seem strange to talk about history when discussing licensing in the "new" field of biotechnology, but the origin of the field and its rapid development in the short period since its beginning explain the significance of biotechnology licensing to universities, teaching hospitals, and research institutes.[1] To a certain extent, universities focused on licensing issues because of the dramatic growth of biotechnology and the opportunities it presented for technology transfer.

In the early development of biotechnology, cutting-edge research (including the identification of potential products) took place in university rather than commercial laboratories. There just was not a critical mass of scientific expertise in industry. Although that situation has changed and university scientists now look to industry as well as other academic institutions when searching for collaborators, a significant portion of cutting-edge research remains in universities. Moving the results of this research into the commercial sector required the development of technology transfer programs and licensing expertise.

As practical applications of biotechnology were identified and their commercial potential was appreciated, new ventures were created that drew heavily on university faculty as founders, consultants, and members of scientific advisory boards. Many of these academic scientists received an equity interest in the new companies. This raised policy concerns for the universities and presented new challenges for the emerging technology transfer programs. Although faculty members had founded companies before and had served as consultants and board members, those who participated in these new biotechnology ventures were basic research scientists, not engineers, and they retained their faculty positions rather than resign to become entrepreneurs or industry employees.

At the same time, there were changes in national policy that resulted in increased university licensing activity. The enactment of U.S. Public Law 96-517[2] meant that by the early 1980s universities and small businesses could easily take ownership of inventions made as the result of federally funded research. Prior to that time, the government owned federally funded inventions and it was necessary for the university to petition the government in order to obtain ownership. Each federal funding agency had its own criteria for determining whether to grant these requests, and thus the process was cumbersome and time consuming at best.

Public Law 96-517 (as amended by Public Law 98-620) represented a turning point for university technology transfer. Since most university research is at least partially supported by federal funds, the legislation gave the university title to most inventions resulting from university-based research. In addition, the law mandated and encouraged the university to establish technology transfer programs to ensure that product development and commercialization of these inventions took place.

Under Public Laws 96-517 and 98-620, the university must have its faculty sign agreements that commit the faculty member to report and to assign ownership of inventions to the university. The university, in turn, is required to report these inventions to the federal government and, within a specified time period, to inform the government whether it will take title and file a patent application. If the university does not wish to file a patent application, the government may do so on its own behalf or, upon request from the inventor, it may release the invention to the inventor. Whether the university retains title or the invention is released to the inventor, the owner has certain obligations: It must

1. Notify the government before abandoning the patent application or patent so that the government may continue prosecution or maintenance in its own name.
2. Submit an "Invention Utilization Report" periodically (no more frequently than annually) providing information on patent status, licensing, and/or development activity, and income.
3. Give preference to domestic small businesses when granting licenses.
4. Diligently seek development and commercialization of the invention.
5. Require an exclusive licensee to manufacture substantially in the U.S. any products to be sold in the United States.

6. Grant the government a royalty-free, nonexclusive license to practice the invention, or have it practiced, for governmental purposes.

In addition, the government reserves the right (a) to require the owner (or its exclusive licensee) to grant licenses on reasonable terms or (b) to grant licenses itself if the owner or its exclusive licensee has not diligently sought to bring the invention into public/commercial use or if necessary to alleviate health or safety needs not being met by the owner or its exclusive licensee.

In practice, so long as the university and its licensee are diligently attempting to develop and commercialize an invention, the government is unlikely to exercise these reserved rights. As a result, most U.S. industry accepts the government's rights without significant concern.

By the mid-1980s, the growth rate for traditional sources of funding for university research, especially the federal government, slowed, and in some cases threatened to or did decline. Thus, the possibilities of new sources of income from licensing to the burgeoning biotechnology industry and research support from that industry were very appealing. In addition, there was continued federal emphasis on technology transfer as an important factor contributing to the health of American industry. This even resulted in funding of programs designed to promote university-industry collaboration.

It is within this context that university attitudes toward and expectations from biotechnology licensing must be considered. The university's approach to biotechnology licensing is colored by issues of academic principle and the university's public interest responsibility, as well as the university's fiduciary responsibility for managing its assets. Because the rise in biotechnology licensing by universities occurred so recently and so quickly, the policies governing it, and university-industry relationships in general, are still being examined by the legislature and the press as well as university faculty and administrators.

B. The Importance of Biotechnology Licensing Today

In recent years, more and more companies have made the strategic decision that acquiring technology through licensing (licensing-in) is an important approach to the development of new or improved products and that a synergy can be created between the company's internal R & D programs and technology the company acquires through licensing. The small biotechnology companies that concentrated on R & D and contract research have now begun to licence products and processes for which they may not have the resources for scale up, regulatory approval, and marketing. These companies are entering into a variety of relationships with larger companies—all the way from ordinary license deals to marketing partnerships or joint ventures. And it is not just the small biotechnology companies who are licensing their technology; major pharmaceutical companies are licensing each other as part of joint ventures and marketing partnerships as well.

At the same time, there has been a marked increase in licensing from universities to industry. Not only are more universities becoming active in identifying and marketing their technological developments, but industry has realized how important these university-originated inventions can be, particularly in biotechnology.

II. LICENSING FROM UNIVERSITIES, TEACHING HOSPITALS, AND RESEARCH INSTITUTES

A. Goals of University Licensing Programs

Because universities have a general commitment to the public interest through their educational and research programs and because research in universities is heavily supported by public funds, the principal goal of the university licensing program is to bring university technology into public use as rapidly and widely as possible. As a secondary goal, the university expects to receive a fair financial return from the companies that commercially exploit university technology. These two goals must be pursued without disrupting the primary missions of the university—education of students and the creation of knowledge and its wide dissemination as well as, in the case of medical institutions, patient care. Thus, the university's desire to foster technology transfer cannot undermine traditional academic values such as the right of the faculty to choose their research topics, to collaborate with other researchers, and to determine where and when to publish the results of their research. Fortunately, it appears that with a good policy framework, it is possible for the university to meet its technology transfer goals through a licensing program without conflicting with its educational and research missions.

The competitiveness of the United States has recently heightened government interest in technology transfer from university and government laboratories to U.S. industry. Congress is now asking whether universities should give preference to U.S. industry when licensing inventions that result from federally supported research. At present, federal law and regulation requires preference for small business and that products to be sold in the United States be manufactured substantially in the United States, but the licensing of foreign corporations is not precluded nor discouraged. Nonetheless, most universities quite naturally seek out U.S. licensees first—they are closer to home and there are fewer cultural and language barriers—but foreign companies are becoming more aggressive in seeking out university technology, and some of these foreign companies appear more willing than their U.S. counterparts to invest in early stage, high-risk technology.

So far, the U.S. biotechnology industry seems to be holding its own with its foreign competition in regard to licensing from universities, but there is a growing concern that this valuable source of biotechnology may increasingly be captured by foreign companies.

Should that occur, the U.S. biotechnology industry will surely suffer but so will U.S. universities because the result would undoubtedly be legislation restricting the university's right to license foreign companies. Thus, to a certain degree, university technology transfer programs have had thrust upon them a third goal—enhancing U.S. competitiveness and thus the U.S. economy.

B. Finding Potential Licensees

Having identified a biotechnology invention and having filed a patent application, the next step in the commercialization process is to identify potential licensees—companies that have the ability and resources to invest in the development of the technology and bring products to market. Most biotechnology inventions resulting from university research are far from being products, and the challenge is to determine how a company can turn the invention into a viable product. For some inventions, this is fairly obvious but for others the inventors and the licensing staff must employ their joint creativity to discover a range of potential uses, some of which may be far afield from the inventor's original research interests. In either case, the task is to discover the product potential of an invention and then find a company or companies that will turn the invention into useful products.

Although the biotechnology industry is finite, there are new companies being formed every day, existing companies change their focus, established companies incorporate biotechnology techniques into their traditional manufacturing methods, and companies fail and disappear. To find an appropriate licensee in this changing environment, university licensing professionals make use of a networking approach that in some ways is similar to that used in finding a new job or tracking down a news story. And since many biotechnology companies are actively looking for technology to license, their search efforts often intersect with those initiated by the university.

Clearly the primary source of leads in this search is the inventor. Even those faculty who are not particularly interested in the commercialization of their inventions often know more about the companies that might be potential licensees than they might initially think. The licensing professional needs to explore the inventor's contacts in industry—which industry scientists have expressed interest in his or her work in the past, for which companies does the scientist consult, which industry scientists have written journal articles the inventor found useful, and, simplistically, which companies does the inventor think might be able to use the invention. Of course, if the invention resulted from research that was funded (even in part) by an industrial sponsor or if the inventor has had industrial funding in the past, those companies are natural candidates. It is not surprising that many licenses turn out to be with companies that have worked with the inventor or were suggested by the inventor.

In addition to companies identified by the inventor and those chosen from the personal contacts of the licensing staff, others can be located through the use of various directories and computer data bases that provide information on the biotechnology industry. A few examples are *The Biotechnology Directory, Corporate Technology Directory* (*Corptech*), and numerous data bases available through Dialogue. Most technology licensing offices have also compiled their own data bases of contacts in the biotechnology industry with whom the staff have interacted in the past.

Ultimately the licensing professional contacts these companies by letter or by telephone to explore interest in the invention. Even if a company has no interest in the particular invention, the contact person can provide referrals to other companies for which the technology might be more suitable as well as information about the types of technology in which the company would have an interest in the future. As personal relationships are developed at more and more companies, the licensing staff can better target these initial contacts and present the technology in a manner that focuses on the particular company's areas of interest.

When a company shows some interest, a brief nonconfidential description of the invention, its advantages, and its patent status is provided along with a nondisclosure agreement to be signed by the company if it wants to review a copy of the patent application or other more detailed, unpublished information. Assuming there is serious interest, company scientists often discuss the invention with the inventor.

Through this networking approach, hopefully one or more companies are identified that are not only interested in the invention but appear to have the capability to develop products and market them.

C. Evaluating Potential Licensees

Having found one or more potential licensees, how does the university decide whether they should be granted licenses? Ideally, each candidate company prepares an R & D/business plan describing the products it intends to develop, the research and development steps it foresees, its timetable and budget for the development process, and the eventual market size it expects. These plans, even if brief, can provide a basis for comparing several companies and whether their expectations are realistic. In addition, the proposed timetable provides a framework for due diligence milestones that can be incorporated in a license agreement.

The primary criterion, of course, is the ability of the company to develop the invention and bring products to market within a reasonable time frame. If several companies have similar abilities, other considerations can be taken into account: the encouragement of U.S. industry, and of small business in particular, as well as the potential financial return to the university (and thus to the inventor).

Of course, in many instances, multiple potential licensees are not readily apparent. When a candidate company is identified that has the capability to develop and bring products to market, the university must decide whether to move directly to negotiating a license or to continue the search for additional or alternate licensees. This is a subjective decision based on the potential value of the invention and the probability that other companies will, indeed, have sufficient interest to compete for a license. Sometimes, it is more important to "strike while the iron is hot" than to delay hoping to find a "better" deal.

When evaluating potential licensees or even deciding where to look for them, there are a number of general options that might be considered. The first is whether a large, established company or smaller start-up company is more appropriate. The large established company offers certain advantages: It is a known quantity, it has the resources to undertake the expense of development (and in the case of pharmaceutical products the regulatory approval process), it has a marketing capability, and it has the assets to indemnify the university and the inventor from product liability claims. On the other hand, there are certain disadvantages. Large, established companies sometimes are slow to make decisions or it is difficult to reach a decision maker within the company. They also have many projects competing for the company's resources, and thus your project may be shelved or delayed, a situation that can be compounded by the "Not Invented Here" attitude. Fortunately, in recent years, many of the larger pharmaceutical and chemical companies have made corporate decisions that licensing-in technology is essential to their future growth and thus efforts are underway to curb the N.I.H. response.

Dealing with small companies and start-ups has the distinct advantage that it is generally easy to reach the actual decision makers and decisions can be made rapidly. Even though these companies may also suffer from the N.I.H. problem, they tend to be receptive to new ideas; at least they lack the large internal bureaucracy which seems to nourish the N.I.H. response. If the invention will become a significant product for a small company, the availability of development resources is reasonably assured; the company's future depends upon it. But if, for example, Food and Drug Administration (FDA) approval is required, a small company may not have the resources to complete that process without a corporate partner. Thus, it is important to know that the small company recognizes this and is prepared to negotiate a realistic arrangement with a larger company. A concern in licensing small companies is the possibility that they will acquired by another company and priorities will change so that the licensed invention will no longer be viewed as important and development will slow or stop. In the case of start-ups established to exploit the technology, one must evaluate the motivation of the investors carefully—are they interested in building a company that will produce products or are they only interested in the short-term potential of taking the company public? Although these need not be mutually exclusive, there are occasional instances where the latter objective is so overpowering that the company does little actual development work but rather expends its effort on creating a market for the company's stock. This can significantly delay the commercialization of the licensed technology and may even kill its chance to reach the marketplace. Another drawback to small companies and start-ups is their rather high failure rate; technology licensed to a company that fails can be tied up in bankruptcy court for years.

The decision of whether to license a large established company, a small business, or a new start-up is one that is made on a case-by-case basis taking into account the pros and cons of each type of company. For some inventions, the choice is clear and for others it is less the type of company that is important than the capability of the particular company, whether large or small, established or new.

Because the establishment of start-ups to exploit university technology has become such a topic of interest, it is important to understand when it should be the approach of choice. Some inventions are unlikely to be fully exploited if licensed by an existing company and a start-up founded specifically for that purpose offers significant advantages. What kinds of inventions benefit? They are generally inventions that have multiple applications and yet are very early stage and thus high risk. Usually it is also important that the technology be sufficiently innovative so that there is not an existing established industry with which the start-up would have to compete. And last, but certainly critical, at least some of the inventors must be willing to actively participate in the company—even if only as consultants or members of a scientific advisory board.

For the right invention, the start-up provides the means to achieve development of a wide range of applications. Lita Nelsen of the Massachusetts Institute of Technology describes this as the "hub and spoke" concept. The start-up devotes its efforts to developing the core technology (the hub) and identifying various applications. It then forms strategic partnerships with a number of companies, each developing a specific application (the spokes) This permits the simultaneous development of multiple products rather than their sequential development. From the university's perspective, this hub and spoke approach is ideal for meeting the university's commitment to ensure the widest possible use of its inventions as rapidly as possible.

Of course, from a financial point of view, there are added advantages to a start-up if the company is a success and the university has taken equity in it—the university benefits from the value of the stock even if its patents do not issue and no royalties are paid. It also benefits if the company expands into new fields not based on the university's technology. To keep this option in perspective, however, one must remember that start-ups are risky business—they can fail for many reasons—and there are conflicts of interest—both real and perceived—that can haunt the university, the involved faculty, and the company. Nonetheless, for the right technology and the right inventors, a start-up can be ideal.

D. Negotiating the License Agreement

Having decided that one or more companies have the capability to develop and market products based on a university invention, the licensing professional and the company then negotiate the terms under which the invention will be licensed. License negotiations require that both the company and the university understand their own objectives and plans for the technology and communicate those objectives and plans to the other party. Although presumably the parties have a mutuality of interest—getting products to market—there are bound to be differences of opinion about how that can best be achieved and under what conditions. The important thing to keep in mind is that ultimate common interest and for each party to remain flexible so that a balance can be found between the needs of the university and those of the company. For a license agreement to be successful, it must be a good deal for both parties.

E. The License Agreement: What Do All Those Clauses Mean and What Are All the Negotiations About?

Perhaps the best way to understand what is involved in license negotiations and what the license agreement is attempting to achieve, is to review a license on a clause-by-clause basis. For each clause or section, I will point out its purpose, explain its meaning, and discuss the types of issues that occur in its negotiation. The license agreement appended to this chapter is a format created by Harvard University for the exclusive licensing of its biotechnology inventions when both patent rights and samples of a biological material are being conveyed to the company. The license agreements used by other universities may look different than this format but are generally similar in their overall treatment of the significant points. And since the example is a format, the terms of actual individual license agreements signed by Harvard differ from it depending on the particular circumstances surrounding the technology and the licensee company.

F. Introduction and Whereas Clauses

Most license agreements begin by identifying the corporate entities entering into the agreement—the organization granting the license to the technology (the licensor) and the organization obtaining the license (the licensee). This is often followed by a series of "whereas" clauses that set forth the context of the agreement. For example, if the technology resulted from research sponsored by the licensee, that background might we recounted in a whereas clause. The university frequently uses these clauses to make note of its policy objectives.

G. Definitions

Most important is to define what is being licensed—in this case, both "patent rights" and "biological materials." Since most university licenses are entered into before patents have issued, patent rights are defined in terms of the initial U.S. patent application as well as any later U.S. filings arising from that initial filing (i.e., continuations, continuations-in-part, and divisional applications) and any foreign applications corresponding to them. Since the definition of patent rights can determine which products will require the payment of royalties, the definition can be very important. Some companies want to limit the definition to issued patents or to "valid claims of issued patents" so that no royalties would be required so long as a patent application remained pending. Given the length of time it can take for patents to issue, this can have a significant impact on income. The existence of a strong pending patent application still provides a licensee with an element of protection, especially for a product that requires significant investment, because competitors will be reluctant to make that investment because when the patent issues they will face a possible infringement suit.

Biological materials must be identified carefully—perhaps in terms of their ATCC deposit number or sequence information. Since the licensee may be using those materials as a starting point to create modifications or derivatives, those modified materials are included in the definition as well.

Licensed products and licensed processes are defined in terms of patent rights and biological materials. Royalties are paid on revenue generated from these defined terms.

The definition of net sales determines the base on which royalties are calculated. Depending on how the licensee conducts its business, this definition may have to be altered; for example, to include income from diagnostic services provided by the licensee. In this rather typical definition, the net sales is essentially any revenue that will actually be retained by the licensee.

In some cases Affiliates are defined not only in terms of companies owned by the licensee but also companies that have a significant ownership interest in the licensee.

H. Grant of Rights

One of the major points of discussion in license negotiations is the scope of the rights to be granted: whether the license will be exclusive, semiexclusive, or nonexclusive; whether it will be limited by field or territory; or whether it will be for a limited period. Although this format is designed for exclusive licenses granted for a limited time period, there are occasions when semiexclusive or nonexclusive licensing might be more appropriate. Technologies that require significant development and expensive regulatory approvals, such as pharmaceutical products, are most commonly licensed exclusively. On the other hand, manufacturing processes that might have application in the production of many products are usually licensed nonexclusively. There are, however, many variations possible to permit the grant of rights to be tailored to a particular technology or industry. For example, when the market is large enough and the investment required is not extraordinary, a limited number of licenses might be granted (semiexclusive) or licenses might be limited by geographic territory. If there are multiple products possible, exclusive licenses for particular products or fields of use could be granted. Although this format limits the length of the license, after which the licensor may grant licenses to others, licenses are frequently for the life of patent rights. The university licensor considers carefully how broad the scope of the license should be for a particular technology; the broader the scope, generally the more rigorous are the requirements imposed on the licensee for diligent effort to develop the technology.

Even in this exclusive license, exclusivity is not total. Because most university inventions are made with support from the federal government, a nonexclusive license must be granted to the government (this is also true for a number of foundations that support university research). In addition, the university itself must retain the right to use the invention in its research and to grant research purposes–only licenses to others. This is important so that the patent would never be used to discourage further research in the field.

The issue of the licensee's diligent effort is dealt with in clauses 2.2 (c), (d), (e), and (f) of the Harvard license agreement, and these clauses are among those that are the most hotly discussed during license negotiations. The first of these clauses is merely an umbrella agreement that the licensee will be diligent. The second is what is often referred to as the "march-in" provision that gives teeth to the requirement for diligence. In this format, if the university determines that the licensee has not made significant efforts after 3 years, it can terminate the license completely or change the license to a nonexclusive one. This general language does not include specific milestones that the licensee is expected to meet because those milestones depend on the type of product being developed and the stage of development of the invention being licensed.

Much of the license negotiations is devoted to determining what milestones should be specified in this march-in provision. Ideally, the R & D plan prepared by the licensee will include a development timetable that identifies key points in the development process. These can then be enumerated with target dates that both the university and the licensee agree are reasonable. If those targets are not met, and there are not acceptable reasons for that failure, the university has an objective basis on which to to terminate the license or convert it to a nonexclusive one. If the technology is so preliminary that a realistic development plan cannot be stipulated, then language such as that in the format must be used and the licensee has to accept the subjective judgment of the university.

When a license is granted for inventions that have multiple uses or where the licensee does not have worldwide marketing capability, sublicensing is a critical means of ensuring rapid and widespread utilization. Hopefully, both the university and the licensee will, in those cases, see sublicensing as being in their joint interest and the rest of the license terms should be designed to encourage that. However, if the licensee desires to hoard the technology until it gets around to exploiting the additional uses or penetrating other geographic markets, clause 2.2 (f) of the Harvard license agreement provides a mechanism for a limited form of march-in—the university can request the licensee to grant sublicenses on reasonable terms and, if the licensee refuses, the university can grant those licenses itself. The university would be restricted from granting licenses if the licensee could show that no greater availability of products would result; i.e., the licensee itself can meet the demand in the additional field or territory.

If the license has been restricted by field (as is the case in this format), clause 2.2 (g) of the Harvard license agreement provides a mechanism for the licensee to obtain rights to additional fields of use. The university is not precluded from licensing these other fields exclusively to third parties, but if it has not done so, the licensee is assured of at least a nonexclusive license on terms as favorable as any granted by the university.

This format permits the university to terminate the license completely at the end of the license period, or to extend the license on an exclusive or nonexclusive basis. For products that have involved a significant investment, the licensee often insists that it be guaranteed at least a nonexclusive license for the life of patent rights. This would still permit the university to grant licenses to other companies, thus benefiting the public by introducing competition on price, quality, and/or service.

Because the requirements imposed by the federal government extend beyond the grant of a nonexclusive license to the government, clause 2.5 of the Harvard license agreement preserves all other rights the government has in inventions made with federal research support. Among those rights are the government's own march-in rights.

I. Royalties

The financial terms of license agreements are perhaps the center of license negotiations. Reaching agreement on the value of the licensed technology is often even more difficult than determining the scope of the license or the march-in terms. This format includes very simple financial terms—an up-front license issue fee, running royalties expressed as a percentage of net sales, a share of nonroyalty sublicense income, and annual license maintenance fees that change to minimum royalties when product sales begin. The final clause in the format is designed to ensure that the licensor receives its full royalty from a foreign licensee. In some countries, most notably Japan, payment of royalties to a foreign licensor is subject to a withholding tax (10% in Japan); this clause ensures that the licensor still receives the full stipulated royalty.

Only recently have universities begun to look on their technology portfolios as an asset that should be managed in much the same way as the university's endowment is managed. However, just as public interest concerns affect the investment of the endowment (divestiture of stock in tobacco companies or in companies doing business in South Africa), so too do public interest concerns take precedence over financial return in licensing. But within the limits of public interest and academic concerns, universities are striving to license so as to receive a favorable financial return—one that is "usual and customary" for equivalent technology licensed under similar terms. Of course, determining what is usual and customary is not easy.

An issue of *Intellectual Property Happenings*[3] contained the following chart prepared by Thomas Kiley (an attorney and former Vice President of Corporate Development at Genetech) showing representative royalty rates for biotechnology licenses:

Recombinant Pharmaceutical Products	Exclusive (%)	Nonexclusive (%)
Early stage or developmental	7–10	3–4
Approvable or approved	12–15	5–8
Monoclonal Antibody	**Exclusive (%)**	**Nonexclusive (%)**
Therapeutic use	5–7	3–4
Diagnostic use	3–4	1–2
Component with significant, high value–added apparatus	1–2	0.5
Enabling or Supplemental Technology	**Exclusive (%)**	**Nonexclusive (%)**
Production method	5–7	0.5–2
Drug delivery component	2–3	0.5–2

Even when attempting to apply these representative rates, many factors must be taken into account: the strength of patent protection (have the patents issued, how strong are the claims, in which countries do patents or applications exist), the amount of development necessary, the profit margin for the particular product(s), the extent of competition, the likelihood of infringers, and what royalties must be paid to others. This latter factor is particularly important in biotechnology; few biotechnology products will be subject to only one royalty. Almost all products require the payment of royalties to two or more licensors: for the gene splicing process, for the protein to be expressed, as well as the delivery system of the drug or the format of the diagnostic. If the university is licensing the protein product or some other technology that is central to the finished product, will it be able to negotiate a "reasonable" royalty in light of the cumulative royalty burden? Because of the university's public interest mandate, it may feel forced to accept a lower royalty so that total royalties do not become so burdensome that a prospective licensee is unwilling to undertake necessary product development. Attempts to solve this problem include full or partial credits for royalties paid to third parties against a portion of royalties due the university. Obviously the university is more receptive to such arrangements if at least some of the other licensors also allow similar credits against their royalties. This stacking of royalties is becoming such a common problem that universities and industry need to begin thinking about fresh solutions.

In an effort to deal creatively with the royalty provisions of biotechnology licenses, universities have employed many approaches. Some of these are described below. None of these approaches should be viewed as the university ideal; rather each is an attempt to respond to a particular situation or type of situation:

1. When there are several alternate technologies available to licensees, the university might set financial terms so that multiple licensees will be encouraged to experiment with the invention, hopefully finding it useful and sufficiently inexpensive so they will not search for an alternative. As an example, the university would charge a small initial fee for a time-limited research license; at the end of that research purposes license period, the company would pay a slightly larger annual fee to continue R & D and to receive an option to obtain a commercial license for a stipulated up-front payment plus a low running royalty.
2. In an effort to respond to the cash-poor small company seeking to bring a licensed product to market, the university might set a low initial license fee and low initial royalty rates. As sales increase and the company's cash flow improves, the royalty rate would increase until it reaches an agreed-upon level.
3. For a research product expected to have a short half-life in the market, the university might set relatively high initial royalties while sales are brisk and

the company's competitive advantage is at its peak. As alternate or improved products become available and sales fall, the royalty rate might decline.

Field of use or product-dependent financial terms provide examples of still other ways to accommodate to the realities of the marketplace. A purified recombinant protein might merit a higher royalty rate when used for a therapeutic product than for a diagnostic product. Within these fields, further royalty differentials might be provided depending on the market for the product—a therapeutic or diagnostic product for a cardiovascular disease could carry a higher royalty than could products for the diagnosis or treatment of a disease primarily occurring in the Third World.

J. Reporting

This section of the license includes three principal reporting requirements. The first is the company's R & D plan that serves as a baseline against which the university can measure progress. The second is an annual report of progress of the licensee's efforts to develop and market products under the license. These progress reports permit the university to determine whether the licensee is meeting the agreed-upon milestones or due diligence requirements specified in the license. The third type of report is the royalty report. This format requires reports of sales and royalties due, along with payment, to be made semiannually. This is perhaps the most common schedule, although quarterly reports and annual reports also are frequently used. A minimum royalty becomes due with the royalty report for the second half of the calendar year in any year in which it is not exceeded by the calculated or running royalty.

Because federal reporting requirements include a report of the date of first commercial sale, the licensee is required to provide that information to the university in addition to the above reports.

K. Record Keeping

In order that the licensor can verify the royalty reports, the licensee must retain records for some period of time (in this case 3 years) and to make them available for audit. Since performing audits involves a cost, the licensee is required to pay that cost if the audit shows underpayment.

L. Domestic and Foreign Patent Filing and Maintenance

Although patent filing and prosecution are the responsibility of the university (as provided in this format), the university is expected to consult with the licensee to obtain its views and the licensee is required to reimburse the university for its costs. Consultation with the licensee is in the university's interest because the licensee should have a better understanding of the commercial situation and thus can make a significant contribution to strengthening claims that are particularly important for the licensee's business. The licensee also is in the best position to make the economic determination of where foreign patent coverage should be sought. Although the university is not required to follow the licensee's advice, so long as that advice is reasonable, it probably will. In those instances where there are differences of view, the university can pursue coverage that the licensee feels is unnecessary or not economically worthwhile and, unless the licensee reimburses the university for those costs, it looses rights under those patent applications and resulting patents.

Other approaches are also possible. The university may permit the licensee to take responsibility for patent prosecution but retain for itself the right to review and approve the licensee's actions as well as the right to take such additional actions as the university feels necessary. Occasionally, universities agree to share the costs of patent prosecution with the licensee or to bear those costs entirely. Generally such arrangements only occur in nonexclusive or semiexclusive licenses or when the eventual product is something like an orphan drug.

M. Infringement

Both the university and the licensee have a common interest in ensuring that licensed patents are respected in the marketplace. Unlicensed users of patented technology reduce the licensee's profits and reduce the royalties paid to the university. Infringers should either become sublicensees or be forced out of the market. Whether or not to sue for patent infringement, however, ultimately becomes an economic decision—is the loss in revenue (currently and in the future) sufficient to justify the expense of the potential litigation? If so, an infringement suit is initiated. Sometimes that is sufficient to resolve the matter—the infringer either takes a license, withdraws from the market, or changes its product so it no longer infringes.

This license format gives the licensee the first opportunity to bring suit for infringement. The university must be consulted (because it will likely be named a co-plaintiff in the suit) and may encourage the licensee to attempt to sublicense the infringer prior to initiating litigation proceedings.

If the licensee proceeds with the suit, this agreement allows the licensee to apply a portion of royalties due the university toward payment of a portion of the expenses of the suit. In this way, the university shares in the financial burden of the suit, but there is still a sufficient cost to the licensee that it will not pursue litigation when it is of questionable long-term economic benefit. This format limits the amount of royalties that can be diverted to supporting the litigation, so that the university continues to receive at least 50% of the royalties due it. Since the university is sharing in the cost of the litigation, it also shares in the recoveries if the suit is won.

N. Termination of Agreement

Although this format is written so that the license extends for a limited period, it is common for licenses to be granted for the life of patent rights and for the license to extend beyond that time period if biological materials that are not generally available have been provided. The license may be terminated for default, upon bankruptcy of the licensee, or upon notice and payment of a termination fee. This latter provision is to permit a licensee to terminate the license voluntarily in the event it is no longer interested in developing or marketing the licensed products. The payment of the termination fee is in recognition that the technology has been tied up for some time and that time is now lost and the university must find another licensee that has to begin the development process again.

O. General

Although the university will warrant that it owns the patent rights and the biological materials being supplied to the licensee, it will not warrant either that the patents are valid or that licenses under other patents are not required. The university feels that as part of the licensee's evaluation of the patent rights being licensed, the licensee should make an investigation of patent validity and whether the licensed patents are dominated by other patents (i.e., if practiced would infringe other patents). Even if the licensed patent rights are not dominated by other patents, it is still possible that the product the licensee plans to produce will require a license under other patents. For example, the university could license a protein to be used as a diagnostic, but the test format the licensee plans to use might be covered by a patent owned by a third party. Even though the licensed protein did not infringe that third-party patent, the planned product would, and the licensee would need a license from the owner of the third-party patent as well as from the university.

The licensor university specifically warns the licensee that no warranties are made regarding the licensed patent rights or biological material. This section is printed in capital letters because that is required by the Uniform Commercial Code in order for such a disclaimer to be effective.

In recent years, universities have become increasingly concerned about the possibility that licensed technology will become part of a product that is subsequently the subject of significant product liability claims. If that occurs, the university fears it may be seen as a "deep pocket" and thus would be named as a defendant along with the manufacturer of the product and the manufacturer's insurance company. In order to protect the university from the costs of defending itself and from the costs of any award that might be made by the courts, the licensee is required to pay any such costs by providing the university with an indemnity and by obtaining insurance to back up that indemnity. The two long clauses of the Harvard license agreement, 9.4 and 9.5, which specify the details of that indemnity and insurance were created by the organization that provides insurance to Harvard University and its affiliated hospitals. That organization concluded that, even though no "bare" licensor (i.e., a licensor of technology who had no further involvement in the development, manufacture, or marketing of the eventual product) had been found liable in a product liability case, trends in case law were sufficient to warrant caution.

Although these clauses appear at first glance to be complicated and onerous, they are really quite straightforward and reasonable. Their purpose is to shift financial risk for claims resulting from the products produced by the licensee (or its sublicensees) from the university (and the inventor) to the licensee. And since the indemnity by the licensee is only as good as there are assets to meet its obligations, the licensee is required to carry insurance sufficient to withstand that financial risk.

There are three parts of these clauses that prospective licensees find troubling, at least initially. The first is the fact that the university is requiring the indemnity *even if the university has been negligent.* Although at first glance that appears unfair, it is important to recognize that the university is granting the licensee the right to use patent rights and biological materials as part of the licensee's efforts to develop a product. The university itself is not designing, testing, or manufacturing that product. Those activities are the function of the licensee, and thus, if there were negligence on the part of the university, it is the licensee's responsibility to discover it and correct it before a product reaches the market.

The second area that concerns some prospective licensees is the requirement that the licensee obtain insurance. But without insurance, how can the university be assured that there will be sufficient funds or assets available for the licensee to meet its obligations to protect the university from financial claims? In particular, start-up companies and small businesses have limited assets and thus their indemnities are only meaningful if adequate insurance coverage is in place. Start-up companies sometimes do not have any insurance coverage and are fearful that the costs will be prohibitive. Although that is not an unrealistic fear, the company itself should be concerned about the possibility of uninsured product liability claims leading to bankruptcy. Most start-ups are relieved when they realize that the requirement for insurance coverage does not begin until the company either initiates clinical trials or sells products. By that time, most start-ups plan to acquire insurance, assuming it is available and economically feasible. Because of this concern about the cost and availability of insurance, we have found it necessary, on occasion, to assist companies is locating coverage to satisfy the requirement in the license.

Large, established companies sometimes react negatively to an insurance requirement, maintaining that they are self-insured and that their assets alone should be sufficient to meet their obligation to indemnify the university. It is for this reason that the clause includes a mechanism for the licensee's self-insurance program to be reviewed, and if adequate, to be substituted for the requirement to obtain insurance coverage. The existence of a self-insurance program is quite different from merely being uninsured—such programs include careful evaluations of risk and the setting aside of reserves to meet those risks and often the obtaining of insurance coverage for extraordinary losses.

The third area of concern is the requirement that insurance coverage be maintained for a time period after the license has terminated. This is, however, necessitated by the fact that the products produced and sold under the license continue to be used by purchasers for some time after sale and damage from the products may not be recognized for some time after use. This, coupled with the fact that most insurance policies only provide coverage for claims made during the period the insurance is in place, means that even though the licensee has stopped selling a product, it (and thus the university) remains at risk of product liability claims for some time.

All of this attention to what would seem a very remote possibility may appear excessive, but a single class-action product liability judgment against the university could be in the millions of dollars and without the licensee's indemnity and insurance to back that indemnity, the university's teaching and research programs could be jeopardized. That is a risk that most universities are unwilling to accept.

Universities are generally not willing for a licensee to advertise their products using the university's name and reputation, and thus any such use requires specific approval. An acknowledgment (perhaps in small print) that the product is protected by patents owned by the university is generally approved, but implying that the university endorses the product is an entirely different matter.

Since the university wants to know the companies it licenses, it is unwilling to permit the licensee to transfer the license to another company without the university's permission. An exception is made in the event the licensee company disposes of the entire business associated with the license, e.g., no approval would be required to transfer the license if the licensee sold the division that manufactured products under the license.

The United States has numerous laws and regulations governing the export of products and information. Some of those regulations specifically deal with the export of biotechnology and genetically engineered microorganisms. The penalties for violation can be quite severe and those penalties can be imposed on the owner or originator of the technology. This clause puts the licensee on notice that it must comply with those export regulations and that, if its actions result in a penalty against the university, the licensee must defend the university and pay any penalties imposed on the university.

Because it is always possible that the licensee and the licensor will disagree about the interpretation of some portion of the agreement or whether one or the other party has met its obligations, an arbitration clause is included. It is, of course, hoped that the parties will resolve their differences without arbitration or litigation, but if they cannot, this binds them to pursue arbitration before litigation. Arbitration is not a simple or necessarily inexpensive method of dispute resolution, but it is more likely to be so than litigation and it is also more likely to be timely.

III. DEVELOPING AND CULTIVATING THE LICENSOR-LICENSEE RELATIONSHIP

A. Principal Differences Between the University and Industry in License Negotiations

As pointed out in the above discussion of the license agreement, license negotiations focus on four principal areas:

1. Scope of the license
2. Value of the technology—the timing and level of payments
3. March-in terms—determination of milestones
4. Indemnity and insurance

In determining the scope of the license, the first issue is the level of exclusivity; will the license be exclusive or nonexclusive. Some technology, which has broad utility, should be licensed nonexclusively, but many biotechnology inventions are the basis of pharmaceutical or diagnostics products that require some element of exclusivity to justify the investment required to achieve FDA approval. Many of these demand a single licensee, but a few may have multiple markets or markets large enough to permit the granting of a limited number of licenses, often referred to as semiexclusive licenses. Exclusivity may be also be limited by geography (an exclusive license might be granted for the United States and Canadian markets, another exclusive license granted for Europe, and still another for Asia and the rest of the world) or by field of use (for example, separate exclusive licenses might be granted for diagnostic purposes and for therapeutic purposes). Even if worldwide exclusivity for all fields of use is granted, the licensor can ensure that an invention with multiple uses will result in multiple products by requiring that, if the licensee does not develop and market one or more products, it will grant sublicenses to others to do so.

Another issue in regard to the scope of the license is whether improvement technology will be added as it is developed by the licensor. The university must be careful not to make commitments regarding improvements that will limit its ability to find sponsors for the inventors' future research or will restrict its freedom to license technology resulting from research by other members of its faculty. Although a prospective licensee often wants a guaranteed license to all improvements and to related technology that arise anywhere within the university, it usually is willing to accept the university's limitations when it realizes that as an exclusive licensee it will be the only party to whom the university can license improvements that fall within the scope of the claims of the originally licensed patents. Negotiations then focus on the financial terms associated with the improvement technology.

Prospective licensees occasionally seek rights to "know-how" as well as to patent rights and biological materials. Many universities do not license (and certainly not exclusively) know-how because it is often equated with confidential information. Universities are unequipped to guarantee the confidentiality of any sort of information because that would be contrary to their mission to disseminate knowledge. That does not mean that there is no undisclosed information (some things are rarely, if ever, the subject of articles, presentations, or even discussion with colleagues because they have little or no scholarly importance) but just that the university cannot commit to keep it secret. If know-how is included in the license, no confidentiality obligation is placed on the university or its faculty; the information would merely be conveyed to the licensee and the licensee would be granted the right to use it.

Valuing licensed technology is never simple and ultimately its value is whatever the licensee and licensor are willing to agree upon. There are some royalty rates that have become generally accepted as reasonable (see discussion above regarding the royalty section of the license agreement). Many have taken the view that when exclusively licensing patent rights on product claims, the royalty should represent approximately 25% of the profits from the sale of the product. In the case of process improvement technology, a similar rule of thumb exists— the licensor should receive 25% of the savings realized from the introduction of the improvement.

Often as important as the royalty rate might be, the licensee expresses more concern about up-front license fees, milestone payments, license maintenance fees, and minimum royalties. The thinking is that running royalties can be passed on to the consumer (so long as they do not put the product at a competitive disadvantage), but these cash payments may come at times when the company does not have the cash flow to support them. Thus, the level of these payments and their timing frequently becomes the focus of negotiation. Although the university might like to receive a sizeable license fee at the outset, even if that means accepting a lower running royalty rate, the licensee may view the technology as unproved and may be unwilling to invest large sums until it is surer that products will result. One approach is to make the license fee payable over a period of time, perhaps linked to the achievement of certain development milestones. Thus, as the technology clears developmental hurdles and thus is that much more likely to result in a product, additional installments of the license fee become due.

Because the university wants to ensure that the licensee is diligently pursuing development of the technology and, if it is not, that the licensee will terminate the license so the university can find another company to do so, the imposition of license maintenance fees can be useful. Not only does this permit the spreading over time of up-front payments, but annual payments, especially if they escalate over time, cause the company to evaluate its continuing interest in the technology. If the company has only a limited interest and the technology is in temporary storage rather than under active development, the need to make a payment can well prompt the company to voluntarily terminate the license. That is certainly more efficient than the university having to march-in under the due diligence requirement in order to impose termination on the licensee.

Minimum royalties serve two functions for the university—they place a lower limit on income once products begin to be sold and they provide another mechanism for ensuring due diligence. Establishing the level of minimum royalties can be difficult, but if the university and licensee can agree upon the level of anticipated sales, minimum royalties can readily be established as the royalties on some percentage of those anticipated sales. The percentage should be low enough that the company would probably find the product undesirable to continue manufacturing and high enough to represent reasonable efforts on the part of the licensee to achieve a reasonable share of the total potential market.

March-in or due diligence terms are essential from the university's perspective—it must be in a position to recapture licensed technology that is not being adequately exploited so that it can license it to a more diligent or capable company. The most effective march-in terms include objective milestones that the company agrees are reasonable. When a company decides it will develop any technology, whether its own or that licensed from a university, it establishes some form of R & D plan with a series of developmental targets and a timetable for reaching them. If it finds that it cannot achieve those targets within the expected time period, it reassesses the technology and whether further development efforts are economically sound. Once the licensee recognizes that the university is not trying to impose standards in excess of those the company would place upon itself, negotiations move more smoothly. By establishing milestones within the license, the university is simply placing some limits on how far the company can deviate from its own R & D plan. If that deviation becomes too great, the university can terminate the license. Of course, the university is unlikely to do so if the company can justify its failures and convince the university that it is making diligent efforts.

Although indemnity and insurance clauses are certainly the subject of considerable negotiation, there is little to add to that included above in the discussion of the clauses themselves.

As you can see from the discussion of the areas that are the subject of most negotiation, the essence of successful license negotiations is the ability to establish common ground between the licensee and the licensor. In most of the above areas, the parties have to find an approach that takes into account the needs of both. To do that, both parties have to know what is really important to them and what is not and they have to be flexible in attempting to satisfy the most significant needs of the other party. The one thing that does the most damage to license negotiations is a lack of flexibility on the part of either party—each has to be looking for new approaches based on his or her understanding of the needs of the other. That does not mean that either party is a pushover, but rather that each is sincerely seeking a solution that both find acceptable.

Successful license negotiations result in an agreement that is good for both parties; one that is based on some understanding of the other's interests and one that recognizes no agreement can adequately deal with every eventuality. Not only must the agreement itself allow for flexibility, it must also deal with the prospect of the deal going sour. Although negotiations may proceed smoothly and the negotiators may have a good rapport, situations and people change, and thus it is critical that when disputes arise, a means for their resolution exists and, if the parties want to end the license, that should be as easy as possible. Going into this "marriage" without providing for the "death of a spouse" or for "divorce" is a big mistake.

B. The Continuing Partnership

Commercialization is only just beginning when the license is signed and in many ways so too is the technology transfer process. Success requires continued efforts on the part of the university and the licensee. Regular and substantive communications are essential—the patent position must be developed and maintained and that requires involvement by both the licensee and the university. In addition, the university must take an interest in the developmental efforts of the licensee and offer at least encouragement if not constructive criticism. The university should never be in the position of suddenly discovering the that licensee did not meet an important milestone and then deciding to terminate the license. If development is getting behind schedule, the university should investigate why before milestones are missed and it should work with the licensee to reevaluate the licensee's commitment to the licensed technology (does the licensee want to terminate the license?) as well as to determine whether future milestones should be changed.

A willingness to renegotiate the license when circumstances change applies to other areas as well. For example, the licensee may encounter unexpected competition and may find the agreed-upon royalty burdensome. Unless the university is willing to renegotiate, the licensee may decide to terminate the license as unprofitable. In many ways, the license is a form of partnership. The university only succeeds if the licensee is able to succeed as well. If the university keeps in mind that its primary goal is ensuring that its technology is widely used, then working *with* the licensee to enhance the licensee's success is obvious. Being flexible and accommodating when a licensee is having difficulty developing or marketing a product does not have to mean letting the licensee get away with unsatisfactory performance or allowing the licensee to reap all the financial benefits from the license.

IV. CONCLUSION

Technology transfer of biotechnology innovations from the university to the commercial sector is now an integral part of the university's mandate. It may not be the university's first priority, but it has become an accepted part of its mission. The commercialization of university-originated technology has important implications for the public welfare and for U.S. industry and it offers the opportunity for the university to realize income that can contribute to its teaching and research goals as well as provide much needed added income for its innovative faculty.

These benefits are not without costs. The university has to make an investment in staff to find, evaluate, protect, and license university technology. If that investment is made solely in anticipation of significant financial gains, it is likely that the university will be disappointed. But if the other benefits are taken into account, including the possibility that industrial research funding and consulting opportunities for faculty may result, then the investment is more than worthwhile.

There is also a cost in oversight effort. Interactions with industry still pose conflicts. The university must be vigilant to ensure (and to reassure the public) that these interactions are not undermining the fabric of the university. To permit technology transfer or industrially supported research to destroy the open research and teaching environment would be tantamount to killing the goose that lays the golden egg. Industry would lose a source of innovation, but more importantly, the nation and the world would lose a place where ideas are so freely exchanged and debated that unsuspected connections are found that contribute to the knowledge base on which science is built.

These costs are manageable and the benefits far outweigh them.

NOTES

1. For the remainder of this chapter, the term *university* is used instead of repeating the list: university, teaching hospital, and research institute.
2. Public Law 96-517, known as the Bayh-Dole Bill, was enacted on December 12, 1980 as an amendment to the patent code. After several attempts, the law was finally

implemented by federal regulations contained in Circular A-124 with an effective date of March 1, 1982, which appeared in the *Federal Register* Vol. 47, No. 34, on February 19, 1982. Public Law 96-517 was subsequently amended by Public Law 98-620 on November 8, 1984. That amendment extended the earlier legislation to the few federal agencies not already covered and removed a number of limitations placed upon the granting of exclusive licenses to large businesses. New implementing regulations appeared in the *Federal Register* Vol. 52, No. 52, March 18, 1987, and these regulations superseded Circular A-124 with an effective date of April 17, 1987.

3. *Intellectual Property Happenings*, April 1990, p. 3.

APPENDIX

License Format
Patent and Biological Materials — April 1990
Exclusive, solely owned by Harvard
Time Limited
Field Limited

LICENSE AGREEMENT

This Agreement is made and entered into between the President and Fellows of Harvard College (hereinafter HARVARD) having offices at the Office for Patents, Copyrights and Licensing, 124 Mt. Auburn Street, Suite 256, Cambridge, Massachusetts, 02138 and ‹company› (hereinafter LICENSEE), a corporation of ‹city, state› having offices at____ .

Whereas HARVARD is the Owner by assignment from ‹inventor› of the entire right, title and interest in United States Patent Application Serial No. ‹s.n.› filed ‹filing date›, entitled ‹invention›, in the foreign patent applications corresponding thereto, and in the inventions described and claimed therein; and

Whereas HARVARD is the Owner by assignment of the BIOLOGICAL MATERIALS as defined below; and

Whereas HARVARD is committed to policy that ideas or creative works produced at HARVARD should be used for the greatest possible public benefit; and

Whereas HARVARD accordingly believes that every reasonable incentive should be provided for the prompt introduction of such ideas into public use, all in a manner consistent with the public interest; and

Whereas LICENSEE is desirous of obtaining an exclusive world-wide license in order to practice the above referenced invention covered by patent rights in the United States and in certain foreign countries, and to manufacture, use and sell in the commercial market the products made in accordance therewith; and

Whereas HARVARD is desirous of granting such a license to LICENSEE in accordance with the terms of this Agreement.

Now therefore, in consideration of the foregoing premises, the parties agree as follows:

ARTICLE I

Definitions

1.1 PATENT RIGHTS shall mean United States patent application Serial No. ‹s.n.› filed ‹filing date›, the inventions described and claimed therein, and any divisions, continuations, continuations-in-part, patents issuing thereon or reissues thereof; and any and all foreign patents and patent applications corresponding thereto; which will be automatically incorporated in and added to this Agreement and shall periodically be added to Appendix A attached to this Agreement and made a part thereof.

1.2 LICENSED PRODUCTS shall mean products claimed in PATENT RIGHTS or products made in accordance with or by means of LICENSED PROCESSES or products made utilizing BIOLOGICAL MATERIALS or incorporating some portion of BIOLOGICAL MATERIALS.

1.3 LICENSED PROCESSES shall mean the processes claimed in PATENT RIGHTS or processes utilizing BIOLOGICAL MATERIALS or some portion thereof.

1.4 BIOLOGICAL MATERIALS shall mean the materials supplied by HARVARD (identified in Appendix B) together with any progeny, mutants, or derivatives thereof.

1.5 NET SALES shall mean the amount billed or invoiced on sales of LICENSED PRODUCTS less:

(a) Customary trade, quantity or cash discounts and non- affiliated brokers' or agents' commissions actually allowed and taken;

(b) Amounts repaid or credited by reason of rejection or return; and/or

(c) To the extent separately stated on purchase orders, invoices or other documents of sale, taxes levied on and/or other governmental charges made as to production, sale, transportation, delivery or use and paid by or on behalf of LICENSEE.

1.6 AFFILIATES shall mean any company, corporation, or business in which LICENSEE owns or controls at least fifty percent (50%) of the voting stock.
1.7 FIELD shall mean__________________________________ .

ARTICLE II

Grant

2.1 HARVARD hereby grants to LICENSEE and LICENSEE accepts, subject to the terms and conditions hereof, a worldwide license in the FIELD under PATENT RIGHTS and a worldwide license to use BIOLOGICAL MATERIALS, to make and have made, to use and have used, to sell and have sold the LICENSED PRODUCTS, and to practice the LICENSED PROCESSES, for a term of __________years. Such license shall include the right to grant sublicenses, subject to HARVARD's approval which approval shall not be unreasonably withheld. In order to provide LICENSEE with a period of exclusivity, HARVARD agrees it will not grant licenses under PATENT RIGHTS to others except as required by HARVARD's obligation in paragraph 2.2 (a) or as permitted in paragraph 2.2 (b) and that it will not provide BIOLOGICAL MATERIALS to others for any commercial purpose. LICENSEE agrees during the period of exclusivity of this license in the United States that any LICENSED PRODUCT produced for sale in the United States will be manufactured substantially in the United States.

2.2 The granting and acceptance of this license is subject to the following conditions:

(a) HARVARD's "Statement of Policy in Regard to Inventions, Patents and Copyrights" dated March 17, 1986, Public Law 96-517, Public Law 98-620 and HARVARD's obligations under agreements with other sponsors of research. Any right granted in this Agreement greater than that permitted under Public Law 96-517 or Public Law 98-620 shall be subject to modification as may be required to conform to the provisions of that statute.

(b) HARVARD shall have the right to make and to use and to grant non-exclusive licenses to make and to use, for research purposes only and not for any commercial purpose, the BIOLOGICAL MATERIALS and the subject matter described and claimed in PATENT RIGHTS.

(c) LICENSEE shall use reasonable effort to effect introduction of the LICENSED PRODUCTS into the commercial market as soon as practicable, consistent with sound and reasonable business practices and judgment; thereafter, until the expiration of this Agreement, LICENSEE shall endeavor to keep LICENSED PRODUCTS reasonably available to the public.

(d) HARVARD shall have the right to terminate or render this license non-exclusive at any time after three (3) years from the date of license if, in HARVARD's reasonable judgment, LICENSEE:

(i) has not put the licensed subject matter into commercial use in the country or countries where licensed, directly or through a sublicense, and is not keeping the licensed subject matter reasonably available to the public, or

(ii) is not demonstrably engaged in research, development, manufacturing, marketing or licensing program, as appropriate, directed toward this end.

In making this determination HARVARD shall take into account the normal course of such programs conducted with sound and reasonable business practices and judgment and shall take into account the reports provided hereunder by LICENSEE.

(e) All sublicenses granted by LICENSEE hereunder shall include a requirement that the sublicensee use its best efforts to bring the subject matter of the sublicense into commercial use as quickly as is reasonably possible and shall bind the sublicensee to meet LICENSEE's obligations to HARVARD under this Agreement. Royalties charged for sublicenses by LICENSEE shall not be in excess of normal trade practice. Copies of all sublicense agreements shall be provided to HARVARD.

(f) In the event that LICENSEE is unable or unwilling to grant sublicenses, either as suggested by HARVARD or a potential sublicensee or otherwise, HARVARD may directly license such potential sublicensee unless LICENSEE reasonably satisfies HARVARD that such sublicense would be contrary to sound and reasonable business practice, and that the granting of such sublicense would not materially increase the availability to the public of products manufactured under this license.

(g) A license in any other field of use in addition to FIELD shall be the subject of a separate agreement and shall require LICENSEE's submission of evidence demonstrating its willingness and ability to develop and commercialize the kinds of products or processes likely to be encompassed in such other field. If HARVARD has non-exclusively licensed in such other field prior to LICENSEE'S request, HARVARD shall grant to LICENSEE a non-exclusive license under terms and conditions at least as favorable as the previous non-exclusive license.

2.3 Upon expiration of the period of exclusivity of this license, HARVARD may at its sole option extend the period of exclusivity, extend the license on a non-exclusive basis, or permit the license to terminate.

2.4 HARVARD hereby grants to LICENSEE the right to extend the licenses granted or to be granted in paragraphs 2.1 and 2.3 to an AFFILIATE subject to the terms and conditions hereof.

2.5 All rights reserved to the United States Government and others under Public Law 96-517 and 98-620 shall remain and shall in no way be affected by this Agreement.

ARTICLE III

Royalties

3.1 LICENSEE shall pay to HARVARD a non-refundable license fee in the sum of ______________________________ upon execution of this Agreement {and a sum of ______________ upon issuance of the U.S. patent}.

3.2 LICENSEE shall pay HARVARD, during the term of the license of paragraph 2.1, a royalty of _____ percent (%) of the NET SALES of all LICENSED PRODUCTS sold by LICENSEE and its AFFILIATES or sublicensees. In the case of sublicenses, LICENSEE shall also pay to HARVARD _____ percent (%) of non-royalty sublicense income (e.g., license issue fees, license maintenance fees, etc.) If this license is converted to a non-exclusive one and if other non-exclusive licenses are granted, this royalty shall not exceed the royalty being paid by other licensees during the term of the non-exclusive license. On sales between LICENSEE and its AFFILIATES or sublicensees for resale, the royalty shall be paid on the resale.

3.3 HARVARD shall have the right to terminate this license in the event that LICENSEE does not pay to HARVARD at least the following amounts in license maintenance fees and/or minimum royalties:

First calendar year	–$
Next calendar year	–$
Next calendar year and each year thereafter.	–$

ARTICLE IV

Reporting

4.1 Prior to signing this Agreement, LICENSEE has provided to HARVARD a written research and development plan under which LICENSEE intends to bring the subject matter of the licenses granted hereunder into commercial use upon execution of this Agreement. Such plan includes projections of sales and proposed marketing efforts.

4.2 LICENSEE shall provide written annual reports within sixty (60) days after June 30 of each calendar year which shall include but not be limited to: reports of progress on research and development, regulatory approvals, manufacturing, sublicensing, marketing and sales during the preceding twelve (12) months as well as plans for the coming year. If progress differs from that anticipated in the plan provided under 4.1, LICENSEE shall explain the reasons for the difference and propose a modified plan for HARVARD's review and approval. LICENSEE shall also provide any reasonable additional data HARVARD requires to evaluate LICENSEE's performance.

4.3 LICENSEE shall report to HARVARD the date of first sale of LICENSED PRODUCTS (or results of LICENSED PROCESSES) in each country within thirty (30) days of occurrence.

4.4 LICENSEE agrees to submit to HARVARD within sixty (60) days after the calendar half years ending June 30 and December 31, reports setting forth for the preceding six (6) month period the amount of the LICENSED PRODUCTS sold by LICENSEE, its AFFILIATES and sublicensees in each country, the NET SALES thereof, and the amount of royalty due thereon and with each such royalty report to pay the amount of royalty due. Such report shall be certified as correct by an officer of LICENSEE and shall include a detailed listing of all deductions from NET SALES, sublicensee income or from royalties as specified herein. If no royalties are due to HARVARD for any reporting period, the written report shall so state. If royalties for any calendar year do not equal or exceed the minimum royalties established in paragraph 3.3, LICENSEE shall include the balance of the minimum royalty with the payment for the half year ending December 31. All royalties due hereunder shall be payable in United States dollars. Conversion of foreign currency to U.S. dollars shall be made at the conversion rate existing in the United States on the date of royalty payments by LICENSEE. All such reports shall be maintained in confidence by HARVARD, except as required by law, including Public Law 96-517 and 98-620. Late payments shall be subject to an interest charge of one and one half percent (1½%) per month.

ARTICLE V

Record Keeping

5.1 LICENSEE shall keep, and shall require its AFFILIATES and sublicensees to keep, accurate and correct records of LICENSED PRODUCTS made, used or sold under this Agreement, appropriate to determine the amount of royalties due hereunder to HARVARD. Such records shall be retained for at least three (3) years following a given reporting period. They shall be available during normal business hours for inspection at the expense of HARVARD by HARVARD's Internal Audit Department or by a Certified Public Accountant selected by HARVARD and approved by LICENSEE for the sole purpose of verifying reports and payments hereunder. Such accountant shall not disclose to HARVARD any information other than information relating to accuracy of reports and payments made under this Agreement. In the event that any such inspection shows an underreporting and underpayment in excess of five percent (5%) for any twelve (12) month period, then LICENSEE shall pay the cost of such examination.

ARTICLE VI

Domestic and Foreign Patent Filing and Maintenance

6.1 LICENSEE shall reimburse HARVARD for all reasonable expenses HARVARD has incurred for the preparation, filing, prosecution and maintenance of PATENT RIGHTS and shall reimburse HARVARD for all such future expenses. HARVARD shall take responsibility for the preparation, filing, prosecution and maintenance of any and all patent applications and patents included in PATENT RIGHTS, provided however that HARVARD shall first consult with LICENSEE as to the preparation, filing, prosecution and maintenance of such patent applications and patents and shall furnish to LICENSEE copies of documents relevant to any such preparation, filing, prosecution or maintenance.

6.2 HARVARD and LICENSEE shall cooperate fully in the preparation, filing, prosecution and maintenance of PATENT RIGHTS and of all patents and patent applications licensed to LICENSEE hereunder, executing all papers and instruments or requiring members of HARVARD to execute such papers and instruments so as to enable HARVARD to apply for, to prosecute and to maintain patent applications and patents in HARVARD's name in any country. Each party shall provide to the other prompt notice as to all matters which come to its attention and which may affect the preparation, filing, prosecution or maintenance of any such patent applications or patents.

6.3 If LICENSEE elects not to pay the expenses of a patent application or patent included within PATENT RIGHTS, LICENSEE shall notify HARVARD not less than sixty (60) days prior to such action and shall thereby surrender its rights under such patent or patent application.

ARTICLE VII

Infringement

7.1 With respect to any PATENT RIGHTS under which LICENSEE is exclusively licensed pursuant to this Agreement, LICENSEE or its sublicensee shall have the right to prosecute in its own name and at its own expense any infringement of such patent, so long as such license is exclusive at the time of the commencement of such action. HARVARD agrees to notify LICENSEE promptly of each infringement of such patents of which HARVARD is or becomes aware. Before LICENSEE or its sublicensees commences an action with respect to any infringement of such patents, LICENSEE shall give careful consideration to the views of HARVARD and to potential effects on the public interest in making its decision whether or not to sue and in the case of a LICENSEE sublicense, shall report such views to the sublicensee.

7.2 If LICENSEE or its sublicensee elects to commence an action as described above and HARVARD is a legally indispensable party to such action, HARVARD shall have the right to assign to LICENSEE all of HARVARD's right, title and interest in each patent which is a part of the PATENT RIGHTS and is the subject of such action (subject to all HARVARD's obligations to the government and others having rights in such patent). In the event that HARVARD makes such an assignment, such assignment shall be irrevocable, and such action by LICENSEE on that patent or patents shall thereafter be brought or continued without HARVARD as a party, if HARVARD is no longer an indispensable party. Notwithstanding any such assignment to LICENSEE by HARVARD and regardless of whether HARVARD is or is not an indispensable party, HARVARD shall cooperate fully with LICENSEE in connection with any such action. In the event that any patent is assigned to LICENSEE by HARVARD, pursuant to this paragraph, such assignment shall require LICENSEE to continue to meet its obligations under this Agreement as if the assigned patent or patent application were still licensed to LICENSEE.

7.3 If LICENSEE or its sublicensee elects to commence an action as described above, LICENSEE may reduce, by up to fifty percent (50%), the royalty due to HARVARD earned under the patent subject to suit by fifty percent (50%) of the amount of the expenses and costs of such action, including attorney fees. In the event that such fifty percent (50%) of such expenses and costs exceed the amount of royalties withheld by LICENSEE for any calendar year, LICENSEE may to that extent reduce the royalties due to HARVARD from LICENSEE in succeeding calendar years, but never by more than fifty percent (50%) of the royalty due in any one year.

7.4 Recoveries or reimbursements from such action shall first be applied to reimburse LICENSEE and HARVARD for litigation costs not paid from royalties (if any) and then to reimburse HARVARD for royalties withheld. Any remaining recoveries or reimbursements shall be shared equally by LICENSEE and HARVARD.

7.5 In the event that LICENSEE and its sublicensee, if any, elect not to exercise their right to prosecute an infringement of the PATENT RIGHTS pursuant to the above paragraphs, HARVARD may do so at its own expense, controlling such action and retaining all recoveries therefrom.

ARTICLE VIII

Termination of Agreement

8.1 This Agreement, unless extended or terminated as provided herein, shall remain in effect for __________ years.

8.2 In the event that one party to this Agreement shall be in default in the performance of any obligations under this Agreement, and if the default has not been remedied within ninety (90) days after the date of notice in writing of such default, the party giving such notice may terminate this Agreement by written notice.

8.3 In the event that LICENSEE shall become insolvent, shall make an assignment for the benefit of creditors, or shall have a petition in bankruptcy filed for or against it, HARVARD shall have the right to terminate this entire Agreement by giving LICENSEE written notice of such termination.

8.4 Any sublicenses granted by LICENSEE under this Agreement shall provide for termination or assignment to HARVARD, at the option of HARVARD, of LICENSEE's interest therein upon termination of this Agreement.

8.5 LICENSEE shall have the right to terminate this Agreement by giving thirty (30) days advance written notice to HARVARD to that effect and paying a termination fee of $__________. Upon termination, a final report shall be submitted and any royalty payments and unreimbursed patent expenses due to HARVARD become immediately payable.

8.6 Sections 8.5, 9.2, 9.4, and 9.5 of this Agreement shall survive termination.

ARTICLE IX

General

9.1 HARVARD represents and warrants that the entire right, title, and interest in the patent applications or patents comprising the PATENT RIGHTS have been assigned to it and that HARVARD has the authority to issue licenses under said PATENT RIGHTS. HARVARD does not warrant the validity of the PATENT RIGHTS licensed hereunder and makes no representations whatsoever with regard to the scope of the licensed PATENT RIGHTS or that such PATENT RIGHTS may be exploited by LICENSEE, an AFFILIATE, or sublicensee without infringing other patents.

9.2 HARVARD EXPRESSLY DISCLAIMS ANY AND ALL IMPLIED OR EXPRESS WARRANTIES AND MAKES NO EXPRESS OR IMPLIED WARRANTIES OF MERCHANTABILITY OR FITNESS FOR ANY PARTICULAR PURPOSE OF THE TECHNOLOGY, LICENSED PROCESSES OR LICENSED PRODUCTS CONTEMPLATED BY THIS AGREEMENT. Further HARVARD has made no investigation and makes no representation that the BIOLOGICAL MATERIALS supplied by it or the methods used in making or using such materials are free from liability for patent infringement.

9.3 LICENSEE shall not distribute or release the BIOLOGICAL MATERIALS to others except to further the purposes of this Agreement.

9.4 (a) LICENSEE shall indemnify, defend and hold harmless HARVARD and its directors, trustees, officers, faculty, medical and professional staff, employees, students, and agents and their respective successors, heirs and assigns (the "Indemnitees"), against any liability, damage, loss or expenses (including reasonable attorneys' fees and expenses of litigation) incurred by or imposed upon the Indemnitees or any one of them in connection with any claims, suits, actions, demands or judgments arising out of the design, production, manufacture, sale, use in commerce, lease, or promotion, by LICENSEE or by a sublicensee, AFFILIATE or agent of LICENSEE, of any product, process or service relating to, or developed pursuant to, this Agreement. LICENSEE's indemnification under this Section shall apply to any liability, damage, loss or expense whether or not it is attributable to the negligent activities of the Indemnitees.

(b) LICENSEE agrees, at its own expense, to provide attorneys reasonably acceptable to HARVARD to defend against any actions brought or filed against any party indemnified hereunder with respect to the subject of the indemnity contained herein, whether or not such actions are rightfully brought.

(c) At such time as any product, process, service relating to, or developed pursuant to, this Agreement is being commercially distributed or sold (other than for the purpose of obtaining regulatory approvals) by LICENSEE or by a sublicensee, AFFILIATE or agent of LICENSEE, LICENSEE shall at its sole cost and expense, procure and maintain comprehensive general liability insurance in amounts not less than $2,000,000 per incident and $2,000,000 annual aggregate and naming the Indemnitees as additional insureds. {During clincial trials of any product, process or service relating to, or developed pursuant to, this Agreement by LICENSEE or by a sublicensee, affiliate or agent of LICENSEE, LICENSEE shall at its sole cost and expense, procure and maintain comprehensive general liability insurance in such equal or lesser amount as HARVARD shall require, naming the Indemnitees as additional insureds.} Such comprehensive general liability insurance shall provide (i) product liability coverage and (ii) broad form contractual liability coverage for LICENSEE's indemnification under this Agreement. If LICENSEE elects to self-insure all or part of the limits described above (including deductibles or retentions which are in excess of $250,000 annual aggregate) such self-insurance program must be acceptable to HARVARD and the Risk Management Foundation. The minimum amounts of insurance coverage required shall not be construed to create a limit of LICENSEE's liability with respect to its indemnification under this Agreement.

(d) LICENSEE shall provide HARVARD with written evidence of such insurance upon request of HARVARD. LICENSEE shall provide HARVARD with written notice at least fifteen (15) days prior to the cancellation, non-renewal or material change in such insurance; if LICENSEE does not obtain replacement insurance providing comparable coverage within such fifteen (15) day period, HARVARD shall have the right to terminate this Agreement effective at the end of such fifteen (15) day period without notice or any additional waiting periods.

(e) LICENSEE shall maintain such comprehensive general liability insurance beyond the expiration or termination of this Agreement during (i) the period that any product, process, or service, relating to, or developed pursuant to, this Agreement is being commercially distributed or sold by LICENSEE or by a sublicensee, AFFILIATE or agent of LICENSEE and (ii) a reasonable period after the period referred to in (e) (i) above which in no event shall be less than fifteen (15) years.

9.5 LICENSEE shall not use HARVARD's name or any adaptation of it in any advertising, promotional or sales literature without the prior written assent of HARVARD.

9.6 Without the prior written approval of HARVARD, the entire license granted pursuant to this Agreement shall not be transferred by LICENSEE to any party other than to a successor to the business interest of LICENSEE relating to the PATENT RIGHTS. This Agreement shall be binding upon the successors, legal representatives and assignees of HARVARD and LICENSEE.

9.7 The interpretation and application of the provisions of this Agreement shall be governed by the laws of the Commonwealth of Massachusetts.

9.8 LICENSEE agrees to comply with all applicable laws and regulations. In particular, it is understood and acknowledged that the transfer of certain commodities and technical data is subject to United States laws and regulations controlling the export of such commodities and technical data, including all Export Administration Regulations of the United States Department of Commerce. These laws and regulations among other things, prohibit or require a license for the export of certain types of technical data to certain specified countries. LICENSEE hereby agrees and gives written assurance that it will comply with all United States laws and regulations controlling the export of commodities and technical data, that it will be solely responsible for any violations of such by LICENSEE or its AFFILIATES or sublicensees, and that it will defend and hold HARVARD harmless in the event of any legal action of any nature occasioned by such violation.

9.9 Written notices required to be given under this Agreement shall be addressed as follows:

If to HARVARD:

Office of Patents, Copyrights and Licensing
Harvard University
124 Mt. Auburn Street, Suite 256
Cambridge, MA 02138

If to LICENSEE:

____

____

____

or such other address as either party may request in writing.

9.10 Should a court of competent jurisdiction later consider any provision of this Agreement to be invalid, illegal, or unenforceable, it shall be considered severed from this Agreement. All other provisions, rights and obligations shall continue without regard to the severed provision, provided that the remaining provisions of this Agreement are in accordance with the intention of the parties.

9.11 In the event of any controversy or claim arising out of or relating to any provision of this Agreement or the breach thereof, the parties shall try to settle such conflicts amicably between themselves. Subject to the limitation stated in the final sentence of this section, any such conflict which the parties are unable to resolve shall be settled through arbitration conducted in accordance with the rules of the American Arbitration Association. The demand for arbitration shall be filed within a reasonable time after the controversy or claim has arisen, and in no event after the date upon which institution of legal proceedings based on such controversy or claim would be barred by the applicable statute of limitation. Such arbitration shall be held in Boston, Massachusetts. The award through arbitration shall be final and binding. Either party may enter any such award in a court having jurisdiction or may make application to such court for judicial acceptance of the award and an order of enforcement, as the case may be. Notwithstanding the foregoing, either party may, without recourse to arbitration, assert against the other party a third-party claim or cross-claim in any action brought by a third party, to which the subject matter of this Agreement may be relevant.

9.12 This Agreement constitutes the entire understanding between the parties and neither party shall be obligated by any condition or representation other than those expressly stated herein or as may be subsequently agreed to by the parties hereto in writing.

IN WITNESS WHEREOF, the parties hereto have caused this Agreement to be executed by their duly authorized representatives.

The effective date of this Agreement is ____________________, 19__

PRESIDENT AND FELLOWS OF
HARVARD COLLEGE

By:______________________________

Name and Title:____________________

‹COMPANY›

By:______________________________

Name and Title:____________________

Index

About the Editor

GALE R. PETERSON is Head of Intellectual Property at Cox & Smith Incorporated, San Antonio, Texas. Serving on the editorial boards of several journals and author of more than 40 professional papers, Mr. Peterson is a former Patent Examiner with the U.S. Patent and Trademark Office and a former Advisor to the Chief Judge of the U.S. Court of Customs and Patent Appeals (now the U.S. Court of Appeals for the Federal Circuit). Mr. Peterson is past Chairman of the Intellectual Property Law Section of the State Bar of Texas, a Fellow and past President of the Society of International Business Fellows, and a member of the U.S. Bar/Japanese Patent Office Liaison Council. He is active in numerous committees of the American Bar Association, the American Intellectual Property Law Association, and the United States Trademark Association, among others. Mr. Peterson received the B.S. degree (1970) in electrical engineering from the University of Nebraska, Lincoln, the J.D. degree (1975) from the University of Baltimore, Maryland, and the L.L.M. degree (1978) in trade regulation law from George Washington University, Washington D.C.